The Effects of Air Pollution on the Built Environment

AIR POLLUTION REVIEWS

Series Editor: Robert L. Maynard
(*Department of Health, Skipton House, London, U.K.*)

Air Pollution Reviews – Vol. 2

The Effects of Air Pollution on the Built Environment

Editor

Peter Brimblecombe
University of East Anglia, UK

ICP

Imperial College Press

Published by

Imperial College Press
57 Shelton Street
Covent Garden
London WC2H 9HE

Distributed by

World Scientific Publishing Co. Pte. Ltd.
5 Toh Tuck Link, Singapore 596224
USA office: Suite 202, 1060 Main Street, River Edge, NJ 07661
UK office: 57 Shelton Street, Covent Garden, London WC2H 9HE

Library of Congress Cataloging-in-Publication Data
The effects of air pollution on the built environment / editor Peter Brimblecombe.
    p. cm. -- (Air pollution reviews)
    ISBN 1-86094-291-1 (alk. paper)
    1. Air--Pollution--Environment aspects. 2. Weathering of buildings. 3. Building
materials--Deterioration. I. Brimblecombe, Peter, 1949– II. Series.

TD883 .E36 2002
690--dc21   2002068691

British Library Cataloguing-in-Publication Data
A catalogue record for this book is available from the British Library.

Printed in Singapore.

CONTENTS

CONTRIBUTORS

Peter Brimblecombe
School of Environmental Sciences
University of East Anglia
Norwich NR4 7TJ, UK

Dario Camuffo
Instituto di Scienze
   dell'Atmosfera e dell'Oceano
   Consiglio Nazionale delle
   Ricerche (CNR), Padova, Italy

J. N. Cape
CEH Edinburgh
Centre for Ecology and
   Hydrology (CEH)
Bush Estate, Penicuik EH26 0QB
Midlothian, Scotland

C. I. Davidson
Department of Civil and
Environmental Engineering
Carnegie Mellon University
Pittsburgh, PA 15213, USA

S. E. Espenhahn
ETSU Health and Safety
   Consultants
Harwell, Oxfordshire
OX11 0RA, UK

V. Etyemezian
Desert Research Institute
755 E Flamingo Road
Las Vegas, NV 89119, USA

S. Finger
Department of Civil and
   Environmental Engineering
Carnegie Mellon University
Pittsburgh, PA 15213, USA

Dave J. Hall
Building Research Establishment
   (BRE)
Garston, Watford
Herts WD2 7JR, UK

Ron Hamilton
School of Health and Social
  Sciences
Middlesex University
Bounds Green Road
London N11 2NQ, UK

Rob Inkpen
Department of Geography
University of Portsmouth
Buckingham Building
Lion Terrace
Portsmouth PO1 3HE, UK

C. Jaiz-Jimenez
Consejo Superior de
  Investigaciones Cientificas
  (CSIC)
Instituto de Recursos Naturales
Y Agrobiologia de Sevilla
Avenida de Reina Mercedes
No. 10, Apartado 1052
41080 Sevilla, España

Vladimir Kucera
Swedish Corrosion Institute
Kräftriket 23 A
SE-104 05 Stockholm, Sweden

V. Kukadia
Building Research Establishment
  (BRE)
Garston, Watford
Herts WD2 7JR, UK

David S. Lee
Technology Leader, Atmospheric
  Processes
The Defence Evaluation and
  Research Agency (DERA)
Pyestock
Air Systems Sector, Propulsion
  Department
Farnborough
Hants GU14 0LS, UK

Johanna Leissner
DG Research DI.5
Commission of European
  Communities
200 rue de la Loi
Brussels B-1049, Belgium

I. D. Leith
CEH Edinburgh
Centre for Ecology and
  Hydrology (CEH)
Bush Estate, Penicuik EH26 0QB
Midlothian, Scotland

P. M. Lewis
Tun Abdul Razak Research
  Centre
Brickendonbury, Hertfordshire,
  UK

Cristina Sabbioni
Institute of Atmospheric
Sciences and Climate (ISAC)
National Research Council (CNR)
Via P. Gobetti 101–40129
Bologna, Italy

S. I. Sherwood
562 Chenango Street
Binghamton NY 13901, USA

Bernard J. Smith
School of Geography
Queen's University Belfast
Belfast BT7 1NN
North Ireland, UK

A. M. Spanton
Building Research Establishment
   (BRE)
Garston, Watford
Herts WD2 7JR, UK

Michael Steiger
Institute of Inorganic and
   Applied Chemistry
University of Hamburg
Martin-Luther-King-Platz 6
20146 Hamburg, Germany

M. F. Striegel
National Center for Preservation
   Technology and Training
NSU Box 5682
Natchitoches, LA 71497, USA

W. Tang
Department of Civil and
   Environmental Engineering
Carnegie Mellon University
Pittsburgh, PA 15213, USA

Johan Tidblad
Swedish Corrosion Institute
Kräftriket 23 A
104 05 Stockholm, Sweden

S. Walker
Building Research Establishment
   (BRE)
Garston, Watford
Herts WD2 7JR, UK

J. Watt
School of Health and Social
   Sciences
Middlesex University
Bounds Green Road
London N11 2NQ, UK

Tim Yates
Building Research Establishment
   (BRE)
Garston, Watford
Herts WD2 7JR, UK

PREFACE

Air Pollution and Our Cultural Heritage

The effects of air pollution on building materials became linked to a wider debate about the protection of cultural heritage. We find ourselves in an age where the links between environment, its pollution and culture seem more obvious than ever. Pollution has become a feature of artistic expression broadening the public interest. The century, just complete, began with the paintings of Monet that rendered impressions of the smoke pollution of London and the novels of Arthur Conan Doyle, had his hero Sherlock Holmes pacing the fog-bound streets of London. In the United States, fictional detectives, such as Raymond Chandler's Philip Marlowe, were driving around Los Angeles in the 1940s when photochemical smog was born. Hardly surprising that Hollywood's detective *film-noir*, ultimately encapsulated a vision of smog, perhaps most intensely seen in the futuristic *Blade Runner* (1982). Here it is always as dark as night, because the orange sun can only cut through the pollution at the tops of buildings. As the century ended, the Ken Saro-Wiwa's literary visions of the environment impelled him to an activism, that became a justification for Nigerian government to order his execution.

The destructive effect of air pollution on our built heritage has long been apparent. Recent interest seems particularly driven by the concerns over the widening effects of acid rain in the 1970s and 1980s. The damage from deposition of sulphur dioxide, often accumulated over centuries, at last found voice. The Convention on Long Range

Transboundary Air Pollution fostered a number of international cooperative programmes in the late 1970s. Among these the international cooperative programme for the effects of air pollution on materials gave rise to a long series of studies at the behest of the United Nations Economic Commission for Europe.

Across the Atlantic, 1980 saw the US Congress authorise the National Acid Precipitation Assessment Program (NAPAP), an inter-agency task force under the auspices of the Council on Environmental Quality. By the early 1990s, NAPAP had produced a series of reports: *Acidic Deposition: State of Science and Technology*. The program was re-authorised through the 1990 Clean Air Act Amendments and examined trends in emissions, deposition and effects, to evaluate the expected benefits of the 1990 legislation. It looked at important construction materials such as metals, paint systems and concrete, and the culturally important materials such as marble, limestone and bronze.

Within Europe, the European Commission promoted studies of the effects of air pollution on cultural heritage, most particularly under the European Union's STEP and Framework programmes from the 1980s. This policy-driven research developed through a period of increasing interest in the concept of sustainability. Sustainability was set to be a central consideration of policy development in the 21st century. Cultural heritage was not explicitly recognised, as a strategic imperative, within seminal documents such as the *Brundtland Report*. Nevertheless, it appeared in official European publications relating to our cultural heritage. The Council of Europe prepared a document about *Sustained Care of Cultural Heritage* and the European Commission's Framework 5 Programme sought tools for the sustainable management of cultural heritage. The move to link sustainability and heritage was motivated by an admirable desire to preserve so much of what we value, although sustainability was not well defined within the heritage context.

Research often focused on dose-response function research and concerns about the mechanisms of damage under laboratory and field conditions. There was much interest in investing decay of well-characterised ambient exposures for standard size samples, as well as for individual structures. The STEP and Framework Programmes in Europe initially focussed on stone, but as with the NAPAP research widened to take an interest in many other materials, e.g. wood, glass and metals. More recently, European research paid greater attention to indoor materials, such as paper, leather and silver.

This research area gave value that went well beyond the field of materials research. There were important spin-offs into other fields, which characterise the research as innovative and show it had escaped the bound imposed by traditional disciplines. One example of this is the use of research on the behaviour of salts on monuments to gain an understanding the moisture requirements of the house-dust mite, implicated in childhood asthma.

The work that contributes to this volume comes from a period when research on damage to cultural heritage was especially active. Times have changed and the future directions of such research is not necessarily clear. The shifting focus of European research under Framework Programme VI seems likely to foster very different projects. In the U.S., similar changes are felt perhaps with less interest in the outdoor dose-response function and increasing attention on the indoor environment.

Nevertheless, buildings remain a most permanent feature of our culture. Whatever the direction of future research, it is clear that an understanding of the way air pollutants affect materials are of practical concern to many involved in the regulation and control of air pollutants. It is hoped that this volume will be of use to the increasing number of scientists, students, conservators and practitioners whose concerns lie at the interface between research and its application.

CHAPTER 1

LONG TERM DAMAGE TO THE BUILT ENVIRONMENT

P. Brimblecombe and D. Camuffo

1. Introduction

Our cultural property is meant to last. We build in stone. Memorials of our heritage are constructed of the most durable materials. It has become common to talk of our crumbling heritage as the victim of air pollution, sea-level rise, urban development, poor taste, etc. At such times, it is easy to see these processes as recent and to forget that the buildings we are so anxious to preserve are a product of continued environmental and social change. These buildings are records both of the forces that created them and those which have altered them. The patina they accrete is part of the heritage they represent. It is part of our attraction to ruins.

This is not to welcome the destruction of our buildings, but to recognise its inevitability. It is wanton or rapid damage, falsification, disrespect, insensibility that characterises true destruction.

It is important to see buildings in their historical context. This is true of their artistic and social history, but here we will be primarily concerned with the environmental history. As with social history, this is partly to help us interpret the current form of the building, but also to aid with an understanding of the physical changes that have taken place. It further offers the opportunity to gain insight into the long term mechanisms of damage. This historical context may need to be

1

several millenia long to cover the span of our built heritage, although a particular focus on the last few centuries can often be appropriate because of the large number of important buildings from this period.

Architects have long recognised environmental considerations in building structures to survive long periods. In the early modern period, they began to believe that damage was caused by time, weather and smoke. Sir Christopher Wren and his colleagues recognised the interaction of sulphur dioxide with stone as there had been recognition of the problems of sulphation through much of the 17th century London where coal was burnt (Brimblecombe, 1992).

This chapter will look at the effects of long term changes in climate and air pollution on damage to materials in the built environment.

2. Changes in Climate

Substantial direct or indirect evidence of climate changes can be found in ancient literature. The oldest climatic scenario, referring to the rise of temperature in the holocene (the climatic period after the end of the last glaciacion, ca. 10,000 years ago), can be found in the Bible, i.e. book of the Genesis. At the very beginning of the book, God is found engaged in an apparently strange action: to separate between them, waters, earth and sky (Genesis, 1, 6–10), as they were all mixed together. In reality, during the warming period, the glaciers on the mountains Zagros and on the other mountains all around the Fertile Crescent melted, and the waters collected in the plain formed marshes and fog. A clear and permanent distinction between earth, waters and sky was made only when the global warming obliged the glaciers to retreat on the top of the mountains and to feed only a few rivers, and the strong sun dried the marshes forming a fertile plain (Issar, 1990). Again, the earth in the beginning was very fruitful, with all kinds of plants which produced abundant fruits (Genesis, 1, 11–13) in a so fortunate situation that Man in that time had the possibility of living without any need of working. This early period in the Hebrew memory, called the Eden, ended when God sent his angel to punish Adam. The punishment was made with a fire sword (Genesis, 3, 23–24), i.e. a climate change, which dried forever all their country, so that Adam and his sons had to work the earth to live (Camuffo, 1990).

Several meteorological observations and the mention of climate changes can be found in the Sumerian, Assyran, Babylonian and Hittite cuneiform tablets and other Far Eastern texts (Camuffo, 1990). The Sumer kingdom ended under the pressure of climatic hazards scouged by storms transporting desert sand and then invaded by other people (Babylonian and Assyran) under pressure of the same climate deterioration. Descriptions of the climatic situation in 2000 BC, characterised by famine, dryness, tornadoes and sandstorms are often found in the collection of tablets edited by Pritchard (1969), e.g. *"the afflicting storm by tears is not adjured; the destructive storm makes the land tremble and quake; like the flood storm it destroys the cities. The land-annihilating storm set up (its) ordinances in the city; the all-destroying storm came doing evil"* (A Sumeran lamentation: lamentation over the destruction of Ur, verses 197–201, in Pritchard (1969); see also item v. 111, v. 173–174, v. 190–192; Hymnal prayer of Enheduanna: the adoration of Inanna in Ur, v. 71; Lamentation over the destruction of Sumer and Ur, v. 61; v. 130–132; v. 391–394; v. 491–493; The curse of Agade, v. 120, v. 123; v. 171–174; The Myth of Zu, v. 31–32; Dumuzi and Inanna: prayer for water and bread). The Hyttite kingdom declined in 1200 BC in a similar climatic context, affected by dryness and storms and in short arrived at its end. In the same period, the war of Troy and the travels of Odysseus occurred in the context of peoples migrations for increasing dryness in the 12th century.

The Bible also mentions a number of climatic events (Camuffo, 1990): e.g. severe drynesses and famines occurred between 1200 and 1025 BC (Ruth, 1, 1); for three years during the David's kingdom (1010–970 BC) (II Samuel, 21, 1–10); for three years again starting in 858 BC (I Kings, 17, 1; 17, 7; 18, 1; II Kings 4, 38); at mid-700s BC with scarce and irregular rain before the harvest time (Amos, 4, 7); in 735 BC when Isaiah complains the scarcity of breads and water (Isaiah 3,1) and slightly later when the Nimrun was dried, the grass and all the green disappeared (Isaiah 15, 6 and Jeremy, 48, 34) and this dryness continued to extend to Lebanon and the mountain highlands (Isaiah, 33, 9). The spring rainfall ceased during the Joiakim kingdom between 609 and 597 so that soon the wells dried and the dry earth cracked (Jeremy 3, 3; 38, 6 and 48, 43); another famine followed in 787 to which contributed the dryness and the siege of Jerusalem

(II Kings, 25, 3); dryness again was found in 520 when the crops were lost, also those cultivated with help of the nocturnal dew (Aggeo 1, 11); another terrible dryness is mentioned in which the hot sun fired grass and trees, and dried up rivers (Joel, 1, 19–20): this is probably the same mentioned by Thucydides (*Pelop. Wars*, 1, 23) during the Peloponnesian wars (431–404 BC) and by Strabo (*Geographia*, I, 3, 2) and Herodotus (*Histories*, 1, 94); a long dryness and famine was found between 45 and 50 AD, which culminated in 48–49 AD (*Acts Apostles* 11, 28) and interested Palestine, Mesopotamia, Greece, Egypt and most of the Roman empire (Joseph Flavius, *Antiquitates Judaicae*, 19, 201 and 20, 101; *Bellum Judaicum*, 2, 204; Svetony, *Vita Claudii*, 18; Cassius Dion, *Historia Romana*, 60, 11; Tacitus, *Annales*, 13, 43).

A history of climate that covers the period of our architectural heritage can be constructed from such descriptive documentary records along with some instrumental records for the most recent centuries. Although other sources of data (e.g. varves, ice cores, tree rings are also important to climatologists), the particular focus of our interest in areas of human occupation makes documentary observations especially important. While excellent records exist in Asia, the climate record has been most extensively studied in Europe.

Climate change was ignored during the 19th century development of scientific meteorology. The late Hubert Lamb argued that these meteorologists had been imbued with classical writers. The climate of the classical Mediterranean was superficially like our own, so Victorian meteorologists found it hard to develop ideas of long term climate change. Although if they read Aristotle's *Meteorologica* more closely they would soon discover that Egypt was seen as becoming continually drier. As for ancient Greece, it was argued the Argive was marshy at the time of the Trojan war, supporting only a small population with Mycenae in excellent condition. By Aristotle's time, Mycenae had become dry and Argos fertile.

Despite the difficulty in accepting long term climatic change and a departure from climatic norms, the gradual increase in temperature became persistent enough to provoke a real interest in long term climatic change by the middle of the 20th century.

2.1. History of Climate

It is simplest to examine some of the broad regional changes that have featured in our views of climate history. The interest in historical climatology fostered by work of men such as Lamb, Flohn, Alexandre, etc. had led to a view that the High Middle ages of Europe were characterised by the Medieval Warm Period. The late medieval period saw a pronounced deterioration of weather. The climate became colder, and led to the early modern period being effectively a *Little Ice Age* (Grove, 1998).

As more reliable evidence is gathered, this simple picture has become less acceptable. It may be that the basis of the Medieval Warm Period is largely biological and phenological and if it existed in a climatological sense it was probably more complex than popularly believed (Ogilvie and Farmer, 1997). A widespread expansion of glaciers was apparant in the *Little Ice Age*, but this event covers different periods in different parts of the world and locations (Bradley and Jones, 1992). There were also times of relative warmth. It may be that rather than regarding this as a period of cold in Europe that stretched from the 16th to 19th centuries, to see it broken into shorter periods of colder weather.

In the middle of the 17th century, a new, growing interest for observing Nature developed in Italy with Galileo Galilei and the Academy of Cimento. This was created and supported by the Granduke of Tuscany Ferdinando II de' Medici and his brother Leopold. They played a leading role in creating the modern science which substituted the Aristotelic dogmas. In this context, the thermometer, the barometer and a hygrometer were invented and applied, creating the first meteorological network, called *Rete Medicea* (1654–1667), composed of six stations in Italy, plus Paris, Innsbruck, Osnabrück and Warsaw. In order to get comparable observations, the instruments were made by the same manufacturer in Forence, and then sent to the ten observing stations. The aim of this network, was to start with an international cooperation to gather climatic data that would have been useful for the future generations. When the meteorological instruments and practice became more mature, in 1723 a new meteorological network, with some medical interest, was proposed by

James Jurin, secretary of the *Royal Society, London* and the gathered data were published in the *Philosophical Transactions* from 1724 to 1735. A third international network (1776–1786) was proposed by Louis Cotte under the auspices of the *Societé Royale de Medicine, Paris.* This network had a brilliant start with 22 observing stations, but terminated under the pressure of the French Revolution. More successful was the fourth international network, proposed in the meantime by the *Societas Meteorologica Palatina* (SMP), Mannheim, and led by the Prince Karl Theodor von Pfalz and his secretary John Jacob Hammer. The SMP network was active in the period 1781–1792 and was composed of 39 stations: 35 in Europe (except the Iberian peninsula and the British islands), one in Russia, one in Greenland and two in the Massachussetts. The observations were published in the *Meteorologischen Ephemeriden* from 1783 to 1795. Giuseppe Toaldo was a contributor from Padova, who funded the *Giornale Astrometeorologico,* in which he published meteorological comments and statistical forecasts, especially aimed to the agriculture. The next milestone was an international agreement between France and Britain, signed in 1860 at Airy, followed by the international Congresses in Leipzig (1872), Vienna (1873) and Rome (1879), in which the basis were established for the birth of the World Meteorological Organisation.

Lamb (1966) suggested that alarms at sudden climate change in Europe at the end of the 18th century may have been a catalyst for the development of early observational networks, in France (1776) and Prussia (1817). However, there were other scientific pressures, and in Britain, a particular interest in climate change over the later part of the 18th century was driven by the effects of climate on agriculture. By the mid-19th century, Koppen attempted to use the instrumental data gathered by H.W. Dove to assess likely global temperature variation since 1750, while in England, Mossman (1897) undertook an analysis of the London's long term climate record.

The late 19th century also marked the beginning of widespread meteorological observations and an increase in northern hemisphere temperature still continuing today (see Fig. 1). The instrumental record suggests some very warm summers at the beginning of World War II. From that point, there was a short cooling and then a continuing rise since the 1970s that is attributed to global warming from increased

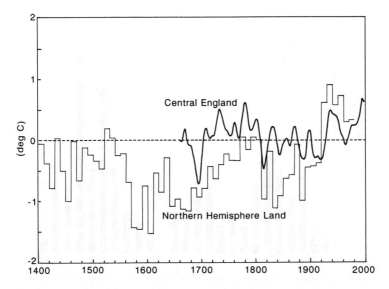

Fig. 1. Northern hemisphere temperature anomalies for the period 1400–1969 compared with the Central England Temperature Record (1660–1989). Anomalies calculated with respect to the period 1861–1960 (Jones and Hulme, 1997).

CO_2 and other greenhouse gases. The changes have been rather subtle and certainly the shifts in annual temperature have been just a few degrees, which would at first suggest only the slightest effect on building materials, but the mechanisms are such that even slight changes can have profound effects. However, we should note that specific locations show different historic changes (Fig. 1).

At a very local level, it is possible to examine climate change at specific buildings, e.g. the Trajan Column after the architect Apollodorus of Damascus built it in Rome in AD 105. The flooding of the River Tiber could be regarded as a climatic indicator of the dominant meteorological features of Rome, at least as a first approximation, and constitutes one of the longest series in the word. In fact, the documentary sources cover an uninterrupted period of 24 centuries because the flooding waters caused severe damage in Rome, so that the number of people killed, the most important damage, the interventions, and the level reached by the waters were carefully recorded. In the cold season, the Tiber floods may be considered as an index of

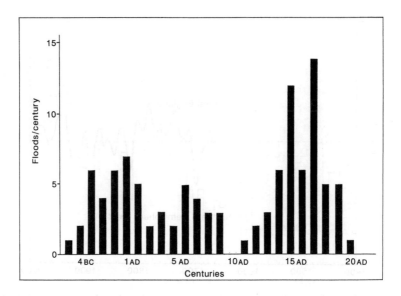

Fig. 2. Secular frequency distribution of major floods in the Tiber.

heavy rainfall, associated with cold inflows over the relatively warm Mediterranean waters. Floods occurred as shown in Fig. 2, with particular severity during some periods: at the beginning of the 2nd century BC, in the 1st century BC, in the 2nd AD, between the years 371 and 860 AD, at the end of the 15th century and in the 17th one. The worst periods were certainly the second half of the 1400s and the whole of the 1600s, two periods when there were various climatic perturbations. Although heavy floodings occurred during the Spörer Minimum of solar activity (AD 1416–1534) all of these anomalies cannot be attributed to the solar forcing, because in other cases the floodings (i.e. the "hydrological effect") occurred for example, before the solar minimum, as for the Maunder Minimum (AD 1645–1714) (Camuffo and Enzi, 1996).

2.2. Freeze-Thaw Cycles

Frost damage is a familiar form of stone weathering in temperate and polar climates. It leads to delamination of material from porous stone which becomes wet. The process can be treated as follows.

Water vapour becomes saturated when its temperature reaches the dew point. However, in micropores, the vapour tension in equilibrium with a curved liquid surface becomes lower and lower with increasing the concavity of the meniscus at the interface between the liquid water and the air. This effect is described by the Kelvin equation which states that the saturation pressure and therefore the relative humidity (RH) in equilibrium with a meniscus is calculated by means of the equation

$$RH(r) = 100 \exp(2\sigma V_m / rRT)$$

where σ is the surface tension of water, V_m is the molar volume of the liquid sorbate (i.e. $V_m = 18$ cm^3 for pure water), r is the radius of curvature of the meniscus ($r < 0$ for concave meniscus, as in the micropores), and R is the gas constant. As the radius of curvature r is the main parameter, the Kelvin formula states that the logarithm of the RH in equilibrium with the meniscus is inversely proportional to the radius of curvature of the water surface. RH levels greater than the above equilibrium value cause condensation, and lower RH evaporation. The effect is important only for very small r, i.e. $r < 0.1$ μm. The problem of surface moisture and condensation is very complex, and depends on the chemico-physical characteristics of both the atmosphere and surface. Total porosity, total pore surface, spatial association of pores that may form pockets and necks, pore size, pore form, pore radii distribution are important variables in the weathering of stones. Stones are characterised by a wide variety of pores and necks, with different shapes and sizes, which range from angstroms to millimetres. The porosity may change with time, especially in the subsurface layer for migration of salts, leaching, dissolution, erosion, and other physical, chemical and biological actions.

When the temperature drops below zero, freezing-thawing cycles can develop, and the pressure exerted by the ice crystals in the pores of materials may have disruptive effects. However, the curvature of the meniscus exerts an important effect between all the phases: the solid, the liquid and the gaseous phase, so that the Kelvin effect in lowering the freezing point in micropores can be calculated. Although a number of equations exist with small empirical corrections (Fagerlund, 1975; Clifford, 1981; Iribarne and Godson, 1986), the chief equation for the Kelvin freezing point depression ΔT_f is

$$\Delta T_f = T_f \ (2 \ M \ \sigma_{sl}) \ / \ (r \ \rho_s \ \Delta H)$$

where the labels s, l refer to the solid and liquid phases, $\sigma_{sl} =$ 17.2 erg/cm^2, M is the molar mass of the substance, ρ_s is the solid density, ΔH is the molar heat of fusion, and $L_f = \Delta H/M$ is the latent heat of fusion. For water, $\Delta H = 18 \ (80 - 0.5 \ \Delta T_f)$ cal/mole and $0.5 \ \Delta T_f = \Delta(c_w - c_i) \ dT$, where c_w and c_i are the specific heats of water and ice at p = const. Experimental observations lie between the data computed with the two surface tensions, i.e. s_{sl} for the solid-liquid interface and σ_{sg} for the solid-gas interface, showing that often the ice is covered with a water film.

The formula derived from the Kelvin equation has some important consequences (for a derivation and further discussion, see Camuffo, 1998). (1) Many micropores are filled with water although the relative humidity in the air is far from saturation. The smaller the pore, the higher the probability of being filled. (2) Condensation first occurs in the finest pores and then in the larger ones. (3) In micropores, the freezing point is lowered by the Kelvin effect and the ambient temperature must fall below zero in order to reach freezing. (4) Freezing occurs first in larger pores and then in the finer ones. However, if the pores are at least partially filled with water and the connections between the largest and finest pores are filled, the ice crystals will start to form in the largest pores; when all the liquid water has changed phase, it is transferred from the smallest pores to feed the growth of the ice crystals in the largest pores, so that the finest pores act as a reservoir of water and are emptied. Should this happen, the freezing-thawing damage preferably occurs in largest pores, especially in the case of stones characterised by the presence of both large and fine pores.

If we make the assumption that frost damage is related to the number of freeze-thaw cycles stone suffers, it is possible to assemble a record of how this stress varies with time. Unbroken instrumental records taken in Rome since 1782 provide good information on the meteorological situation over the last two centuries. The freeze-thaw cycles are not as frequent today as in the past (it reached the maximum frequency in the mid-19th century).

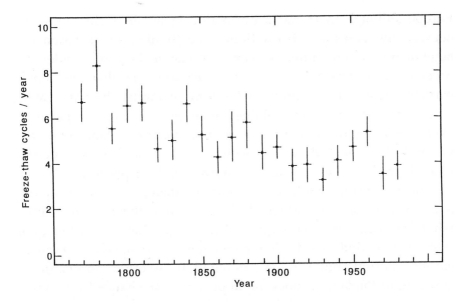

Fig. 3. Decadal means of the freeze-thaw cycles in Central England.

The number of freeze-thaw cycles each year is a sensitive function of average temperature and varies across Europe. At warmer locations, there are no freezing events, while at more moderate climates it increases, although under very cold conditions it is reduced because it remains very cold most of the year. Indeed, under polar conditions there may be no thaw events at all, but such climate regimes have not traditionally been rich in built heritage. The change in freeze-thaw cycles with time is shown for Central England in Fig. 3. We see a substantial increase in events before 1800. Interestingly, this marks an important period or urban development in many European cities and climatic considerations may well have influenced the way in which durable stones were chosen for buildings.

2.3. Storms and Precipitation

Long term trends in precipitation are less well understood than temperature, partly because of the high spatial and interannual variability. Over the long duration, it may have varied only slightly in

Europe although there was a dry period in much of the continent between the 1780s and 1810s (Jones and Bradley, 1992). Given the large degree of interannual variability, it may be best to consider the occurrence of extremely wet or extremely dry years that can impart particular stress on buildings, by lowering or raising the water table.

The time-of-wetness of buildings often emerges as a key factor in likely dissolution of stone surfaces. We have some knowledge of evaporation back to the 17th century which could be used to estimate the potential for changes in time-of-wetness over long periods.

A parameter that may be more important than precipitation amount is the wind-driven rain which drives water into the building fabric. Wind-driven rain is often expressed as the product of precipitation amount and wind speed. This parameter is not measured even in modern period, but can be estimated from other observations. However, in England, it does not appear to have changed very much in the last century or so.

Camuffo *et al.* (1999) have estimated the change in storminess along some Mediterranean coasts from descriptive records. It is well known that in coastal areas, buildings deteriorate more rapidly than at inland sites. The damage is chiefly attributed to the presence of sea salt, which is generated and transported inland by offshore winds, especially when they are strong. Sea spray deposits on buildings, trees and other obstacles, but its concentration decreases rapidly with distance from the coast.

Many million of tons of salt are annually transferred from the oceans to the atmosphere each year. In normal conditions, marine aerosols mainly originate from bubbles in whitecaps, i.e. from the disintegration of the film that forms the cap of the bubble, or from the central jet that forms when the cavity of the bubble collapses and the jet drops seem to be the main source of the so-called giant salt nuclei. During storms, the gale force wind is much more effective in breaking waves, transferring large amounts of seawater to the atmosphere as sea spray. In coastal areas, the deposition rate of sea salt has been evaluated in the range between 21 and 3200 μg cm^{-2} yr^{-1}. Extreme events can drive rain deep into building materials or carry salt onto the upper parts of structures in maritime locations. The salt can contribute to salt damage under humidity changes or pollutant deposition.

The wind itself can cause direct damage to structures. Lightning is also capable of damaging buildings. Historical changes in frequency of these are known at a number of locations and there may be hints of increases in frequency at urban locations (e.g. Mossman, 1897). A list of historical buildings hit and ignited by lightning in Venice since 1388 shows the total number of strikes is relatively low [24, although in three cases more than one bell tower was hit during the same thunderstorm (Camuffo *et al.*, 1999)]. In the course of centuries, the S. Marco bell tower was hit 11 times, as it is the tallest bell tower (i.e. 100 metres; all the other bell towers are much lower and were hit only once or so). The first conclusion is that only the taller bell towers were at risk, not the buildings of normal height. In 18 May 1776, a lightning rod was applied to the S. Marco bell tower, and then successively to the other relatively tall buildings as well. After the introduction of lightning roads to the taller towers, this risk practically disappeared. This pinpoints once more the importance of preventive measures.

2.4. Biological Factors

There are secondary effects of climate on buildings via biological interactions. We are unsure of the ways in which temperature and humidity changes may affect the growth of microbiological organisms on the surface of buildings. Although this may be important, it is possible that air pollution is an even more important factor in mediating biological attack on building materials (Urzi *et al.*, 1999).

3. Changes in Air Pollution

The effects of air pollution on buildings have been so important in the 20th century, that there has been much less interest in the effect of long term changes in air pollution on the rate of deterioration of building stone. Nevertheless, European architecture has suffered enormously from centuries of exposure to atmospheric pollution. By the 17th century, there was sufficient understanding of the damaging effects of smoke that sporadic attempts were made to solve the problems it caused.

3.1. History of Air Pollution

Urban air pollution was known from classical times. Rome, with a population in excess of one million, consumed enormous amounts of wood, such that temples were blackened by soot and the health of sensitive individuals affected. Citizens had some redress under Roman law if their property rights suffered interference and legal decisions against the production of smoke are known (Brimblecombe, 1987a).

Patterns of fuel use, and more particularly, changing fuel types seemed to trigger concern about urban air pollution in the past. Especially noticeable was the transition to coal as a fuel. There are important differences in the historical development of fuel use, control policies and the climate of various European cities. The changes to differences in the evolution of air pollution experienced.

It is possible to estimate the changes in pollutant concentration occurring within a city over the centuries by modelling the fuel use. This has been done for London (Brimblecombe, 1977b), York (Hipkins and Watts, 1996) or American cities such as Boston (Sherwood and Bambaru, 1991) or Pittsburgh (Davidson, 1979). Alternatively, there been studies of the exposure of individual monuments such as York Minster (Bowler and Brimblecombe, 1991) or Trajan's Column (Camuffo, 1993). Studies of the long term change in air pollution may be based on the history of fuel use and examine civic and legal records, travellers' descriptions and medical writings. In some cities, scientific measurements of air pollutants began in the late 19th century.

London has an interesting air pollution history because the shift from wood to coal, as a fuel, began so early. While the transition there began in the 13th century, it was as late as the 20th century in some parts of Europe. When fuel is changed, it may be accompanied by a noticeable change in the quality of the smoke. The unfamiliar smell of coal smoke led to early fears about the health risk, through the belief that disease was carried in malodorous air (miasmas). Coal in 13th century London was mostly used for industrial processes such as lime-mortar production and it was not until the late 16th century, with the widespread construction of household chimneys, that the fuel began to be used domestically (Brimblecombe, 1987b). At this time, we can find early observations of damage. John Thornbrough, the Dean of York Minster, suggested that coal should be desulphurised to make it less offensive. In the early years of the 17th century, King James I complained of the terrible damage done to the fabric of St. Paul's cathedral and Archbishop Laud fined the brewers upwind for damaging the same building. The diarist John Evelyn described the effects of smoke in his book Fumifugium of 1661, which told how air pollution made churches and palaces look old, clothes and furnishings foul, and paintings yellow.

3.2. Early Acid Rain and Dry Fogs

Lucius Annaeus Seneca (4 BC?–AD 65) gives us a clear evidence about air pollution and acid rain, discussing the reasons why water has a different taste in different places: "The first is the type of soil from where it comes. The second depends on the soil, whether it arises from its transformation. [He believed that the springs were the result of a direct transformation of the soil into water.] The third from the air which is transformed into water. The fourth due to the corruption of the water when it is contaminated by polluting agents" (*Nat. Quaest.* III, 20, 1). The observation that dripping droplets or running water eroded stones (Lucretius, *De Rerum Natura* 1, 314; Ovid, *Ars Amatoria* 1, 475–476) was another indirect mention of acid water in ancient Roman times.

However, the early scientific observations of atmospheric acidity (Camuffo, 1992) can be found only after the development of the science in the 1600s which followed the example by Galileo and the *Accademia del Cimento*. The physicist Honoratus Fabri published in 1670, an extensive treatise of physical and meteorological sciences in which he devoted a whole chapter to the problem of impure water. Of course, several types of impure rainfall are considered, from the rain which seems like blood from the inclusion of reddish lime, to a type of corrosive rain that damages the agriculture: "A certain type of rain causes spot damage to the fruit, and sometimes the whole fruit is burned, e.g. the rain that falls during the hottest hours in the middle of the day, with rarer but larger drops; as it evaporates rapidly leaves some acid particles on the fruit. Due to their great capacity to acidify, the fruit is burned, dessiccated, or undergoes spot lesions, or some other damage" (Fabri, 1670). Of course, at that time, and in the hot season, when domestic heating was absent, we cannot attribute this atmospheric acidity to anthropogenic activity (Camuffo and Daffara, 1999). The only possible cause for background or widespread pollution was attributable to natural factors, and volcanoes were the only possible source of powerful emissions. In fact, a major eruption of the volcano Etna (38°N, 14°E) occurred in 1669, the year preceding the publication of the treatise by Fabri.

Sixty years later, Hermannus Boerhaave, wrote in 1731, and published in Venice in 1749, an important treatise containing the basic ideas concerning air pollution, atmospheric scavenging and acid rain. A sample from the text, dating more than two and half centuries ago, reads as: "The soot deposited by the smoke at the top of very high chimneys by the combustion of plants, and in chemical distillation it is transformed.... These fossil compounds called sulphurs each time they are burnt, are so completely dispersed in the air that they disappear: the saline and acid parts are transformed into suffocating emanations.... The same sulphur, if it is alone, is transported away by the air, once it has been transformed by heat into minute particles.... Also the winds themselves, when they transport the air with all it contains from place to place, always move materials from the places where they originated, and gradually change the composition of the air, continually removing substances belonging to one place and giving back always those which have been taken up a short time before.... The rain could be called a real and proper agent that washes the atmosphere, and that collects all types of substances that are suspended in the air. However, the rain that falls after long periods of dryness is very different from that which falls during the rainy period. The water of storms is different from other waters, without speaking of the effects of the winds, which carry atmospheric water from place to place; for this reason, with persistent winds which always blow violently from one zone only, the rain is full of emissions belonging to that remote zone. The winds perturb these substances, they mix them with the rains, they disperse them, carrying them from distant places and sometimes form a mixture which is so wonderful that it is a joy when such rains irrigate the plants and the fields. The frequent observations of the rains falling from the sky during the hot season, and collected in vases which have been carefully cleaned and then observed for a long time, shows that it putrifies spontaneously being transformed into a fetid and putrid lattice.... The rain which falls in the hot seasons, with strong winds, in urban areas, in low lying [or humble] places consists of very impure water, where putrification occurs, where animal and vegetable products, and products of any other nature are dispersed in the air by man in great quantities and in various ways. And the rain coming from such places will be even more impure if the air has been very nebulous, dense and fetid, such as to carry a horrendous odour to the nostrils, air and emissions that are harmful to the lungs" (Boerhaave, 1749). The same author also had a clear idea about the role of the volcanoes as agents of acidification, and of the long-range transport of volcanic aerosols: "We should remember that the winds carry in the air, like waves, the sand from Egypt and Lybia, and transport the ashes of Etna over long distances. The dusts of Vesuvius were dispersed in the air for over one hundred miles.... The ashes erupting from the volcano (Vesuvius, 41°N, 14°E) were found to have been transported over one hundred miles away" (Boerhaave, 1749).

Besides acidic rain, volcanic emissions also caused *dry fogs* (Stothers and Rampino, 1983) which consist of a mist composed of gases and aerosols, foul smelling, and characterised by a reddish colour. These reddish, maloderous fogs which do not wet surfaces (for this reason they are called "dry"), appeared at dawn and sunset, sometimes persisting even into the middle of the day. They appeared especially in the summer (instead of the cold season) and were often accompanied by red dusks, weak sun, solar and lunar haloes, which caused damage to the vegetation and brought in their wake, epidemics. The most famous event, which affected for months the whole of Europe, occurred in 1783 for the eruption of the Laki volcano, Iceland (Thodarson and Self, 1993; Grattan and Charman, 1994; Grattan and Brayshay, 1995).

The damage caused by these fogs on vegetation were so important that they were described in treatises on agricultural meteorology: "There are 'fogs which wet' bodies they touch and they are in the majority; but there are also dry fogs, whereby the hygrometer moves towards 'dry'.... Among the different species of fogs we can, for agricultural purposes, reduce them to two only — the damp fogs and the dry fogs. The damp fogs favour agriculture, as long as the plants are not near maturation. They bring, like the dew, humidity and substances in solution.... The dry fogs, form a haze through which the sun appears to be a bright red, not a real fog. These hazes.... spread over Europe in 1783, where they lasted for months" (Orlandini, 1853). In Italy, the most important dry fogs occurred in the years: 1374, 1465, 1499, 1587, 1592, 1648, 1682, 1710, 1734, 1735, 1775, 1780, 1783, 1785, 1786, 1791, 1794, 1802, 1803, 1805, 1812, 1814, 1816, 1819, 1821, 1824, 1831, 1866, 1869, 1883/4, 1886 when chroniclers and scientists described fouling, reddish fog, leading to damage to vegetation, people illness, weather anomalies, and other unpleasant effects (Camuffo and Enzi, 1995).

3.3. Early Descriptions of Damage

Damage to buildings in the past can be established from descriptions of the state of the exterior, most particularly those of architects from the beginning of the early modern period. Nevertheless, classical descriptions survive and Horace complained of smoke blackening to the temples in Rome. In religious buildings, smoke encrustations might be seen as desecration, but secular buildings seem to require an appreciation of architecture as an art form. In 17th century England, there was a predominance of interest in the damage to religious buildings, but this should not be taken to mean that it was primarily a religious concern. The complaints of King James I about smoke damage to St. Paul's Cathedral came from someone who also complained about the besooted state of English kitchens.

There was a clear recognition that coal smoke was a most important cause of damage. Keepe (1683) tells us that at Westminster Abbey you see "the skeleton of a church than any great comeliness in her appearance, being so shrivelled and parcht by the continual blasts of the northern winds, to which she stands exposed, as also the continual smoke of the sea-coal which are of a corroding and fretting quality...."

At York Minster, many houses whose smoke damaged the fabric, were demolished under the guidance of the Dean during the early 1700s. At the same time, smoke and scurff [an archaic word for a saline or sulphurous encrustation] were removed from the stonework during cleaning operations (Drake, 1736). The aim of the repair work was to "stop up all the Cracks, Flaws and Perishing of the Stones, with excellent Cement and Mortar, that.... this Fabrick might yet bid Defiance to Time and Weather for many succeeding Generations". In a similar vein, Nicholas Hawksmoor concluded his repairs at Westminster "will stand 1000 years" (Westlake, 1863). Architects involved in such large scale restorations had yet to realised the impermanence of their repairs. Interest gradually spread to dwellings, and in London of the 18th century, the rate of darkening to the paintwork of some houses was so rapid that repainting needed to be frequent (Malcolm, 1770).

3.4. Industrial Development and Pollution

The late 18th and 19th centuries saw an enormous amount of industrialisation across Europe. The development of the steam engine and the intense urbanisation created very special problems. In England, civic administrators soon became aware that their old legislative frameworks were inadequate to cope with pollution on this much enlarged scale (see Manchester's problems in Bowler and Brimblecombe, 2000a). The problems continued throughout the 19th century with the development of much sanitary legislation concerned with the health of urban populations. These beginnings to modern pollution environmental legislation reflect a desire for cleaner air, but it was often thwarted by weak laws that were erratically applied and poor control technology. The regulations usually failed to achieve substantial abatement of smoke in the 19th century. However, the framework initiated pressure on industry, either their workers had to be trained or they had to be replaced by automated stoking. Gradually, the pall of smoke over cities lessened, though it was soon to be replaced by different pollutants originating from mobile sources.

3.5. Victorian Approaches to Damage

The cities of the Victorian period were so badly polluted that there was much awareness of the need to choose building stones carefully for new structures. Major building works often began with the formation of a select committee of "experts" — architects, chemists and stonemasons, who could then deliberate over the right choice of stone to resist the polluted atmosphere. The Palace of Westminster (1837–1870), constructed after much thought on the ideal stone, began to manifest signs of stone decay in 1847. This showed that "expert" opinion on such matters did not always guarantee that buildings would not succumb to the enhanced weathering (Scott, 1861). At Westminster Abbey, a committee of architects and chemists reported on the best means of preserving the stonework (Carpenter, 1966). They used a scientific approach and concluded that disintegration resulted from acid vapours, particularly sulphuric acids, although they also saw hydrochloric acid as a contributor.

Where thoughtful choice of stone failed to provide adequate protection against the polluted atmospheres, there was some enthusiasm with induratating the stone with protective solutions. Materials such as linseed oil, beeswax, paraffin, gums and resins were all tried, but these techniques were ultimately discredited. Inorganic materials were also applied to the stone. Baryta-water (barium hydroxide) was thought to prevent damage (Jackson, 1901; Church, 1901). Professor Desch also recommended "magnesium silicofluoride", and for the more porous sections of the stonework, "double silicofluoride which contains zinc" to prevent the growth of vegetation at the Abbey (Desch, 1921). These techniques also failed and the early 20th century applications of silicates had produced grey patchy discolouration pitted with small spalls.

On hindsight, it is now accepted that the stone indurators of the period 1850s–1920s had often wreaked more havoc than centuries of decay and weathering (Ashurst, 1985). Writers of the Victorian period had many theories to explain stone decay so prevalent in their cities. The erratic classical scholar Paley (1878) argued that the blackening of St. Paul's cathedral was really due to the presence of lichens and should be treated with the biocide copper sulphate. While at York, the vulcanologist, Tempest Anderson, wrote that stone decayed from both air pollution and biological attack (Anderson, 1910). However, increasingly smoke abatement was seen as the most appropriate solution, and was strongly advocated by ecclesiastical officials and architects alike.

3.6. Architectural Responses

The Victorians wrote that coal smoke was imparting a "new colour" to towns, but aesthetic opinions about this were divided. Some felt *Time* and *Nature* effected a "softening" appearance on urban buildings and accepted the effects of smoke, while others claimed that all that was required to bring colour back to London was to regularly wash buildings with water. Yet others argued that architects should aim at designs that would become "beautifully coloured by Nature charged with smoke" (e.g. Ricardo, 1896).

The problem was deeper than the simple mechanical response hinted at above. By late Victorian times, the issue of air pollution encompassed many aspects of the design of urban buildings. Architects recognised that within cities there were problems of overcrowding, access to light, air pollution, stone discolouration, indoor pollution and inappropriate building styles. The urban environment needed to be an aspect of their designs. At the end of the 19th century, the range of responses included (Bowler and Brimblecombe, 2000b):

(1) abandonment of gothic and the promotion of less detailed classical styles not so suceptible to air pollutants,
(2) careful choice of pollution-resistant stone and glazed materials,
(3) extensive provision of clear window glass to let in more light,
(4) adoption of electric light and air filtration to lower pollution indoors, and
(5) development of dust excluding interior fittings to protect books, etc.

3.7. The Twentieth Century

The present century has seen a gradual decline in the use of coal in cities of Europe and North America. This has been replaced by oil, gas or electricity (often generated outside the urban area). These changes have led to the emergence of an entirely new kind of air pollution — photochemical smog. This arises when volatile organic compounds (primarily from liquid fuels), sunlight and nitrogen oxides interact. Ozone is the most characteristic compound found in these smogs. Photochemical smog first began to be noticed in Los Angeles during the Second World War and was actively studied in the 1950s when the mechanism of their formation was finally established. At first, it was assumed that these types of smogs would be restricted to warm sunny climates, but soon they began to be found at most industrialised locations.

Ozone is harmful to organic materials, especially those with double bonds. However the oxidising atmosphere of photochemical smog is also able to oxidise nitrogen oxides to nitric acid, which can damage metals and stone. Ozone may also enhance sulphate formation on building surfaces.

In many cities, the potential for damage by sulphur-derived acidifying substances appears to have lessened in recent decades. Vehicle fuels (particularly now low sulphur diesel oils are available) are not as significant a source of sulphur as coal. Many had thought that the improved urban air quality would be matched by proportionately lower rates of degradation to historic buildings. The rates of stone damage may not have declined as much as hoped, although at St. Paul's in London, limestone weathering seems to occur at a fairly constant rate (Trudgill *et al.*, 1991). In Britain, ferrous metals appear to be less rapidly corroded than in the earlier part of the century (Butlin *et al.*, 1992).

At the end of the century, many cities find that diesel soot is an important source of soiling (see Chapter 10) reflecting yet another fuel change of the 20th century.

3.8. Economic Analysis

Economic materials provide a further potential source of evidence regarding historical changes in building damage and are especially relevant as we are very much concerned about the costs that air pollution imposes on contemporary buildings. Many cathedrals have kept detailed accounts of their expenditure for many centuries. Their fabric accounts record money spent in both construction and repair of cathedral property. In some cases, these accounts begin at the foundation of the building and may be detailed enough, in later periods, to preserve all the original vouchers accounting for the expenditure on individual items.

In England, cathedral income was rarely sufficient to undertake all desirable repairs and was often dependent on large gifts from wealthy patrons to restore dilapidated parts of the building. In more recent times, religious authorities have been faced with great structural change. Under an 1840 Act of Parliament, cathedrals lost wealthy benefactors as stipends were abolished. Since then, there was a frequent need to appeal for money from the Ecclesiastical Commission. Currently, cathedrals have a broad approach to funding maintenance. The importance of public involvement was stressed with the formation of societies, such as the Friends of Norwich Cathedral or Friends of York Minster. These have widened the base for support, interest and concern for financial problems. Casual visitors to cathedrals are usually reminded of the great running costs and encouraged to make donations during their visit or attracted to purchase souvenirs in cathedral shops.

The annual expenditures from the fabric fund of four English cathedrals (York Minster, Westminster Abbey, Norwich Cathedral and Exeter Cathedral) illustrate the increases in expenditure across the centuries (Brimblecombe *et al.*, 1992). The amounts have been corrected for inflation using the prices of consumables established by Phelps-Brown and Hopkins. Figure 4 shows expenses corrected to the 1447 value using their indices. We see that at York, there were declines in fabric expenditure from the 14th through to the 16th centuries as the Minster was finished. However, in recent centuries the existing records suggest that the rate of inflation in the price of consumables is insufficient to explain the increasing costs in maintaining the fabric. This observation is similar at other sites. An analysis of this data suggests, rather surprisingly, that the rate of increase was somewhat greater over the period 1700–1850 (especially

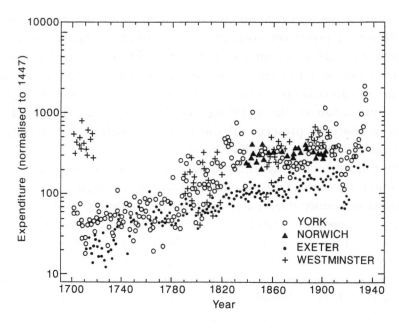

Fig. 4. Annual expenditure from the fabric funds of four English Cathedrals corrected for inflation.

for York Minster) than for 1850–1930. The lack of a rapid rise in expenditure in the late 19th century, a time of rapidly increasing air pollution in UK cities, is rather unexpected.

With the exception of the early data from Westminster, the changes over time show a remarkable similarity in the rise in fabric expenditure at all the sites. The most immediate explanation for this would be to assume that the pressure on fabric expenditure was rather similar across the nation. However, the air pollution histories of these locations are quite different. Westminster, and to a lesser extent York, have been coal-burning cities for many centuries. Norwich and Exeter have been less polluted, so their cathedrals would have been expected to show lower levels of expenditure.

These differences are not evident, so it may be that we cannot attribute the enormous costs in mantaining cathedral fabric simply to air pollution. There are social interactions that make expenditure a good deal more complex than this. Catastrophic costs may be driven by the importance now given to culturally significant buildings. Restoration is no longer work for the local jobbing builder or the mason and now requires specialist skills and expertise far beyond that traditionally expected of local artisans. This is so important that some cathedrals give evidence of a rebirth of local skills in their workshops. Possibly "major restoration programmes", with considerable profile and media interest, have also increased public relations costs, consultant fees and other items not traditionally requiring high expenditure.

We have to remember that "Cathedrals are getting older". Air pollutants clearly cause great damage to materials outdoors. With this few would disagree, but it may be more difficult to relate the catastrophic increase in costs over the last century or so to air pollution alone. If the vast costs are related to factors other than air pollution, it is likely that these buildings will continue to require high expenditure despite the improvements in the corrosive nature of urban air.

As Peter Foster (1985), surveyor to the fabric of Westminster Abbey for many years wrote: "I have endeavoured to take the best advice and to act upon it. To blame everything on acid rain, atomic waste, aerosols or whatever is the latest 'Green' view of the world, is surely 'simplistic'...."

3.9. Archeometric Sources of Information

The historical analysis above has restricted itself to relatively conventional approaches using documentary data to establish changes in air pollution over time. Documents can be supplemented by pictorial sources of information. Roger Lefevre has shown that paintings of buildings can often record the patterns of blackening. Viles in her studies of building damage in Oxford (Viles, 1996) found early photographs useful, and Tang *et al.* (Chapter 11), also used photographic methods.

Modelling is a useful approach to establishing past pollution concentrations (Brimblecombe, 1977b). Where trace gas concentration can be estimated over a long time period, it can be used to estimate the deposition of pollutants to buildings. Figures 5(a) and (b) show

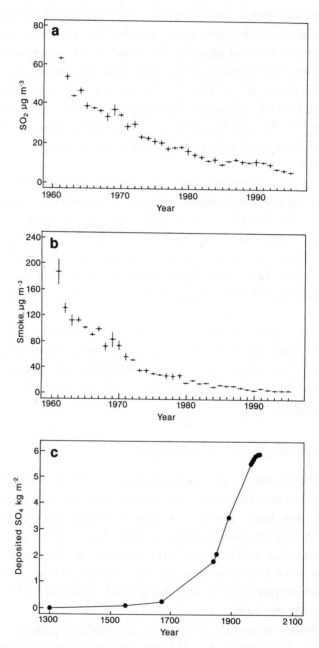

Fig. 5. (a) Measured sulphur dioxide concentrations in York (sites 4–9 normalised to site 4) and (b) smoke concentrations. (c) Cumulative long-term deposition of sulphate to the minster assuming a deposition velocity of 1 cm/s.

measured pollution concentrations in York. Figure 5(c) shows the cumulative deposition of sulphate to a building such as York Minster assuming a deposition velocity of 1 cm/s (using Hipkins and Watts, 1996). These calculations suggest that much of the pollutant deposition had already occurred before the 20th century and that the current rate of accumulation is relatively small.

Pittaluga (1994) and Cesarino and Pittaluga (1995) studied the past pollution by means of a stratigraphic analysis of the gases which have been absorbed and the particles which are sticking on the several layers of paint coatings found on the historical buildings in Genova. Knowing the date of each painting layer, it was possible to date the intervals of time in which each deposit of pollutants was formed. Ausset *et al.* (1998) have studied the history of air pollution and building damage through a direct examination of old crusts on stone. Ancient crusts of buildings have revealed unburnt wood fragments along side microspherules from wood combustion. Within crusts from the Arch of Constantine is possible to trace a decline in gypsum concentration as one penetrates from the recent crust surface into older layers. This is taken to represent an increase in the sulphur burden on Rome's air with time.

4. Recent Changes in Modern Pollutants and Materials

Our view so far has considered the changes of the past. Although it is often suggested that we study history to learn about the future, we should be cautious in our view of parallels for the future. There has been an enormous change in the types of materials used for buildings. We should not only consider structural elements, but recollect the vast range of polymers now in use as coatings and sealants. Changes in the types of air pollution have been equally large as we have moved to atmospheres that are often controlled by photochemical oxidation. Broader atmospheric changes have also thinned the ozone layer, so ultraviolet fluxes have also increased.

One should note that polymers are often sensitive to ozone. Although the changes in UV flux are small, many polymers are sensitive to light and the photodegradation can be enhanced by some polyaromatic hydrocarbons present in the atmosphere.

We should also reflect on the suitability of modern architecture in relation to the urban environment in which it is found. Its simple lines, limited surface detail and a preoccupation with glass, tile, metal and generally resistant materials may be easy to clean. However, sealants are needed that resist degradation, and the deposits of diesel soot from the contemporary is also particularly visible on the light-coloured surfaces that are much in favour.

References

Anderson T. (1910) The decay of stone antiquities. *Museums J., York.*

Ashurst J. (1985) *Archit. J.* (21/28 August), 40–43.

Ausset P., Bannery F., Del Monte M. and Lefevre R.A. (1998) Recording of pre-industrial atmospheric environment by ancient crusts on stone monuments. *Atmos. Env.* **32**, 2859–2863.

Boerhaave H. (1749) *Elementa Chemiae.* Coleti, Venice, Vol. 1.

Bowler C. and Brimblecombe P. (1991) Long term stone decay at York Minster. *Europ. Cult. Herit. Newslett. Res.* **5**, 47–57.

Bowler C. and Brimblecombe P. (2000a) Control of air pollution in Manchester prior to the Public Health Act, 1875. *Env. Hist.* **6**, 71–98.

Bowler C. and Brimblecombe P. (2000b) Environmental pressures on the design of Manchester's John Rylands Library. *J. Design Hist.* **13**, 175–191.

Bradley R.S. and Jones P.D. (1992) Climate since 1500. In *Climate Since 1500*, eds. Bradley R.S. and Jones P.D. Routledge London, pp. 1–16.

Brimblecombe P. (1977) London air pollution 1500–1900. *Atmos. Env.* **11**, 1157–1162.

Brimblecombe P. (1987a) The antiquity of "smokeless zones". *Atmos. Env.* **21**, 2485.

Brimblecombe P. (1987b) *The Big Smoke.* Methuen, London.

Brimblecombe P. (1992) A brief history of grime. Accumulation and removal of soot deposits on buildings since the 17th century. In *Stone Cleaning*, ed. Webster R.G.M. Donhead, London, pp. 53–62.

Butlin R.N., Coote A.T., Devenish M., Hughes I.S.C., Hutchens C.M., Irwin J.G., Lloyd G.O., Massey S.W., Webb A.H. and Yates T.J.S. (1992) Preliminary results from the analysis of metal samples from the national materials exposure program (NMEP). *Atmos. Env.* **26B**, 199–206.

Camuffo D. (1990) *Clima e Uomo.* Garzanti, Milano.

Camuffo D. (1992) Acid rain and deterioration of monuments: how old is the phenomenon? *Atmos. Env.* **26B**, 241–247.

Camuffo D. (1993) Reconstructing the climate and the air-pollution of Rome during the life of the Trajan Column. *Sci. Total Env.* **128**, 205–226.

Camuffo D. and Enzi S. (1995) Impact of clouds of volcanic aerosols in Italy in the past centuries. *Natural Hazards* 11, 135–161.

Camuffo D. and Enzi S. (1996) The analysis of two bi-millenary series: tiber and po river floods. In *Climatic Variations and Forcing Mechanisms of the Last 2000 Years*, NATO ASI Series, Series I: Global Environmental Change, Vol. 41, eds. Jones P.D., Bradley R.S. and Jouzel J. Springer Verlag, Stuttgart, pp. 443–450.

Camuffo D. (1998) *Microclimate for Cultural Heritage*. Elsevier, Dordrecht, p. 415.

Camuffo D. and Daffara C. (1999) Archaeometry of air pollution: urban emissions in Italy during the 17th century. *J. Archaeol. Sci.*, in press.

Camuffo D., Daffara C. and Secco C. (1999) Fire risk in Venice, Italy, during the last millennium, in preparation.

Camuffo D., Secco C., Brimblecombe P. and Martin-Vide J. (1999) Sea storms in the Adriatic Sea and the Western Mediterranean during the last millennium. *Climatic Change*, submitted.

Carpenter E. (1966) *A House of Kings*. John Barker, Wiltshire.

Cesarino A. and Pittaluga D. (1995) Superfici intonacate: inquinamento, degrado e pulitura nei secoli XVI-XIX. *Scienza e Beni Culturali* 11, 127–138.

Church A.H. (1901) *Chapter House Westminster. Prof. Church's Report on the Treatment of Decayed Stone*. Westminster Abbey Library, City of Westminster, 28 May 1901.

Clifford J. (1981) Properties of water in capillary and thin films. In *Water: A Comprehensive Treatise, Vol. 5, Water in Disperse Systems*, ed. Franks F. Plenum, New York.

Davidson C.I. (1970) Air pollution in Pittsburgh: a historical perspective. *J. Air Pollut. Control Assoc.* 29, 1035–1041.

Desche C. (1901) *Note on the Preservation of Exterior Stonework of Westminster Abbey*. Westminster Abbey Library, City of Westminster, 25 July 1921.

Drake F. (1736) *Eboracum*. London.

Fabri H. (1670) *Physica, id est Scientia Rerum Corporearum*. Anisson, Lyon, Vol. 3.

Fagerlund G. (1975) Determinations of pore-size distribution from freezing point depression. *Materiasux et Constructions* 6, 215–225.

Grattan J. and Charman D.J. (1994) Non-climatic factors and the environmental impact of volcanic volatiles: implications of the Laki fissure eruption of AD1783. *The Holocene* 4(1), 101–106.

Grattan J. and Brayshay M. (1995) An amazing and portentous summer: environmental and social responses in Britain in the 1783 eruption of an Iceland volcano. *Geograph. J.* 161(2), 10.

Grove J.M. (1998) *The Little Ice Age*. Methuen, London.

Hipkins S. and Watts S.F. (1996) Estimates of air pollution in York, 1381–1891. *Env. Hist.* 2, 337–345.

Iribarne J.V. and Godson W.L. (1986) *Atmospheric Thermodynamics.* Reidel, Dordrecht.

Issar A.I. (1990) *Water Shall Flow from the Rock.* Springer-Verlag, Berlin.

Jackson H. *Mr Jackson's Report Upon the Examination of the Stone Work of Westminster Abbey.* Westminster Abbey Library, City of Westminster, May 1901.

Jones P.D. and Bradley R.S. (1992) In *Climate Since A.D.1500,* eds. Bradley R.S. and Jones P.D. Routledge, London, 1992 pp. 649–665.

Keepe H. (1683) *Monumenta Westmonasteriensa.* London.

Lamb H.H. (1966) Our changing climate past and present. In *The Changing Climate,* ed. Lamb H.H. Methuen, London, pp. 1–20.

Malcolm J.P. (1770) *Anecdotes of the Manners and Customs of London During the 18th Century.* London.

Mossman R.C. (1897). The non-instrumental meteorology of London 1713–1896. *Quart. J. Roy. Meteorol. Soc.* **23**, 287–298.

Ogilvie A. and Farmer G. (1997) Documenting the medieval climate. In *Climate of the British Isles, Past Present and Future,* eds. Hulme M. and Barrow E., Routledge, London, pp. 112–133.

Orlandini O. (1853) *Trattato completo di Meteorologia Agricola.* Garinei, Florence, p. 299.

Paley F.A. (1878) *Science Gossip* **14**, 170–171.

Pittaluga D. (1994) Nuovi sviluppi interdisciplinari nell'analisi stratigrafica delle superfici intoinacate: esperienze e prospettive. *Scienza e Beni Culturali* **10**, 123–133.

Pritchard J.B. (1969) *Ancient Near Eastern Texts Relating to the Old Testament.* Princeton University Press, Princeton, N.J., p. 710.

Scott G.G. (1861) *Gleanings from Westminster Abbey.* J.H. and Jas. Parker, London.

Sherwood S.I. and Bumbaru D. (1991) Historical urban SO_2 levels. *APT Bull.* **23**, 72.

Stothers R.B. and Rampino M.R. (1983) Volcanic eruptions in the Mediterranean before A.D.630 from written and archaeological sources. *J. Geophys. Res.* **88**(B8), 6357–6371.

Thodarson Th. and Self T. (1993) The Laki (Skaftar Fires) and Grimsvoton eruptions in 1783–1785. *Bull. Vulcanol.* **55**, 233–263.

Trudgill S.T., Viles H.A., Cooke R.U., Inkpen R.J., Heathwaite A.L. and Houston J. (1991) Trends in stone weathering and atmospheric pollution at St. Pauls Cathedral, London, 1980–1990. *Atmos. Env.* **25**, 2851–2853.

Urzi C., Garcia-Valles M., Vendrell M. and Pernice A. (1999) Biomineralization processes on rock and monument surfaces observed in field and in laboratory conditions. *Geomicrobiol. J.* **16**, 39–54.

Viles H. (1996) Air pollution and building decay in Oxford. *Env. His.* **2**, 359–372.

Westlake H.F. (1863) *Westminster Abbey.* Bell and Hyman, London.

CHAPTER 2

BACKGROUND CONTROLS ON URBAN STONE DECAY: LESSONS FROM NATURAL ROCK WEATHERING

B.J. Smith

1. Introduction

Despite the aspirations of architects, the best intentions of builders and the fervent wishes of building owners, stone is not immutable. Sometimes, as in many dense limestones, the rate of decay may be gradual and, given constant climatic conditions, largely predictable. Even then, solution rates can vary in response to fluctuations in pollution levels and changes in the surface morphology of stonework. However, there are also many commonly used building stones which do not decay gradually, but instead experience episodic and some-times catastrophic breakdown. This occurs as strength is exceeded by the cumulative effects of internal decay and/or when stones are subjected to extreme environmental stresses associated with, for example, a severe frost. Included in these stone types are the quartz sandstones that are widely used across northwestern Europe. Characteristically, such stones are immune to all but the most limited solutional erosion, but are particularly prone to disruption by the effects of accumulated salts which produce effects such as granular disintegration, contour scaling and flaking (Smith *et al.*, 1994). In view of these observations and a long acceptance of natural rock weathering, why does the search for immutability in building stones persist?

In part, this search is conditioned by a widespread assumption that once stone is placed in a building, it is immunised from natural processes of weathering. An assumption that is compounded by a lack of public awareness and poor understanding of these weathering processes, the controls on them and their consequences. This belief that decay can be switched on or off is reflected in the commonly held view that where rapid decay is observed, it is in some way accidental and attributable to human intervention. Usually, this entails the creation of an atmospheric environment peculiarly aggressive to stone and the corollary that once polluted conditions are removed, decay will cease. This quest for immutability is further fuelled by references to ancient structures that have survived, recognisably intact for thousands of years. Such references conveniently ignore, however, the multitude of other structures that have not survived. Similarly, earlier phases of conservation and replacement can be forgotten or ignored. Thus, for example, we often neglect the fact that many of the great "Medieval" Cathedrals of England underwent major programmes of stone replacement and re-dressing in the 18th and 19th centuries. Finally, underpinning these observations and conclusions is the inevitable, though rarely acknowledged, assumption that processes of urban stone decay are distinct from and only loosely related to those responsible for the weathering of natural rock outcrops.

2. The Origins of Misconceptions

The roots of the relationships between natural weathering and urban stone decay lie in a mixture of the practical, the theoretical and the philosophical. At the heart of this is a *de facto* separation between researchers and practitioners in the two disciplines. There are sufficient points of contact to have established the various problems as being multidisciplinary, but to-date insufficient discourse to promote interdisciplinary solutions.

2.1. Decisions Governing Choice of Materials

Builders and architects are rightly concerned with the practicalities of stone properties and behaviour. Selection of stone is invariably driven by questions of personal preference and experience, availability, cost and aesthetics, whereby building design can drive choice rather than "fitness for purpose". Where "fitness" is a consideration, it is invariably based upon performance in standard durability tests. Unfortunately, these tests need not relate directly to conditions experienced by the stone in use nor long-term exposure to a succession or combination of processes operating at a range of magnitudes and frequencies. Alternatively, selection may be based upon initial, but rapidly changed, properties including strength, porosity, permeability and colour. This is particularly the case for replacement stone where matches are sought for existing stonework and where rapid decisions are often required for the choice of conservation strategy. As a consequence most conservation measures inevitably treat symptoms of decay rather than causes. Rarely, if ever, are time, expertise or resources available to research individual conservation projects. Exceptions can be made for particularly prestigious buildings and stone decay research is dominated by studies of high profile municipal and ecclesiastical structures at the expense of more mundane, but more numerous domestic, secular and industrial buildings. In some areas, this bias against scientific investigation can become almost institutionalised so that, for example, public funding for renovation through schemes such as the European Union Structural Fund and United Kingdom National Lottery is either not directly available for research or specifically excludes it. Inevitably, therefore, what research there is into urban stone decay is dominated by short-term, problem-based case studies. Pure research has been centred mainly in government-funded laboratories. But, especially in Europe, even this is under threat as these facilities are increasingly required to be self-financing and hence to concentrate upon commercial applications of their research.

2.2. Research Bias and Accessibility

Because of the above factors, pure research into urban stone decay, especially into decay processes and the long-term behaviour of stone, has historically been very difficult to pursue. This difficulty is compounded by the tendency for the results of government-funded research to appear as internal documents, government reports (BERG, 1989) and other forms of "grey literature". Rarely are these documents readily available to the wider research community and, if produced as commercial reports, they may be restricted by client confidentiality. As a by-product of this, there is little demand for and hence little provision of international journals dedicated to or even willing to accept, this kind of research. Instead, publication is dominated by professional magazines and conference proceedings. This has two effects: first it tends to reinforce the insularity of the stone decay/conservation community and, second, it actively discourages participation by university-based researchers. In these days of journal impact factors and subject-based research assessments, few academics can afford to publish outside of international journals or beyond their peer group subject.

There are clearly numerous exceptions to this intellectual isolation and classic texts such as that by Winkler (1973) made a cogent case for a more geological approach to stone decay. More recently, Cooke (1989) and the excellent review of salt weathering processes and products by Goudie and Viles (1998) have successfully brought together a wealth of information from across a wide number of disciplines. Similarly, earlier forays into stone decay by geologists/geographers such as Cooke and Gibbs (1993) produced many perceptive analogies between natural and urban environments, but even this was published in a format largely inaccessible to both research communities, let alone the wider public.

Despite these points of contact, geomorphological and geological research has continued to concentrate on rock weathering in natural environments and has been separated from urban stone decay by its journal and book base, separate terminologies and an absence of funding opportunities. This is exemplified in the UK by the support of environmental and geological research through the Natural Environment Research Council. Thus, for example, salt decay experiments are frequently published as studies in desert geomorphology (Smith and McGreevy, 1988) and studies of fire damage to stone are examined in the context of bush fires (Allison and Goudie, 1994).

This seeming denial of the wider implications of such studies is matched by a suspicion amongst architects and conservators of "academics", who by definition are concerned with the "theoretical and speculative" rather than the "technical, applied and professional" (Collins Concise Dictionary, 2001). In doing this, however, valuable insights into decay processes and alternative perspectives are ignored or missed entirely and work is often duplicated. So that, even in excellent texts on stone decay (Amoroso and Fassina, 1983) there are few, if any, references to parallel studies in natural environments. Likewise, until very recently, standard texts on weathering (Ollier, 1984) made little or no reference to building materials.

At a more fundamental level, the lack of a clearly defined and generally recognised academic base, together with an emphasis upon problem solving, has left urban stone decay without a unifying conceptual framework. The absence of a framework is particularly telling when questions of spatial and temporal variability of decay arise. In comparison, such questions are central to geomorphological and geological research, and have traditionally provided core paradigms within which studies of weathering can be set and their implications projected. It is this ability to provide a structure to decay studies that may eventually be the most significant contribution to be made by natural weathering research. In addition, however, there are many direct lessons to be learnt from the interactions between natural weathering processes and the very special conditions created once stone is taken, shaped and placed in buildings within the built environment.

3. Process Interactions

In recent decades, geomorphological and geological research into weathering has been preoccupied with the detailed explanation of processes that operate, frequently to the detriment of understanding the broader issues of resultant weathering forms and materials. Nonetheless, it has resulted in a large corpus of literature on physical and chemical processes and the factors that control them. Central to this has been the recognition of complicated feedback relationships between processes, fresh and weathered rock, surface morphology and the weathering environment experienced at the rock surface. Change in one variable inevitably triggers a chain of responses in the others and weathering is invariably discontinuous in space and time and highly dependent upon the breaching of stability thresholds within weathering systems. Interactions are numerous, and include such classic salt weathering forms as honeycombs and cavernous hollows (Fig. 1), which once initiated create humid micro-environments suitable for the retention of any deposited salt (Smith and McAlister, 1986). In such interactions a distinction can be drawn between mechanisms responsible for initiating change and those responsible

Fig. 1. The accumulation of a salt efflorescence within the interiors of honeycombs sheltered from rainwash (Salamanca, Spain).

for exploiting changed conditions. The most striking example within urban environments is that of granite. Of itself, "fresh" granite is a highly durable stone, relatively immune to attack from, for example, salts and chemical alteration. However, if microfractures are present or are initiated by other mechanisms (dilatation induced by pressure release during quarrying or directional loading within a building (Fookes *et al.*, 1988) then these can in turn be exploited by a range of other mechanisms (Neill and Smith, 1996). Interactive relationships extend beyond the roles of precursors to include catalytic effects such as enhanced silica solution in the presence of salts and the combined effects of mechanisms/processes acting in sequence related to fluctuating environmental conditions. So that, for example, porous stone containing suitable salts may be subject at different times to expansive forces associated with differential thermal expansion, hydration/dehydration and solution/crystallisation. In addition to which, the same stone may experience periodic freeze-thaw and thus be subject to a stress regime of highly variable magnitude and frequency.

The most widely accepted example within building stone decay literature of the role of natural processes is to be found in studies of limestone solution (see review by Inkpen *et al.*, 1994). Solution loss in polluted environments is a product of the effects of sulphur-rich "acid rain" and dry (gaseous) deposition and naturally carbonated rainfall — so-called karstic solution. There now exist a range of physico-chemical models which aim to identify the relative contribution of each solution process (Livingstone, 1992a and b; Webb *et al.*, 1992). Unfortunately, however, the emphasis appears to be on apportioning individual blame, rather than investigating interactions linked to variations in the chemical composition of precipitation over time, and possible synergisms associated with the chemically complex conditions of one rainfall event.

The processes and mechanisms highlighted above are in no way unique to the built environment. The range of salts present may be different from other salt-rich environments, such as coasts and deserts, but salt damage is not the preserve of any particular salt. Similarly, phenomena such as saline groundwater and groundwater rise are not restricted to buildings with defective damp courses. The effects of this overlap are most graphically illustrated where urbanisation occurs within naturally salt-rich environments. Thus, there now exists an extensive literature on damage to stone and concrete structures in areas such as the Middle East (Cooke *et al.*, 1982; Fookes, 1996) and Central Asia (Cooke, 1994) and coastal areas of the Mediterranean (Zezza, 1996). One of the most comprehensive overviews of process interactions was provided by Fookes *et al.* (1988). In this they identified the range of naturally occurring weathering processes associated with built environments and, more significantly, demonstrated that, although chemical weathering may be of significance in primarily hot, wet climates, many physical weathering processes are not restricted to what are perceived as aggressive climates. Moreover, they also demonstrated, through a survey of weathering rates from outcrops and structures in non-polluted environments, that significant natural alteration and surface loss can occur within the expected life spans of stone buildings or, over what they term as "engineering time".

4. Climatic Controls on Stone Decay

Most weathering studies (including that by Fookes *et al.*, 1988) rely upon macroclimatic distinctions when seeking to differentiate between environmental controls. These distinctions are normally drawn from meteorological observations collected in such a way as to rule out local variability and then averaged over varying periods to smooth out or remove fluctuations in time. It seems reasonable to assume, however, that these observations do not represent conditions experienced at the exposed stone surface of a building. Stone temperatures, for example, differ significantly from those of the overlying air and vary in response to many additional intrinsic and extrinsic variables. In turn, variations in temperature are the driving force behind most of the physical and many of the chemical processes responsible for stone decay. These include not only the obvious cases of freeze-thaw and thermal expansion/contraction, but also any processes influenced by wetting and drying. Because of this, geomorphologists with interests in desert landscapes have made significant contributions to understanding urban weathering environments. This is not simply a consequence of the presence of salts in both environments, but has much to do with a wider commonality, including large expanses of bare, unshaded stone/rock exposed to periodic wetting, high temperature ranges and absolute temperatures and rapid drying out assisted by strong winds (Table 1). Benefits have, however, accrued by the exchange of ideas in both directions and, in particular, processes of moisture absorption and loss from stone have been more systematically investigated on building stones than in relation to the far less

Table 1. A comparison of weathering in hot deserts and urban environments (modified after Winkler, 1973).

	Location/origins/timing	
Underlying causes	*Hot deserts*	*Urban environments*
Temperature contrasts	High	High (on walls)
Moisture from ground	High	High
Moisture from fog	Winter	Frequent
Moisture from condensation	Frequent	Frequent/winter
Salts	Desert floor/residual	Groundwater, stone weathering, polluted air, mortars
Damage	*Hot deserts*	*Urban environments*
Abrasion by wind	Strong	Some near street level
Frost action	Occasional/seasonal	High
Flaking by heat and moisture	Very strong	Moderate (stone-dependent)
Effloresence	Strong	Moderate but common
Sub-florescence	Strong	Moderate to strong
Case hardening/boxwork	Common, hollowed pebbles	Common between mortared joints
Rock coatings	Common, hard brown/black crusts	Light brown stains
Solution	Very slow	Very rapid in polluted atmospheres

predictable rains of desert regions. Thus, desert geomorphologists have much to learn from the work of building stone researchers in relation to water absorption and loss (see Camuffo, 1998 for a review).

4.1. Temperature Controls

Insolation, or temperature-induced weathering, was a preoccupation of desert geomorphology for much of the 20th century and no self-respecting review of desert weathering is complete without a table of rock surface temperatures (Cooke *et al.*, 1993; Goudie, 1997; Smith, 1994). There is debate as to the representativeness of these measurements and the extent to which they are biased by a pursuit of record high temperatures, but it does seem ironic that through them we probably know more about rock surface temperatures in the least populated parts of the world than in densely populated urban areas.

The process of insolation weathering was first generally invoked by returning desert explorers (Goudie, 1997), unsuccessfully simulated in early, somewhat crude experiments (Griggs, 1936) and finally reproduced as a series of minor modifications to stone surfaces (e.g. reflectance and microhardness) under carefully contrived laboratory conditions (Aires-Barros *et al.*, 1975; Aires Barros, 1977). However, the factor that is ignored or missed in these studies is that even under the supposedly sterile conditions found in deserts, temperature cycling of rock never operates in isolation from other processes or in the absence of other factors influencing the condition of the rock. Contrary to popular conception, water in small amounts is frequently available at rock surfaces via dew and fog, salts are often present in abundance and all rock outcrops and debris carry with them stress legacies reflecting, for example, their history of formation and exposure, past climatic regimes and, in the case of rock debris, possible transportation from areas of different climate as well as modification during transport (see Smith, 1994 for a review). The possibilities for these and other complicating factors are multiplied greatly for stone that is transposed to a built environment.

Fig. 2. Cavernous hollows in the Valley of Fire, Nevada, showing the concentration of active weathering in the shaded interiors of the hollows.

In addition to the hunt for proof of insolation weathering, the bulk of rock temperature data from deserts also reflects beliefs as to which temperature characteristics are most significant in controlling physical breakdown. Namely, the highest absolute temperature achieved, normally during summer, and the greatest diurnal temperature range. These values have in turn been those traditionally used in experiments designed to replicate not just insolation weathering, but also salt weathering and the relative durabilities of different rocks. More recently, however, alternative views on what constitute significant temperature controls have emerged.

(1) Firstly, it is readily apparent (Fig. 2) that most active weathering by salts within hot deserts (and on most buildings) takes place within cavernous forms (honeycombs and tafoni) and on shadow surfaces with greatly reduced maximum surface temperatures and diurnal regimes (Dragovich, 1967 and 1981; Rögner, 1987; Turkington, 1998). Within these caverns, moisture availability seems

to be the critical control on weathering rather than extreme temperatures. This view that was also expressed by Schattner (1961) working in the Sinai, who noted that granular disintegration of granites was more rapid and penetrated more deeply, where rocks were not subjected to the strongest insolation. Instead, disintegration was greatest on north- and west-facing slopes exposed to rain-bearing winds.

(2) Despite the perception of hot deserts as very homogeneous macroclimatic areas, in detail absolute temperatures and patterns of change over time are highly influenced by factors such as latitude, altitude, aspect (Smith, 1977) and time of year (Cooke *et al.*, 1993). With respect to the latter, environments that are perceived to be dominated by, for example, high temperatures and salt weathering may also be subject to processes such as frost weathering during winter months. It seems prudent to assume, therefore that resultant decay is the product of more than one process.

(3) The advent of thermistors and data loggers that allow a continuous record of temperature change, and infra-red thermometers that permit accurate measurement of surface temperatures, has focused attention on rapid fluctuation in surface and near-surface temperatures over minutes or tens of minutes. Jenkins and Smith (1990) showed, for example, that in the arid southwest of Tenerife, sandstone surface temperatures could fall up to 15°C within three to 15 minutes if insolation was blocked by cloud cover. The rates of cooling were in excess of 2°C per minute and were steeper than those experienced by the same test blocks during diurnal cooling at the end of the day. The significance of this is that, although such short-term fluctuations may only penetrate a few millimetres, they are concentrated in the zone where very important effects such as granular disintegration and multiple flaking occur. Moreover, there is the probability that fluctuations can occur many times during each day and that they would be particularly effective during winter months and/or at high altitudes where low air temperatures can lead to rapid cooling. These are times and places where, in the absence of high air temperatures, many thermally driven processes have been

presumed to "switch off". The identification of an additional level of thermal cycling also has possibly profound implications for the appropriateness and interpretation of weathering simulation experiments and building stone durability tests. In most cases, these use diurnal temperature cycles which penetrate several centimetres into a stone, but are designed to replicate damage and to measure material loss from grains within a millimetre or so of the surface.

(4) In the same way that it is dangerous to define any complex environment by a single thermal regime, it is misleading to impose it uniformly on the variety of rock types found there. Yet, this is precisely the strategy adopted in oven-based simulation and durability tests. There are good field observations to show that in the same natural environment, rock surface and sub-surface temperatures vary according to a range of thermal properties including thermal conductivity, heat capacity and, in particular, albedo (Kerr *et al.*, 1984). More recently, a combination of field exposure, heating under infra-red lamps and cycling in a conductive oven of four different rock types has confirmed this effect, and highlighted the fact that in nature, rocks such as basalt experience much more extreme temperatures than, for example, lighter coloured limestones and sandstones (Warke and Smith, 1998). The forcing of different rock types through the same heating/cooling cycles in durability tests may also go some way towards explaining discrepancies between laboratory-based ratings of weathering effectiveness and stone durability on buildings exposed to natural climatic conditions.

4.2. Moisture Controls

As noted earlier, students of natural rock weathering have much to learn from the many detailed studies of moisture absorption, movement and loss that have been conducted in the course of research into urban stone decay. Understandably, such studies, at least of building exteriors, have tended to concentrate on damage associated with prolonged dampness or with moisture in excess. Hence, the emphasis on solution loss and damage caused by rising groundwater. However, studies of natural weathering in hot deserts have understandably tended to focus attention on constraints exerted by limited moisture availability, which may provide insights into the effects of limited quantities of water on stonework. Salt weathering, for example is principally concentrated on those parts of structures protected from direct rainwash and subject to direct precipitation from dew, fog and frost — in much the same way that many desert areas derive most of their moisture from these sources (Peel, 1975; Smith, 1994; Goudie, 1997).

One apparent consequence of frequent moistening with limited quantities of moisture in salt-rich environments is the development of a surface efflorescence that is unlikely to be dissolved except under exceptional circumstances of a severe storm or driven rain. It has been claimed, therefore, that under these circumstances salts crystallised near the surface might act as passive pore fillers (Smith and McAlister, 1986), preventing moisture penetration to any depth and effectively restricting any solution/crystallisation to an immediate surface zone of granular disintegration. Similar efflorescences appear on stone samples undergoing sodium sulphate durability tests and it seems evident that their development must affect the nature of these tests as they develop. Some preliminary experiments have shown, for example, that even after two or three cycles of wetting and drying with a weak salt solution (10%), the pattern of moisture loss from test blocks is altered. Solutions which initially penetrated both sandstone and Portland limestone blocks were seen to pond on the surface, leading to a rapid initial moisture loss by evaporation on heating, followed by a much slower loss of moisture sealed beneath a surface zone of salt-filled pores (Smith and Kennedy, 1999). The further significance of this observation is that tests which initially set out to quantify one aspect of stone durability may rapidly transform into a very different trial. So-called salt crystallisation tests might therefore become salt hydration tests or, in extreme instances of surface sealing, a differential thermal expansion test. Similarly, as successive cycles re-distribute salt within stone, the critical threshold for decay might not be the number and magnitude of the cyclic stresses to which it has been subjected, but the number of cycles required to concentrate salts in sufficient quantities within a critical zone needed to trigger effects such as scaling and flaking.

An additional effect noted in hot desert environments is that occasional, intense rainfall falling on bare rock surfaces is rapidly lost as overland flow. Similar effects have been described by Camuffo (1991) in Southern Europe, who noted the complete drying out of external and internal surfaces of monuments due to warming by strong solar radiation. As a consequence, rainfall from typical afternoon showers: "cannot initially penetrate into the pores and capillaries of the stone, because these are not lined by a monolayer of water molecules in the solid state that would allow for the liquid water to run over it" (p. 55). The combined effects of such drying out and possible passive pore filling, is to allow the development of a hydrophobicity in stonework that has not been investigated or appreciated in more temperate environments. Yet, its existence would require an important revision of the number and nature of the environmental cycles to which stone is deemed to be subject.

A final result of the regular deposition of moisture in small amounts on rock surfaces in deserts is the widespread development of iron- and manganese-rich rock coatings or varnishes (Oberlander, 1994). The wider range of surface coatings has been recently reviewed by Dorn (1998) who extends the classification to include lithobiontic coatings, carbonate crusts, case-hardening, heavy metal skins, silica glazes and oxalate crusts, whilst stressing that such crusts are not unique to hot desert environments. The study of such varnishes is relevant to stone decay at a number of levels. For example, data on the temperatures of varnished surfaces in deserts provide a range of insights into effects of a lowering of albedo and the impact this has on the thermal regimes to which soiled stones are subject (Warke *et al.*, 1996). Of equal importance, however, is the value of a large, parallel geological/ geomorphological literature as a reference source for the growing number of observations of "urban" coatings, especially oxalate crusts (Del Monte *et al.*, 1987). The origins and implications of which remain largely unexplored. Literature on the effects of case-hardened layers might also help us to understand the consequences of the outward migration of iron in sandstones, both slowly under natural conditions and when accelerated by chemical cleaning. The breaching of these crusts and the hollowing out of weakened sub-surface layers is directly parallel to the formation of many cavernous hollows that are widespread across hot deserts and other salt-rich areas (Turkington, 1998).

5. The Direct Consequences of Placing Stone Within a Building

In urban environments, the act of placing a stone within a building exposes it to a range of additional stresses, most notably a range of potentially aggressive atmospheric chemicals. These can directly influence decay through reacting with the stone, or indirectly through either providing the ingredients (salts) for, or creating conditions amenable to, physical disruption. The latter can include the weakening of the stone through the selective dissolution of selected constituents or the creation of a secondary porosity that increases susceptibility to processes such as freeze/thaw. More significantly, and with less dependence upon local atmospheric pollution, placing a stone in any structure creates a wide range of physical and chemical conditions that make it more or less prone to damage from what would be considered as natural processes of weathering affecting an, albeit, somewhat unnatural rock outcrop. This includes modification of the climatic environment by the geometry of the structure and its relation to surrounding structures (see Ashton and Sereda, 1982), as well as modifications to the stone as a part of the construction process and the chemical and physical stresses exerted by surrounding materials. These factors can be grouped into three overlapping categories.

(1) *Temperature.* Building geometry can result in high temperatures on bare, exposed surfaces, but reduced temperatures and temperature ranges on shadow surfaces. Albedo can be lowered by soiling, leading to higher surface temperatures, increased diurnal ranges and higher internal temperature gradients. Urban heat island effects may reduce frost frequency and intensity.

(2) *Moisture.* Locally increased evaporation on surfaces exposed to insolation and where building geometry increases surface wind. Reduced evaporation in shadow areas with time of wetness also dependent on aspect. Selective wetting of surfaces determined by geometry, prevailing wind/rain direction and runoff pattern over

building surface. The possibility of rising damp if a damp course is absent or faulty and exposure to splash from road salts.

(3) *Construction effects.* Emplacement in a building can subject stone to compressive, shear and in some instances tensile stressing, but most importantly it typically constrains individual blocks during expansion caused by chemical alteration, expansion of interstitial salts and thermal and moisture cycling. Three dimensional geometry provides opportunities for "convergent" weathering as moisture, temperature and other environmental gradients interact across corners. The use of different stone types in combination can trigger decay where, for example, calcium sulphate salts produced by the reaction of atmospheric sulphur with carbonate stones can wash into adjacent, non-calcareous stonework (Cooper *et al.*, 1991). Similar contamination of non-calcareous stone can occur from surrounding mortars, whilst the use of "hard" mortars in combination with weaker stonework can lead to disruption of the stone during expansion, producing a honeycomb or "boxwork" effect as mortars are left proud of the retreating stone (Fig. 3). Finally, the surface dressing of the stone

Fig. 3. A boxwork effect produced by the rapid weathering of sandstone blocks to leave a framework of harder and more resistant mortar (Durham Cathedral Close).

influences pollution deposition and patterns of decay. Roughened stonework provides a greater specific surface to react with the environment and sheltered hollows in which aerosols and salts can accumulate protected from rainwash. Smooth stone has reduced opportunities for particulate retention, but repeated environmental cycling normal to the exposed smooth surfaces increases the susceptibility of salt contaminated porous stone to effects such as contour scaling.

6. Rates and Patterns of Decay

Studies of natural weathering, whether by geologists or geomorphologists, take place within established conceptual frameworks. These may not necessarily be directly applicable to conditions experienced on buildings, but they do allow a systematic evaluation of long-term weathering variability in space and time that is beyond reactive studies of stone decay on often unique structures. Thus, because most studies of urban stone decay are dictated by the need to solve a specific problem, there is little underlying conformity in methods of study. Nowhere is the absence of theories and theorists more acutely felt than in consideration of rates of stone decay.

6.1. Temporal Variability

In the majority of cases where rates of decay have been interpolated between observations, or extrapolated from a short-term monitoring period, practitioners have assumed a steady and progressive change. This is loosely based upon geological principles of uniformitarianism, first developed in the 18th and 19th centuries, in which geological processes of erosion and deposition are perceived to occur at a steady, uniform rate and the processes responsible for present-day change are assumed to be the same as those that acted in the past to create similar effects. Such a rationale had the virtue, at the time, of moving geology away from the invocation of unspecified catastrophies to explain change, whilst drawing it closer to mainstream scientific ideas championing concepts of gradual evolution. Unfortunately, the strict application of uniformitarian principles to stone decay ignores many clearly observed characteristics of actual breakdown and loss. Invariably, damage is a response to discrete meteorological events such as a storm or a particularly severe frost and consequently much decay occurs episodically. Exposed stone is in fact generally characterised by long periods of quiescence or very gradual alteration interspersed with periods of rapid change, when internal resistance thresholds are exceeded and sometimes catastrophic damage occurs (Smith, 1996). These thresholds can be breached either by sudden exposure to severe conditions, as with frost, or they may represent the end-product of the gradual accumulation of stress within stonework. A common cause of this stress increase is the internal build up of pollution-derived salts such as gypsum. Indeed, it is generally accepted that it is pressure exerted by repeated crystallisation, hydration and/or differential thermal expansion of interstitial salts that is the main cause of urban stone decay in non-calcareous stones. An example of a hypothetical sequence of episodic decay is given in Fig. 4. This shows the high degree of unpredictability associated with decay and illustrates the difficulties of forecasting the lifetime behaviour of building stone from short-term observations of decay. Clearly, measurements taken during a period of quiescence will over-predict durability, whereas observations during a period of rapid change (disequilibrium) will produce an underestimate. Most importantly, however, Fig. 4 demonstrates the complexity of stone decay, its interactions with environmental factors and the way in which it invariably results from the synergistic interaction of more than one weathering mechanism. In this way, stone can reach a number of critical points during its lifetime when, once rapid decay is initiated, it either stabilises or accelerates until often complete blocks of stone are consumed by processes such as granular disintegration and multiple flaking (Fig. 5). At present, there is no clear understanding of what triggers the positive feedbacks within stone decay systems that lead to this progressive, catastrophic decay, nor the negative feedbacks required to stabilise stonework once rapid decay is initiated. Negative feedbacks may be related to factors such as a limited depth of weakened stone that is rapidly removed to expose a more resistant interior or the loss of salts concentrated near the stone surface. Whereas a

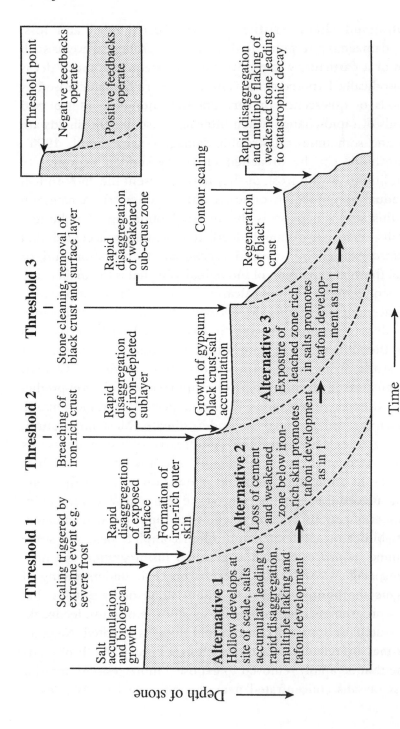

Fig. 4. Hypothetical schematic diagram illustrating the episodic decay of non-calcareous quartz sandstones in a polluted, salt-rich environment, with inset showing feedback options under threshold conditions (modified from Smith *et al*, 1994).

Fig. 5. Rapid, catastrophic decay of a sandstone block by a combination of granular disintegration and multiple flaking (Belfast).

positive feedback may be triggered by the creation of a humid microclimate in hollowed-out stonework that encourages a range of weathering processes. Whatever the mechanisms involved, it is likely that they are complex, interactive and most probably site and stone-type specific.

Whilst such episodic models of decay are most obviously applicable to stones such as sandstones, that are prone to effects such as sudden contour scaling (Fig. 6) and to processes such as salt weathering, they do have a wider applicability. Solution loss is, for example, by definition episodic — only occurring during and after precipitation.

Fig. 6. Contour scaling of a 19th century sandstone headstone to reveal a gypsum crypto-florescence (Belfast). For approximately 100 years, the stone would have exhibited little overt evidence of decay before the instantaneous loss of the contour scale.

Moreover, recognition of its episodic character allows a clearer appreciation of changes that occur between rain falls. This includes surface sulphation that would enhance subsequent solution loss and the possibility of physical weathering.

6.2. Spatial Variability

Assumptions of uniformity have also governed views on the spatial distribution of decay on buildings, yet it is self-evident that stone decay is spatially concentrated. This is often in relation to quite subtle environmental differences (exposure to rain, wind or sun), but also variations in materials properties such as porosity. This can be clearly seen on any complex building where, for example, black gypsum crusts accumulate only in areas protected from rainwash to produce an unappealing "patchwork" effect (Fig. 7). However, even on a flat surface of the same stone of the same age, individual blocks weather at different rates and exhibit different patterns of breakdown. This natural variability often confounds architects and owners of buildings

Fig. 7. The highly variable formation of gypsum black crusts on a church in the city of Bath, related to microclimatic conditions created by the geometry of the building.

who cling to the ideal of a building that "mellows" uniformly. It also means that prediction of decay over a whole building is very difficult, especially when durability predictions that dictate the initial choice of stone are often based on tests of a few small samples under unnaturally aggressive laboratory conditions (Smith, 1996).

7. Inheritance Effects

Natural weathering and erosion operate over such long time spans that it is normal to assume that rock outcrops and debris have long, complex and individual stress histories (Smith and Warke, 1997). This could be produced by climate change, spatial re-location (for example, from mountains characterised by frost weathering to lowlands dominated by chemical processes), or by changing conditions as rocks formed at depth are gradually exposed at the earth's surface. It should not be a surprise, therefore, that apparently identical building stones decay in different ways and at different rates because of different environmental and stress histories. Some of this legacy would have been put in place before the stone was quarried and/or prepared and used in construction (Table 2), but other effects reflect subsequent exposure to, for example, pollution. The most obvious instance of the latter is where pollution-derived salts accumulate within stone over a period of time, but it can include changes in stone properties such as porosity which exert a strong influence on salt weathering and solution loss. Some examples of pre- and post-emplacement effects are given in Table 2. Their greatest significance is that their influence can be unrelated to present-day conditions. Thus, removing atmospheric pollution may not halt decay that is driven by pollutants and weaknesses which accumulated in the past under "dirtier" conditions. This so-called "memory effect" must be an important consideration when projecting the likely success of any strategies for stone conservation or cleaning. In particular, it is the principal reason why replacement stone should not necessarily be sought as a match to the original, unweathered stone, but that it should be chosen to be visually, chemically and structurally compatible with the weathered stone.

Table 2. Possible inheritance effects found in urban stonework (from Warke, 1996).

Causative factors	Inheritance effects
Pre-emplacement factors	
Dilatation effects	Removal of overburden leads to compressional stress release in stone causing gradual opening of joint systems and microfracturing.
Microfracturing caused by quarrying	Quarry blasting can initiate microfracture development and/or propagate existing fracture lines.
Cutting and dressing of stone	May cause roughening of stone surfaces creating potential sites for subsequent accumulation of moisture, salts and organic matter.
Post-emplacement factors	
Thermally-induced microfracturing	Differential thermal expansion and contraction of surface mineral grains and interstitial salt deposits in response to long- and short-term temperature fluctuations may eventually lead to microfracture development.
Frost-induced microfracturing	Repeated freezing of moisture in pore spaces and microfractures may eventually lead to shattering of stone and loss of material.
Chemical dissolution	Dissolution of stone fabric alters surface pore dimensions and may facilitate the subsequent ingress of salt and moisture.
Soiling of stone surface	Soiling of stone surface by particulate deposition changes albedo, increasing absorption of solar radiation and hence surface/subsurface temperature conditions.
Surface crust development	Crusts act as salt reservoirs and can contribute to a decrease in substrate strength as material is leached out.
Salt accumulation and deposition	Interstitial deposition of salt contributes to stone decay through the mechanisms of intergranular crystallisation, hydration/dehydration and thermal expansion/contraction.

Table 2. (*Continued*)

Causative factors	Inheritance effects
Changes in surface morphology	Prolonged exposure to weathering processes leads to increased stone surface roughening allowing accumulation of moisture, salts and general particulate material which facilitate surface weathering processes.
Cleaning	Removal of crusts may expose a weakened substrate to attack by weathering mechanisms resulting in accelerated decay. Additionally, high pressure washing may drive salts deep into the stone fabric and application of biocides may damage individual grain structure and/or intergranular bonds.
Conservation treatments	Application of stone consolidants and other surface preparations alter surface characteristics, and may influence stone response to weathering processes.

8. Concluding Observations and Implications for Stone Conservation

Increased urban pollution undoubtedly accelerates building stone decay, especially on limestones where loss is primarily related to dissolution. Generally, however, it is salts derived from reactions between pollutants and stonework that are the most damaging agents of decay. The precise roles of pollution-derived salts in causing decay are however, very complicated. Firstly, salts can also originate from other sources ranging from groundwater and road salting to natural marine aerosols. It is therefore often difficult to dissociate pollution damage from underlying rates of weathering. Secondly, salt weathering rarely produces a gradual, widespread loss of material. It tends to attack certain blocks or surfaces and for many years there may be few overt signs of damage as salts accumulate and stone is weakened internally. This relative quiescence can be interrupted by a dramatic loss of material, which may eventually be halted as stone achieves a new equilibrium, but can also result in the complete destruction of stone blocks. This sudden breakdown can be triggered by an extreme event such as a severe frost or it can represent the culmination of a long buildup of stress within the stone which finally exceeds the stone's strength to resist. If the existence and nature of episodic decay is not understood, it is very tempting to ascribe accelerated decay simply to increases in pollution levels. This ignores decay that results from the long accumulation of pollution effects and is an overall function of the age of the structure. In crude terms, the older the building the more susceptible it becomes to damage and the more rapidly it weathers.

This willingness to explain accelerated decay solely in terms of increased atmospheric pollution, is also symptomatic of an attitude amongst many people that once stone is placed in a building it should last for ever. This forgets that all stone weathers naturally and that, even without the added complications of pollution, placing stones within a building exposes them to greater stresses than they would experience in natural outcrops. These can result from concentrated loading by other stones, contact with chemically aggressive mortars and renderings and placement next to other, chemically incompatible stone types.

The fact that building stones can carry a "memory" of past pollution and treatment is frequently ignored or discounted when stone is cleaned. This applies especially to structures that do not arouse public interest and thus do not generate specific research into their condition. Under these circumstances many of our "lesser buildings" are prone to "cleaning by formula". This may successfully remove superficial grime, but possibly at the expense of also removing a layer of underlying stone, the mobilisation of salts into areas where they can cause damage or the exposure of weakened sub-surface layers. Without careful consolidation of the newly exposed stonework, decay may therefore proceed more rapidly than on the original, grimy surface.

Much cleaning is also being carried out in the belief (or promise) that pollution legislation has successfully created urban environments in which it is once again safe to expose clean stonework. Local emission-based regulation cannot, however, legislate for continued

Fig. 8. The rapid re-soiling of the Portland Limestone of Belfast Technical College, some seven years after cleaning in the mid-1980s. Note the decrease in soiling away from road level.

transregional inputs of pollution. Similarly, most legislation is retrospective in that it is the reaction to a particular problem, filtered by the time that national and international legislatures need to agree on an important strategy. By the time that one pollution source is countered (for example, domestic burning of sulphur-rich coal), the problem is invariably superseded by a new pollution source. In the UK, we are seeing this effect on our buildings as soiling is increasingly concentrated near ground level in response to vehicle emissions. Thus it is now common to see buildings that were cleaned a few years ago — in the belief that urban smogs were a thing of the past — beginning to show the re-growth of black stains and crusts (Fig. 8).

Finally, much of society's attitude towards building stone decay, not the least those of owners, is that once constructed, buildings should remain pristine. It may be more realistic and ultimately much less expensive, if we come to accept that buildings naturally change and decay over time. Because of variations in microclimate and materials, different parts of buildings will weather in different ways and at different rates. Moreover, the gradual accumulation of natural and anthropogenic stresses within stonework is likely to mean that rates of decay will accelerate over time. Thus, whilst striving not to add unduly, through pollution, to the stress that structures are subject to, we have to accept that they have a finite service life.

Acknowledgements

The writer is indebted to the drawing office of the School of Geosciences, Queen's University for preparing the figures and to Dr. Patricia Warke for her constructive comments on an earlier version of the paper.

References

Aires-Barros L. *et al.* (1975) *Eng. Geol.* **9**, 249–265.

Aires-Barros L. (1977) *Eng. Geol.* **11**, 227–238.

Allison R.J. and Goudie A.S. (1994) In *Rock Weathering and Landform Evolution*, eds. Robinson D.A. and Williams R.B.G. Wiley, Chichester, pp. 41–56.

Amoroso G.G. and Fassina V. (1983). *Stone Decay and Conservation* Elsevier, Amsterdam.

Ashton H.E. and Sereda P.J. (1982) *Durab. Build. Mater.* **1**, 49–65.

BERG (Building Effects Review Group) (1989) *Deposition on Buildings and Building Materials in the United Kingdom.* HMSO, London.

Camuffo D. (1991) In *Weathering and Air Pollution, Course Notes, Venice/Milan, September 1991*, ed. Zezza F. Community of Mediterranean Universities, Bari, pp. 27–42.

Camuffo D. (1998) *Microclimate for Cultural Heritage.* Elsevier, Amsterdam.

Collins Concise Dictionary (2001) *Definition of "Academic".* Harper Collins Publishers, London, p. 7.

Cooke R.U. Halics (1982) *Urban Geomorphology in Drylands.* Oxford University Press, Oxford.

Cooke R.U. (1989) *Geograph J.* **155**, 361–366.

Cooke R.U. and Gibbs G.B. (1993) *Crumbling Heritage: Studies of Stone Weathering in Polluted Atmospheres.* Cathedrals Fabric Commission/National Power plc and PowerGen plc, London.

Cooke R.U. (1993) *Desert Geomorphology.* UCL Press, London.

Cooke R.U. (1994) In *Rock Weathering and Landform Evolution,* eds. Robinson D.A. and Williams R.B.G. Wiley, Chichester, pp. 193–207.

Cooper T.P. *et al.* (1991) In *Science, Technology and European Cultural Heritage,* eds. Baer N.S., Sabbioni C. and Sors A.I. Butterworth-Heineman, Oxford, pp. 456–459.

Del Monte M. *et al.* (1987) *Sci. Total Env.* **67**, 17–39.

Dorn R.I., (1998) *Rock Coatings.* Elsevier, Amsterdam.

Dragovich D., (1967) *Geol. Soc. Am. Bull.* **78**, 801–804.

Dragovich D., (1981) *Stud. Cons.* **26**, 143–149.

Fookes P.G. *et al.* (1988) *Q. J. Eng. Geol.* **21**, 33–57.

Fookes P.G., (1996) In *Proceedings of Geological Society Conference on Engineering Problems, November 1996.* Geological Society, London, pp. 7–23.

Goudie A.S. (1997) In *Arid Zone Geomorphology,* ed. Thomas D.S.G. Wiley, Chichester, pp. 25–40.

Goudie A.S. and Viles H.A. (1998) *Salt Weathering Hazards.* Wiley, Chichester.

Griggs D.T. (1936) *J. Geol.* **74**, 733–796.

Inkpen R.J. *et al.* (1994) In *Rock Weathering and Landform Evolution,* eds. Robinson D.A. and Williams R.B.G. Wiley, Chichester, pp. 119–130.

Jenkins K.A. and Smith B.J. (1990) *Catena* **17**, 449–459.

Kerr A. *et al.* (1984) *Geology* **12**, 306–309.

Livingstone R.A. (1992a) In *Proceedings of the 7th International Congress on Deterioration and Conservation of Stone,* eds. Rodrigues J.D., Henriques F. and Jeremias F.T. Lisbon, pp. 375–386.

Livingstone R.A. (1992b) in *Stone Cleaning,* ed. Webster R.G.M. Donhead, London, pp. 166–179.

Neill H.L. and Smith B.J. (1996) In *Processes of Urban Stone Decay,* eds. Smith B.J. and Warke P.A. Donhead, London, pp. 113–124.

Oberlander T.M. (1994) In *Geomorphology of Desert Environments,* eds. Abrahams A.D. and Parsons A.J. Chapman and Hall, London, pp. 106–119.

Ollier C. (1984) *Weathering.* Longman, London.

Peel R.F. (1975) In *Processes in Physical and Human Geography,* eds. Peel R.F., Chisolm M. and Haggett P. Heineman, London, pp. 110–129.

Rögner K. (1987) In *International Geomorphology,* ed. Gardiner V. Wiley, Chichester, pp. 1271–1287.

Schattner I. (1961) *Bull. Res. Counc. Israel* **109**, 247–266.

Smith B.J. (1977) *Catena* **41**, 41–63.

Smith B.J. and McAlister J.J. (1986) *Zeit. für Geom.* **30**, 445–460.

Smith B.J. and McGreevy J.P. (1988) *Earth Surf. Proc. Landforms* **13**, 697–706.

Smith B.J. (1994) In *Geomorphology of Desert Environments*, eds. Abrahams A.D. and Parsons A.J. Chapman and Hall, London, pp. 39–63.

Smith B.J. *et al.* (1994) In *Rock Weathering and Landform Evolution*, eds. Robinson D.A. and Williams R.B.G. Wiley, Chichester, pp. 131–150.

Smith B.J. (1996) In *Processes of Urban Stone Decay*, eds. Smith B.J. and Warke P.A. Donhead, London, pp. 3–18.

Smith B.J. and Warke P.A. (1997) In *Arid Zone Geomorphology*, ed. Thomas D.S.G. Wiley, Chichester, pp. 41–54.

Smith B.J. and Kennedy E.M. (1999) In *Aspects of Stone Weathering, Decay and Conservation*, eds. Jones M.S. and Wakefield R.D. Imperial College Press, London, pp. 55–64.

Turkington A.V. (1998) *Q. J. Eng. Geol.* **31**, 375–383.

Warke P.A. (1996) In *Processes of Urban Stone Decay*, eds. Smith B.J. and Warke P.A. Donhead, London, pp. 32–43.

Warke P.A. *et al.* (1996) *Earth Surf. Proc. Landforms* **21**, 295–306.

Warke P.A. and Smith B.J. (1998) *Geomorphology* **22**, 347–357.

Webb A.H. *et al.* (1992) *Atmos. Env.* **26B**, 165–182.

Winkler E.M. (1973) *Stone: Properties, Durability in Man's Environment.* Springer, Vienna.

Zezza F., ed. (1996) *Origin, Mechanisms and Effects of Salts on Degradation of Monuments in Marine and Continental Environments, Proceedings of EC Conference, 25–27 March 1996.* Bari.

CHAPTER 3

MECHANISMS OF AIR POLLUTION DAMAGE TO STONE

C. Sabbioni

1. Introduction

Stones constitute one of the most widely used materials in historic monuments and buildings. Therefore, the study of the effects of air pollution on these materials was one of the first to be undertaken with the aim of understanding the physico/chemical, and more recently, biological mechanisms leading to the degradation of the cultural heritage. There is general consensus and wide evidence (Fig. 1), that the damage encountered on building stones in urban environments is predominantly due to air pollution (Winkler, 1973; Amoroso and Fassina, 1983).

While the main qualitative processes involved in pollution-related stone degradation are generally accepted to be gypsum formation and carbonate dissolution, knowledge on the quantitative relationship between pollutants and stone damage remains sparse and continues to generate debate.

The damaging effects of gypsum arise from its far greater solubility in water compared to calcium carbonate. During wet periods, gypsum can dissolve and be transported deep into the pore system of stone. When the stone dries, gypsum re-deposits on or inside the stone, where it produces further damage, due to its higher specific volume compared to calcite, through cycles of dissolution-crystallisation and

(a)   (b)

Fig. 1. Stone decay in urban environments has rapidly increased over the last hundred years. The two photographs show a marble bas-relief dated 1377: **(a)** taken in the early years of the 20th century (Poppi collection, Historical Archives of the Cassa di Risparmio di Bologna), and **(b)** as it is today.

hydration-dehydration (Arnold, 1981; Lewin, 1982). On the basis of S isotope ratios, Pye and Schiavon (1989) demonstrated that the sulphur found in gypsum crusts on building stones always originates from an atmospheric source. Therefore the formation of calcium sulphate on calcareous stone occurs through a reaction with the sulphur species contained in the air (dry deposition) or dissolved in rainwater (wet deposition).

Stone decay is also attributed to the acidity of rainwater, caused by contaminants such as sulphur and nitrogen oxides, as well as carbon dioxide. "Acid rain" corrodes the stone by penetrating the pore structure and reacting with the materials. In particular, calcium salts, which often serve to bind the crystal grains, are highly susceptible to dissolution by acid solutions. Evidence of such processes has been reported in Europe and America (Martinez and Martinez, 1991).

Numerous studies have addressed this very serious problem; in general, they have proceeded in three major directions: (1) studies performed directly on monuments and historic buildings with the aim of identifying the different components due to atmospheric pollution responsible for stone damage, (2) chamber tests performed in the laboratory to examine the interaction between pollutants and stones, and (3) field exposure tests to study the damage produced on stones by different microclimatic and pollution conditions. The main results of these investigations are summarised below.

2. Damage on Stone Buildings and Monuments

Henley (1967) showed that the calcium carbonate (calcite) in stones may be almost completely replaced by gypsum or by a hard black carbonaceous deposit. This pioneering study showed how gypsum formation does not affect the stone homogeneously; the matrix material is almost completely transformed, while the fossil contained in the limestone has little or no deposit, a discrepancy that permits the estimation of the rate of erosion.

Since then a number of studies have been performed on monuments and historic buildings with the aim of identifying the different typologies of damage layers, their composition and origin. In spite of this, the availability of quantitative data remains relatively limited.

In studies on damage, the classification of stone types needs to be related more clearly to degradation mechanisms than geological criteria (genesis, grain size and mineralogical composition). The key properties of stone with reference to deterioration are: (a) chemical composition and (b) porosity (Viles *et al.*, 1997). For this reason, the papers reported in the literature are reviewed here according to the following categories: (1) marble and limestone, (2) stone with high porosity, such as sandstone, and (3) granite.

2.1. Marble and Limestone

The earliest and most extensively studied stones are marble and limestone, largely because they have been widely used internationally in statuary and buildings. At the same time, they represent a more simple system than other stones thanks to their chemical homogeneity, being almost entirely composed of calcium (or, more rarely Ca and Mg) carbonates, and low porosity which permits the identification of an interaction interface between the atmosphere and stone.

The damage typologies observed on monuments built in marble and limestone are related to the way rainwater wets the surface. *White areas* are found where rainwater runoff dominates, producing the dissolution of the carbonate rock so that the original colour of the stone is evidenced. *Black areas* are observed on stone surfaces wetted by rainwater but sheltered from intensive runoff, being the areas where atmospheric pollutants accumulate, along with the products of reactions between stones and atmospheric gas and aerosol. Protection from leaching by rain washout favours the formation of crusts, which are black in colour due to the atmospheric particles they embed during their growth. Water from fog or dew is in general insufficient to "activate" any dry deposit on stone surfaces or to dissolve the calcite (Camuffo *et al.*, 1982).

White areas

To determine the material loss of stone buildings and monuments incurred by wet deposition, typical of white areas, several methods have been employed.

Microerosion measurements have been made of the difference in depth of erodable material using fossil fragments that protrude above the present-day surface as reference level; the fossil fragments were considered to be weathering resistant and to represent the original surface of the stone. In a similar way, lead plugs filling holes have also been used as a reference plane for the original stone surface (Sharp *et al.*, 1982). Using these methods, the rate of surface recession of Portland limestone measured on St. Paul's Cathedral over the period 1980–1985 was on average 0.06 mm/y (Trudgill *et al.*, 1990).

A number of studies have been performed on tombstones. Baer and Berman (1983) measured the depth and thickness of inscriptions on tombstones in marbles from Vermont, Georgia, Massachusetts and Mississippi. They found recession rates ranging from 3.6 mm/100y (Philadelphia, PA) to 2.8 mm/100y (Cypress Hills Cemetery, NY) for urban areas and the value of 1.7 mm/100y for a suburban site (Long Island Cemetery, NY). They concluded that finer grained marbles are eroded more rapidly than coarse-medium grained stones or dolomitic marbles, and that local sources dominate the deposition of pollutants.

Another way of evaluating the wet material loss of a stone is to collect and analyse runoff water, i.e. rain water that has flowed over its surface. This method also provides insight into the leachable reaction products forming during the rainfall event (Leysen *et al.*, 1987; Roekens and van Grieken, 1989; La Iglesia *et al.*, 1994). Measurements performed on St. Paul's Cathedral, London, gave a rate of limestone erosion of 0.27 mm/y, much higher than the values obtained from microerosion meter measurements. The relative standard deviation, found to be 115%, reflects the complex nature of the erosion process (Butlin *et al.*, 1985).

At Gettysburg (Pennsylvania, USA), runoff from small, well-defined catchment areas washed almost entirely by direct precipitation, was collected from two Carrara marble statues, a Carrara obelisk and a Pennsylvania blue marble obelisk (Dolske, 1995). Current meteorological and air quality data were collected to compute the dry deposition of sulphur dioxide (SO_2), nitric acid and sulphate and nitrate aerosol between the rain intervals. Thus both wet and dry deposition of atmospheric pollutants was examined. The data indicate that marble loss from a complex carved surface is several times greater than that from flat vertical surfaces of comparable petrography and exposure history. The author points to the need for adjusting weathering rates based on simple geometry when applied to monuments and statues with more complex surface geometry. The sulphur flux calculations indicate that the bulk of the sulphur is deposited when the marble is dried.

Black areas

Black crusts, studied extensively by numerous authors on monuments all over the world, are found to be composed mainly of calcium sulphate (identified as gypsum) and carbonaceous particles (soot). Most of the literature gives evidence of these components on the basis on analyses performed by optical microscopy, scanning electron microscopy and X-ray diffractometry; examples are reported for monuments in Turkey (Caner *et al.*, 1988), Poland (Haber *et al.*, 1988), Spain (Carretero and Galan, 1996), Greece (Chabas and Lefevre, 1996) and Italy (Garcia-Valles *et al.*, 1997).

The gypsum crust presents a crystalline habit, generally laminar and only rarely globular or acicular, depending on the growth rate of the crystals (Lewin and Charola, 1978; Alessandrini, 1992) and can vary in thickness between 10 μm and 10 mm (Del Monte *et al.*, 1981; Margolis *et al.*, 1988).

The origin of sulphur in gypsum crust formation on monuments has recently been investigated by means of sulphur isotope measurements (Torfs *et al.*, 1997). Relatively large variations of sulphur have been found in gypsum crusts on monuments situated in the centre of Antwerp (Belgium), its suburbs and the surrounding area. More heavy sulphur is encountered in the centre of Antwerp. The variation in the sulphur (S) isotopic composition of the crusts coincides geographically with the variation of SO_2 in the atmosphere, although generally the crusts present a higher $^{34}S/^{32}S$ ratio than the surrounding atmosphere. The petrochemical industry in the harbour is thought to be the main SO_2 emitter in the north of Antwerp, while in the southern region, the local brickyard industry influences the isotopic composition of SO_2 and the damage layers on the stone. The authors conclude that SO_2 makes a greater contribution to gypsum crust formation on stone than the sulphate (SO_4^{2-}) from rainwater.

The atmospheric particles embedded within the gypsum crusts have been analysed and, on the basis of their morphology and elemental composition, different types of particles have been identified, which can be summarised under three general categories: porous carbonaceous particles (soot), smooth aluminosilicate particles, and metal particles mainly composed of iron (Del Monte *et al.*, 1981; Esbert *et. al.*, 1996; Derbez and Lefevre, 1996). The pollution sources emitting these particles have been identified as fuel oil combustion in domestic heating systems and electric power plants, coal combustion, gas oil emission (Sabbioni, 1995), vehicle exhaust (Rodriguez-Navarro and Sebastian, 1996) and biomass combustion (Ausset *et al.*, 1992).

Whether atmospheric particles simply incur aesthetic damage or play a role in the actual formation of the gypsum crust has been the subject of debate. Carbonaceous particles have been cited as damaging to stones since they are responsible for the blackening of the patina, seriously impairing the appearance of monuments, and are catalysts of the SO_2 oxidation process on stone surfaces (Amoroso and Fassina, 1983).

Few quantitative data are reported on the concentrations of the main components present in the black crusts (Table 1). The amount of sulphate is generally over 30% and among minor components, nitrate and chlorides are present in amounts around or above 1%.

When gaseous and aerosol pollutants were measured at different heights on the Column of Marcus Aurelius in Rome, the composition of the black crust was found to relate to the concentration in the air of SO_2 and total suspended particles mainly emitted by domestic oil heating and traffic, particularly diesel engines. Although the high NO_x concentration measured in air is not reflected in the presence of nitrates, the hypothesis is advanced that nitrogen dioxide (NO_2) can enhance the oxidation of SO_2, especially in the presence of metal and carbon particles, which act as catalysts (Brocco *et al.*, 1988). These data are in agreement with Livingston (1985), whose review of nitrate measurements on carbonate stones (marble and limestone) indicated that the nitrite (NO_3^-) concentration never exceeds 3% in weight, which is a factor of ten or more lower than that of SO_4^{2-}.

Table 1. Soluble salt concentrations (%) measured in the black crusts.

	Brocco *et al.* (1988)	Fassina (1988)	Sabbioni and Zappia (1992a and b)	Torfs and van Grieken (1997)	Moropoulou *et al.* (1998)
SO_4^{2-}	41.3	43.7–69.7	31.5–48.5 [*]	31–40 [+]	39.4–60.8
NO_3^-	0.17	0.01–0.25	—	0–0.5 [+]	0–1.36
Cl^-	0.05	0.07–0.52	0.02–0.7	0.1–0.9 [+]	0.13–0.76

Note: [*] Calculated from $CaSO_4.2H_2O$ concentrations.

[+] Only data referring to black crusts are reported.

The sulphation process, that is the transformation of calcium carbonate into calcium sulphate dihydrate, can proceed via two principal reactions:

(1) SO_2 is adsorbed on the stone and reacts with calcium carbonate and water to form calcium sulphite as an intermediate product, which is then further oxidised to gypsum;

(2) SO_2 can also be transformed to sulphuric acid (H_2SO_4), either directly on the stone or in the air, and the subsequent attack on the carbonate leads to the formation of a gypsum crust.

To better understand the importance of the first reaction pathway, Roekens et al. (1989) measured the sulphite (SO_3^{2-}) and SO_4^{2-} content on the limestone of two historic buildings in Belgium using ion chromatography (IC) and spectrophotometry. A mean concentration of SO_3^{2-} of 0.05% was found, far lower than the SO_4^{2-} concentration, which ranged from 2 to 37%; the SO_3^{2-}/SO_4^{2-} is on average 0.73.

Quantitative data on sulphite in damage layers are also provided by Gobbi et al. (1998) in their presentation of a new methodology for the simultaneous identification and quantification of sulphite and main anions present in the black crusts. The data obtained, ranging from 210 to 2600 ppm, confirm the formation of sulphite as an intermediate stage in the mechanism of interaction between SO_2 and carbonate stones. Bacci et al. (1997) propose the principal component analysis of diffuse reflectance near-infrared spectroscopy (NIR) as a non-destructive methodology for discriminating $CaSO_3 \cdot 0.5H_2O$ and $CaSO_4 \cdot 2H_2O$ in calcareous samples.

The carbon compounds present in the damage layers on building materials may have four different origins: (1) calcium carbonate, deriving almost exclusively from the stone (Zappia et al., 1993); (2) deposition of atmospheric particles containing elemental and organic carbon, as primary and secondary pollutants (Saiz-Jimenez, 1993; Turpin and Huntizcker, 1995); (3) biological weathering due to the action of micro organisms such as fungi and lichens, one of the major products of which is oxalic acid, which reacts with the underlying materials, leading to the formation of calcium oxalates (Sabbioni and Zappia, 1991; Saiz-Jimenez, 1995); and (4) surface treatments (oils, waxes, proteins, etc.) frequently used in the past to protect monuments and historic buildings (Rossi Manaresi, 1996).

The total carbon (C_t) present in the black crusts can be considered as being composed of two main fractions:

$$C_t = C_c + C_{nc} \tag{1}$$

with C_{nc}, composed by:

$$C_{nc} = C_e + C_o \tag{2}$$

where C_c is the carbonate carbon, basically due to the stone, and C_{nc} is the non-carbonate carbon, which includes the organic carbon C_o of biogenic and anthropic origin, while the elemental carbon C_e is predominantly a product of combustion processes. The measurements performed on black crusts from stone monuments and buildings in Bologna and Rome were found to contain mostly non-carbonate carbon (90% of Ct), showing concentrations ranging between 2.69 and 1.48%, of which 40% is formed by elemental carbon, a parameter which is linked to the soot embedded within the damage layers. The remaining organic fraction is mainly composed of oxalates, along with other organic compounds, such as formates, acetates n-alkanoic acids, polycyclic aromatic hydrocarbons (PAH) and ethyl esters (Sabbioni *et al.*, 1996a). These organic constituents are tracers of specific anthropogenic sources typical of urban areas, such as vehicle exhaust, road dust, industrial combustion and domestic heating systems (Rogge *et al.*, 1993a and b).

Among organic compounds, calcium oxalates have been widely encountered on stone building surfaces and have been identified as whewellite and weddellite, the mono- and dihydrate calcium oxalate, respectively (Cipriani and Franchi, 1958; Franzini *et al.*, 1984; Guidobaldi *et al.*, 1985; Alessandrini *et al.*, 1989). Whether their origin is linked to protective treatments, biological weathering or pollutant deposition is the subject of controversy, and at present the data are generally insufficient to exclude any of the origins suggested (Realini and Toniolo, 1996), except in the case of studies specifically performed on single monuments.

A number of metals have been measured within the black crusts and some average concentrations found in the damage layers sampled

Table 2. Average concentrations (ppm) of heavy metals measured in the black crusts on stone monuments in different urban sites.

	V	Mn	Fe	Ni	Cu	Zn	Pb
Marble and limestone							
Milan[*]	72	236	18796	13	50	258	883
Venice[*]	29	58	3779	38	11	59	123
Rome[*]	44	133	8061	22	195	118	532
Bologna[*]	58	113	14090	<10	49	177	427
Eleusis[†]	20	200	9000	40	40	300	300
Sandstone and calcarenite							
Brussels[‡]	186	120	16116	13	105	873	516
Bologna[¶]	34	152	653	12	20	48	160
Granada[§]	18		1655	13	12	40	40

[*] Sabbioni and Zappia (1992a)
[†] Torfs and van Grieken (1997)
[‡] Leysen *et al.* (1990)
[¶] Sabbioni and Zappia (1992b)
[§] Rogriguez Navarro and Sebastian (1994)

on marble and limestone monuments in towns of the Mediterranean Basin are reported in Table 2. It can be observed that iron is the most abundant metal at all the sites and the concentrations of remaining metals show a similar order of abundance, indicating that the components contributing to black crust formation have a common origin.

The enrichment factor (EF) of various elements with respect to the carbonate rock have been calculated to identify the component due to the deposition of atmospheric gas and aerosols on the stone surfaces. The average elemental composition of carbonate rock reported by Mason (1966) is used and Ti has been assumed as indicator element:

$$EF_{carb.}(X) = \frac{(X/Ti)\,Black\,crust}{(X/Ti)\,Carb\,rock} \tag{3}$$

where X is the concentration of the investigated element and Ti the concentration of the indicator element in the black crusts and carbonate rock, respectively. By convention, a "cut-off" EF value of 5, a purely arbitrary value, is used to distinguish the origin of the elements. An EF < 5 was taken as an indication that the elements have a significant stone origin and they are considered non-enriched. An EF > 5 is assumed to indicate that a significant proportion has a non-stone origin and is enriched due to atmospheric deposition. At various Italian urban sites, Mg, Al, Si, K, V, Fe, Ni, Mn and Sr were not enriched and the elements showing a non-stone origin were Na, S, Cl, Cu, Zn, Br and Pb (Sabbioni and Zappia, 1992a and b). It should be noted that although iron presents the highest concentration among the heavy metals, it is not due to atmospheric deposition. In maritime sites of the Mediterranean Basin, Torfs and van Grieken (1997) found enrichments for Cl and Na in all sites, while S and Pb were found in some cases; for the other elements the EF were very small, indicating almost no deposition.

2.2. Sandstone

Porous limestones have been widely used in buildings and monuments, thanks to the ease with which they are quarried and worked. Less attention has been focused on damage to stones of high porosity, such as sandstones and calcarenites (Stambolov and van Asperen de Boer, 1972; Rosvall, 1988). Sandstones are sedimentary rocks composed mostly of mineral and rock fragments within the sand-size range (2 to 0.06 mm) and a minimum of 60% free silica, cemented by various materials, including carbonates, while calcarenites are limestone composed predominantly of clastic sand-size grains of calcite (ASTM, 1989).

In the literature, the main degradation mechanisms proposed for these materials are:

(1) The formation of damage layers mainly composed of gypsum on the surface, weakly linked to the underlying rock and therefore subject to easy exfoliation (Rossi Manaresi, 1975; Andersson, 1985; Weber H, 1985; Twilley and Podany, 1986; Mirwald *et al.*, 1988; Fobe *et al.*, 1993). The process is due to the sulphation of the carbonate component of sandstone owing to atmospheric deposition, with the formation of a gypsum patina embedding carbonaceous particles and soil dust (Sramek and Eckert, 1986;

Leysen *et al.*, 1989, Pavia Santamaria *et al.*, 1996). The structure of the stone controls the rate of gypsum formation, which is claimed to alter the original porosity of the stone (Kozlowski *et al.*, 1990).

(2) The dissolution and transport of soluble salts by water circulating within the stone (Winkler, 1982; Subbaraman, 1985; Blaeuer, 1985); the subsequent recrystallisation of these salts produces mechanical stress which can disaggregate the material (Lewin and Charola, 1978).

(3) The formation of calcium oxalate layers (Jones and Wilson, 1985), identified as whewellite and weddellite by Alaimo *et al.* (1986). These alteration patinas have been linked to biological weathering due to the action of micro organisms, such as fungi and lichens, at the stone surface (Krumbein, 1988; Warscheid *et al.*, 1990).

(4) Mechanical erosion due to the action of sand and marine salts transported by the wind (Lal, 1978). This kind of damage produces the segregation of grains, the formation of step profiles and leaves a rough, honeycombed surface (Agrawal *et al.*, 1986).

Only the first two mechanisms are considered to be related to the effects of air pollution, while the third is the subject of discussion, as already mentioned for marble, and the fourth is a typical effect of natural weathering.

The damage patterns on sandstone or calcarenite present entirely different visual features from those described for marble and limestone, the surfaces being blackened homogeneously, regardless of their geometry (Sabbioni and Zappia, 1992b; Camuffo, 1998). The diverse damage feature is ascribed to the different mechanisms of deposition and resuspension of atmospheric particles (particularly carbonaceous particles, responsible for the colour of the crusts) occurring at the surface of the monument. Such stones present high surface roughness due to their intrinsic porosity and mineralogical dishomogeneity, which prevent smoothing and polishing. Furthermore, the wetness of a surface is highly favoured by the presence of pores and capillaries, since condensation occurs at relative humidity below 100%. Surface roughness and high porosity facilitate the deposition of gas and particles, at the same time reducing the removal of particles and damage products by resuspension and washout. Thus on porous stones, the mechanisms of particle deposition and capture are more efficient, while those mechanisms tending to remove particles from the surface after their deposition are less efficient. These synergistic effects produce the homogeneous blackening typical of historic buildings and monuments built in porous stones.

The distribution of damage patterns has been examined on churches built in sandstones in the West Midlands, England, where the total blackened stone was found to be less than 10% in rural environments, while reaching values greater than 40% in urban areas. The use of coal as the dominant domestic and industrial fuel appears to have been an important source for much of the blackening encountered (Halsey *et al.*, 1996).

On sandstone and calcarenite, two damage layers of surface damage have been distinguished: (1) a surface layer (A) of a few millimetres thickness with a composition similar to the black crusts analysed on marble and limestone and (2) a disaggregate layer (B) of the order of one centimetre, where the dissolution of the carbonate matrix occurs due to atmospheric acid deposition, which produces the decohesion of the sandy grains. The detachment of layer A also causes the complete loss of layer B, exposing the underlying undamaged stone to a new cycle of damage (Sabbioni and Zappia, 1992b). The multiplication of decay layers, progressing from the outer surface of the exposed stone toward the interior was first described by Lewin (1982).

Layers A found in Bologna presented average concentrations of 64.4% of gypsum and 1.1% of non-carbonate carbon (C_{nc}), which is assumed as a quantitative marker of the amount of embedded carbonaceous particles (essentially soot) deriving from fossil fuel combustion (Sabbioni and Zappia, 1992b). Concentrations of heavy metals measured on sandstone and calcarenite monuments in European towns are reported in Table 2.

The deposition of atmospheric particles on the sandstone surface is indicated to increase the rate of gypsum formation. Black soiling is also claimed to increase stone deterioration by physical stresses due to a lowering of the albedo of the stone, which increases the absorption of incident solar radiation. As a consequence, the blackened surface layer experiences increased heating/cooling cycles, wetting/drying cycles and thermal expansion: gypsum hydration/dehydration cycles exacerbate stress, producing the detachment of the surface layer A of the stone (Halsey *et al.*, 1996).

2.3. Granite

Granitoid rocks have received scant attention although they have been widely employed because of their good mechanical properties and durability (Lazzarini, 1993).

Two types of damage layers, of different composition and origin, have been found on granite:

(1) Gypsum crusts, where almost all constituents are allochthonous and are due to the deposition of air pollutants; and
(2) Clay-calcitic layers, whose constituents are due to the original rock and must be considered as weathering layers, being a natural evolution of granite (Jeannette, 1980 and 1981).

Pollutants and particularly pollution-derived salts, such as gypsum, are initially confined to the surface stone, but they represent potential agents of degradation as they produce a widespread microfracture network, which confers a much higher porosity than would normally be encountered on such rocks. The secondary microporosity allows the penetration of pollution-derived salts which produce mechanical and chemical change in the rock mass (Neill and Smith, 1996). Alongside gypsum, calcium chloride and calcium nitrate have also been found (Urquhart *et al.*, 1996). The presence of roadside dust and fly ash particles has been shown on granite buildings in urban environments and the fly ash is claimed to be a possible source of the gypsum, which forms a thin black crust (Smith *et al.*, 1993). In some cases, many successive layers of degradation have been recognised, all of them similar in character, with thicknesses ranging from a fraction of a millimetre to 1 or 2 mm (Lewin, 1982).

3. Chamber Tests

In dry deposition, sulphur may either adsorb and react directly with the stone surface, or may first react with the water present in the stone, and subsequently, with the calcareous material. The rate of SO_2

deposition depends on a number of factors. The concentration of SO_2 is obviously important, since the number of molecules colliding with the surface in a unit of time is linearly dependent on it. The presence of oxidants, such as NO_2 or ozone, may strongly enhance the process, since they catalyse the oxidation of the adsorbed species containing four-valent sulphur or calcium sulphite to sulphate. Humidity also plays a major role in the process. Finally, the nature of the stone itself affects the rate of SO_2 deposition to a considerable degree. Polished marbles have been found to adsorb less SO_2 than rough and porous limestone surfaces. The susceptibility of various stones to the corrosive action of SO_2 was found to depend strongly on their pore structure, which in turn can control moisture equilibrium in the stones (Kozlowski *et al.*, 1992). So far, very little is known on how stone-corroding pollutants are distributed in the materials (Mirwald, 1991).

One approach towards defining the key physical and chemical parameters is to conduct controlled laboratory studies set up to determine their impact. The literature reports laboratory tests on building stones performed primarily with the aim of studying the effects of exposure to SO_2.

Simulation tests on various stones, notably marbles, limestones and sandstones, have been performed in the laboratory utilising both static and flow climatic chambers for the control of temperature and relative humidity with different gas concentrations (mainly SO_2, NO_x and O_3) and particles.

SO_2 deposition was determined indirectly by measuring the decrease in the gas concentration in a chamber after exposing the test specimens (Gilardi, 1966; Braun and Wilson, 1970). This indirect method may be prone to significant experimental error owing to deposition on to the chamber walls. The alternative method of directly measuring the amount of material deposited on specimens requires long exposure periods or abnormally high SO_2 concentrations in the chamber to increase the amount of deposit and facilitate direct analysis (Johansson *et al.*, 1988; Gauri *et al.*, 1989). Radioactive sulphur ($^{35}SO_2$) has been used in chamber studies and the deposition on material surfaces measured directly by beta-counting techniques (Spedding, 1969a; Serra and Starace, 1973; Fuzzi and Vittori, 1975). The effects of wet deposition of SO_2, NO_x and ozone (O_3) were investigated in a flow chamber, which allowed the rinsing of the specimens with different kinds of water solution, in analogy to field exposure tests, where the runoff is collected and chemically analysed (Dickinson et al., 1988).

Accelerated tests have generally been designed in order to: (a) investigate the degradation process; (b) study the durability of different building materials, and (c) evaluate the efficiency of a specific preservation treatment (Gauri et al., 1973; Rossi Manaresi et al., 1979). This section will briefly address only point (a).

The results achieved using accelerated weathering tests can be grouped according to three lines of research:

(1) The influence of various parameters with potential impact on the degradation process.
(2) The study of the synergistic action of such parameters.
(3) The consideration of the way in which the transformation takes place at the stone surface.

However, the current reality is that the quantity of detailed results on stone materials of relevance to monument conservation is both very limited and heterogeneous with respect to the above-mentioned research topics. In view of this, it is hardly surprising that modelling in the field is in a very unsatisfactory state. As yet, there is no state equation available for stones that is valid in ambient conditions, and only mostly descriptive damage functions have been proposed. Such difficulties explain why no reliable quantitative predictions exist for stone damage.

Radioactive labelling of SO_2 ($^{35}SO_2$) has shown that the rate of absorption and transformation of SO_2 to the oxidised state depends on relative humidity and surface characteristics.

Spedding (1969b) reported that the saturation of a stone surface with SO_2 is reached rapidly, with the rate of uptake depending on relative humidity (SO_2 was taken up only by the matrix material and not by the fossil). Measurements of SO_2 uptake in laboratory exposure apparatus showed that a humidity of around 90% is essential for the progress of the corrosion reaction (Serra and Starace, 1978). These early results indicated that it is the presence of water which is the key factor in the conversion of calcite to gypsum, and thus, in the corrosive action of an SO_2-containing atmosphere on the stone.

Spiker *et al.* (1992 and 1995) found that the surface reaction process is controlled below a critical relative humidity, which is 70% for a limestone (Indiana) and 95% for a marble (Vermont), confirming the crucial role of pore size distribution in determining the degree of damage.

According to the results achieved in simulation chamber experiments (Gauri and Holdren, 1981; Mangio and Johansson, 1989; Gauri and Gwinn, 1982; Johansson *et al.*, 1988; Sabbioni *et al.*, 1996b), the sulphation of stones takes place through two basic mechanisms: (1) the direct formation of gypsum and (2) the formation of calcium sulphite, as an intermediate stage, followed by oxidation into calcium sulphate dihydrate. The reactions occurring at the stone surface are:

$$SO_2 + H_2O \Rightarrow SO_2 \cdot H_2O \tag{4}$$

$$CaCO_3 + SO_2 \cdot H_2O \Rightarrow CaSO_3 \cdot 0.5H_2O + CO_2 + 0.5H_2O$$

$$2CaSO_3 \cdot 0.5H_2O + O_2 + 3H_2O \Rightarrow 2CaSO_4 \cdot 2H_2O$$

$$CaCO_3 + H_2SO_4 + H_2O \Rightarrow CaSO_4 \cdot 2H_2O + CO_2$$

Analyses of the crystalline species forming on the samples agree in identifying sulphate in the form of calcium sulphate dihydrate (gypsum), while sulphite is generally found as calcium sulphite hemihydrate (Gauri *et al.*, 1982; Kozlowski *et al.*, 1992; Johansson *et al.*, 1988). In particular, Gauri and Gwinn (1982) revealed that moisture in the liquid phase was essential for oxidising calcium sulphite into gypsum, while at relative humidity (RH) < 80% no change in the sulphite took place. In the early literature, calcium sulphite dihydrate ($CaSO_3 \cdot 2H_2O$) was reported as an intermediate product of the interaction between SO_2 and calcium carbonate ($CaCO_3$), as mentioned by Amoroso and Fassina (1983), although this failed to be confirmed in later studies. The possible formation of a tetrahydrate, $CaSO_3 \cdot 4H_2O$, or a double salt, $Ca_3 (SO_4) (SO_3)_2 \cdot 12H_2O$ is advanced by Mangio *et al.* (1991), but no further data substantiate their hypothesis.

Whether the SO_2-carbonate interaction is enhanced or not by the presence of NO_2 is the subject of debate, as investigators have produced controversial results. Gauri and Gwinn (1982), for instance, did not detect any calcium nitrate in marble exposed in a flow chamber to levels of 6 and 12 ppm NO_2, but X-ray diffraction measurements are unable to detect low levels of crystalline phases. Livingston (1985) suggests that no intermediate product is formed in the NO_x-$CaCO_3$ interaction. Consequently, NO_x appears to be only slightly physically adsorbed into the carbonate structure, leading to much lower kinetics for the formation of calcium nitrate, while SO_2 is strongly chemisorbed.

Johanson et al. (1988) found that the SO_2-carbonate interaction is enhanced by the presence of NO_2 and stated that, when deposited on a stone surface, SO_2 is first reversibly adsorbed and then reacts to form the relatively stable $CaSO_3 \cdot 0.5H_2O$ phase, which is eventually oxidised to gypsum $CaSO_4 \cdot 2H_2O$. Due to the saturation of the surface with SO_2 (ads), the deposition and corrosion rates soon drop to low levels. In humid atmospheres (90% RH), NO_2 catalytically oxidises SO_2 (ads), thereby promoting the adsorption of more SO_2 molecules. Further, NO_2 promotes the formation of gypsum from $CaSO_3 \cdot 0.5H_2O$. However, at moderate values of RH (i.e. 50%), NO_2 does not influence the SO_2 deposition rate on calcareous stones.

The following reaction scheme for the interaction of SO_2 with marble is proposed:

$$SO_2(g) \overset{\text{rapid}}{\Leftrightarrow} SO_2(ads) \overset{\text{slow}}{\Leftrightarrow} CaSO_3 \cdot 0.5H_2O(s) \overset{\text{slow}(*)}{\Leftrightarrow} CaSO_4 \cdot 2H_2O(s) \quad (5)$$

The reaction marked with an asterisk (*) is rapid in the presence of NO_2 at high humidity. This scheme indicates that the oxidation process may become the rate-limiting step.

Johnson et al. (1990) and Hanef et al. (1992) confirm that a wet stone surface revealed greatest reactivity with SO_2, as compared to nitrogen monoxide (NO) and NO_2. Umierski (1995) found that when adding NO_2 as second pollutant, no sulphate was detected on the stone samples, and no real increase in total sulphur deposition was found, comparing single SO_2 exposure with a gas mix of SO_2 and NO_2.

The deposition of SO_2 on marble is strongly enhanced by the presence of ozone, especially at high RH: O_3 acts as an oxiliser on the loosely bound sulphurous species on the surface, increasing the rate of conversion of these species. Ozone also increases sulphate production by oxidising crystalline calcium sulphite (Mangio and Johansson, 1989).

The deposition of NO, NO_2, HONO, HNO_3 and PAN on Ihrlerstein sandstone was studied in the laboratory by Behlen *et al.* (1996). The reactivity of the gases differed considerably and the following order of succession was found $HNO_3 > HONO > NO_2 > PAN > NO$.

The effects of HF and/or SO_2 on calcareous stones and sandstones were investigated by Vales and Martin (1986a and b). In the case of calcite, the reaction with HF primarily leads to an increase in porosity and the reaction occurring is represented by the equation:

$$2HF + CaCO_3 \Rightarrow CaF_2 + CO_2 + H_2O \qquad (6)$$

The formation of CO_2 in this reaction is indicated to be another important alteration factor, particularly in conditions of high humidity, as it causes the transformation of the original calcium carbonate into bicarbonate, a much more soluble compound. In sandstones, exposure to an atmosphere of HF in saturation humidity conditions causes a greater change in silicates, the dissolution of silica, and hinders the formation of crystalline phases; the reaction products are probably gaseous or volatile, such as silicon tetrafluoride or even fluosilicic acid, and are thus released from the surface of the samples. The synergistic action of HF and SO_2, leading to the formation of fluorite and gypsum, produces more marked effects than those observed when each pollutant acts alone, such as cracks.

The role of the gaseous pollutants HCl and SO_2 was investigated by Hutchinson *et al.* (1992a), who found that HCl reacts rapidly with calcium carbonate to produce a soluble product, $CaCl_2$. Unlike the relatively insoluble $CaSO_4 \cdot 2H_2O$, it is readily washed away, allowing a continuous reaction which gives rise to crust development.

A number of tests have recently been performed on the interaction between stones, SO_2 and aerosol. Cheng *et al.* (1987) show that fly ash covering marble in a simulation chamber with high SO_2 concentration (100 ppm) favours gypsum formation. Hutchinson *et al.* (1992b) performed experiments on carbonate stones using fly ash produced by fuel oil and coal combustion, and concluded that fly ash particles play no active role in the sulphation process and actually screen the surface, providing a certain degree of protection. This screening role is confirmed by Ausset *et al.* (1996) for soot particles produced by light fuel oil combustion in diesel motors, while fly ash produced by heavy fuel oil combustion is found to play a role in fixing sulphur compounds on stone.

In agreement with these results, carbonaceous particles, emitted by oil fuelled power plants and domestic heating systems, are found to be catalysts of the SO_2 oxidation process on stone surfaces, mainly due to their heavy metal content (Sabbioni *et al.*, 1996b). The role in sulphation of particles from vehicle exhaust confirms a close relationship between the composition of the particulate matter and sulphate on the stone surface: diesel engine exhaust, primarily comprising soot and metallic particles composed mainly of Fe and Cr, Ni, Cu and Mn as trace elements, plays the most important part in the catalytic oxidation of SO_2; emissions from gasoline engines, composed of small quantities of soot and high concentrations of Pb and Br result in less gypsum formation (Rodriguez Navarro and Sebastian, 1996).

Boke *et al.* (1996), during experiments using clay minerals, active carbon and metal oxide, found that the presence of particulate matter does not significantly change the extent of sulphation, but affects the ratio between calcium sulphite hemihydrate and gypsum, showing that particles produce a considerable acceleration of the oxidation process at the carbonate surface; these data underline the importance of quantifying both sulphite and sulphate formed by the interaction between SO_2 and material surfaces. The results are confirmed by Umierski (1995) with reference to stone samples treated with solutions of transition metals, i.e. Fe, Mn, Cu, Pt and V. The total sulphur uptake was only slightly increased by the catalysts, but the sulphites are lower than on untreated samples or completely absent, with Cu and Mn showing the highest catalytic effect.

Defining the role of carbonaceous particles in crust formation is particularly important as gypsum can be autonomously nucleated by soot (Del Monte *et al.*, 1984). In addition, gypsum has been found on materials containing no calcium, such as bronze (Riederer, 1973), a result that induced some researchers to ascribe all gypsum found on monuments to atmospheric deposition alone, without any contribution from the stone (Rossi Manaresi, 1975). Laboratory tests have shown that the gypsum nucleated by carbonaceous particles is negligible (no more than 0.04%), compared to the total gypsum forming by the interaction between SO_2 and stones during the experiment (Sabbioni *et al.*, 1996b).

The dry deposition of pollutants to surfaces is typically expressed as a flux; the relationship between mass flux per unit time per unit area to the surface F (g cm^2 s^{-1}) and the airborne pollutant concentration C (g cm^{-3}) in the vicinity of a structure yields a coefficient referred to as the dry deposition velocity in cm s^{-1}:

$$v_d = F/ [C]. \tag{7}$$

The inverse of the deposition velocity is the resistance to mass transfer described in terms of electrical analogy following Slinn *et al.* (1978) as:

$$v_d = (r_a + r_b + r_c)^{-1} \tag{8}$$

where r_a is the aerodynamic resistance, which is a function of wind speed and air turbulence, r_b is the boundary layer resistance depending on molecular diffusivity and turbulence, and r_c is the surface uptake resistance, i.e. a function of surface chemistry (material composition), surface roughness, porosity and surface moisture. The dry deposition rate differs for different materials by the magnitude of r_c: materials with some acid-buffering capacity (carbonate minerals for example) tend to have low surface resistance and, consequently, a relatively high SO_2 deposition velocity.

A compilation from the literature of SO_2 dry deposition velocity data for various stones estimated in laboratory tests is reported in Tables 3 and 4. For nitrogen compounds, the deposition velocities measured by Behlen *et al.* (1996) for NO, NO_2, HONO and HNO_3 in the chamber at 80% RH were respectively 1.8, 0.12, 0.02, 0.006 and < 0.001 cm s^{-1}.

Few published estimates of r_c for building materials are available: Lipfert (1989a) estimated r_c in the range of 2.12–3.85 s cm^{-1} for limestone at about 85% RH, while for dry limestone and sandstone the value was 2–2.5 s cm^{-1}. Spiker *et al.* (1992) measured r_c values of 1.3 and 34 s cm^{-1} for fresh limestone and fresh marble, respectively, at 75% RH, 26°C and 50 ppb SO_2.

Table 3. Deposition velocity (v_d) of SO_2 measured in chamber tests.

Deposition surface	Conc. (mg m^{-3})	Flow (m s^{-1})	RH (%)	v_d (cm s^{-1})	Reference
Marble (dry)	1.1	low	50;60	0.03	Calc. from Gilardi (1966)
Marble (wet)		low	100	0.13	Calc. from Gilardi (1966)
Marble	10 ppm		100	0.08	Coburn et al. (1993)
Carrara marble	0.09 ppm		50	0.006	Mangio and Johansson (1989)
Carrara marble	0.09 ppm		70	0.015	Mangio and Johansson (1989)
Carrara marble	0.09 ppm		90	0.02	Mangio and Johansson (1989)
Carrara marble	1.44 ppm		50	0.001	Mangio and Johansson (1989)
Carrara marble	1.44 ppm		70	0.002	Mangio and Johansson (1989)
Carrara marble	1.44 ppm		90	0.02	Mangio and Johansson (1989)
Carrara marble	0.4	0.001	90	0.058	Henriksen (1995)
Carrara marble	3 ppm		90	0.02	Calc. from Sabbioni et al. (1996)
Pentelic marble	0.4	0.001	90	0.069	Henriksen (1995)
Pentelic marble	0.3 ppm		90	0.049	Calc. from Sabbioni et al. (1996)
Oolitic marble	0.36		11	0.16	Calc from Spedding (1969b)
Oolitic marble	0.28		13	0.09	Calc from Spedding (1969b)
Oolitic marble	0.1		79	0.83	Calc from Spedding (1969b)
Oolitic marble	0.37		81	1.26	Calc from Spedding (1969b)
Dolomite	10 ppm		100	0.05	Coburn et al. (1993)
Limestone	1–5	low	85	0.097	Calc. from Braun and Wilson (1970)
Limestone	1.1	six times than low	85	0.967	Calc. from Braun and Wilson (1970)
Limestone (wet)		low	100	0.21	Gilardi (1966)

Table 3. (*Continued*)

Deposition surface	Conc. (mg m^{-3})	Flow (m s^{-1})	RH (%)	v_d (cm s^{-1})	Reference
Travertine limestone	3 ppm		90	0.01	Calc. from Sabbioni *et al.* (1996)
Trani limestone	3 ppm		90	0.01	Calc. from Sabbioni *et al.* (1996)
Portland limestone	0.3 ppm		90	0.231	Calc. from Sabbioni *et al.* (1996)
Vicenza limestone	0.4	0.001	90	0.1	Henriksen (1995)
Sandstone (dry)	1.1	low	50;60	0.09	Calc. from Gilardi (1966)
Serena sandstone	0.4	0.001	90	0.093	Henriksen (1995)
Sandstone (wet)		low	100	0.2	Calc. from Gilardi (1966)
Baumberger sandstone	2.66		95	0.0336	Umierski (1995)
Granite (dry)	1.1	low	50;60	0.01	Calc. from Gilardi (1966)
Granite (wet)		low	100	0.002	Calc. from Gilardi (1966)
Granite red and grey	0.4	0.001	90	0.038	Henriksen (1995)
Monzonite blue	0.4	0.001	90	0.009–0.018	Henriksen (1995)
Monzonite green	0.4	0.001	90	0.027	Henriksen (1995)
Soapstone	0.4	0.001	90	0.007–0.04	Henriksen (1995)
Syenite red	0.4	0.001	90	0.027	Henriksen (1995)

Table 4. Deposition velocity (v_d) of SO_2 measured in the chamber in the presence of other gases.

Deposition surface	SO_2 (mg m^{-3})	NO_2 (mg m^{-3})	RH (%)	v_d (cm s^{-1})	Reference
Carrara marble	0.4	0.4	70	0.052	Henriksen (1995)
Carrara marble	0.4	0.4	80	0.084	Henriksen (1995)
Carrara marble	0.4	0.4	90	0.135	Henriksen (1995)
Pentelic marble	0.4	0.4	70	0.094	Henriksen (1995)
Pentelic marble	0.4	0.4	80	0.115	Henriksen (1995)
Pentelic marble	0.4	0.4	90	0.136	Henriksen (1995)
Vicenza limestone	0.4	0.4	70	0.115	Henriksen (1995)
Vicenza limestone	0.4	0.4	80	0.120	Henriksen (1995)
Vicenza limestone	0.4	0.4	90	0.125	Henriksen (1995)
Serena sandstone	0.4	0.4	70	0.090	Henriksen (1995)
Serena sandstone	0.4	0.4	80	0.100	Henriksen (1995)
Serena sandstone	0.4	0.4	90	0.109	Henriksen (1995)
	SO_2 (ppm)	O_3 (ppm)			
Carrara marble	1.44	0.25	50	0.002	Mangio and Johansson (1989)
Carrara marble	1.44	0.25	70	0.015	Mangio and Johansson (1989)
Carrara marble	1.44	0.25	90	0.070	Mangio and Johansson (1989)
Carrara marble	0.09	0.25	50	0.035	Mangio and Johansson (1989)
Carrara marble	0.09	0.25	70	0.080	Mangio and Johansson (1989)
Carrara marble	0.09	0.25	90	0.240	Mangio and Johansson (1989)

Simulation chamber tests have also been employed to predict the thickness of the gypsum crust forming on samples exposed for up to 4500 hours in the laboratory (100% RH and 10 ppm SO_2), and the following equations have been proposed:

$$t = C_{Bo}/ k_s C_{As} \cdot \delta l \tag{9}$$

$$t = C_{Bo}/ k_s C_{As} \cdot \delta l + C_{Bo}/ C_{As} D_{eg} \cdot \delta l^2 \tag{10}$$

where t is the time of exposure in seconds, δl is the thickness of the crust, C_{as} and C_{Bo} are the SO_2 and calcite concentrations expressed as moles/cc, respectively, and k_s and D_{eg} are the surface rate constant and effective diffusion coefficient calculated as 4356 cm/h and 0.193 cm_2 h^{-1}, respectively (for Georgia marble). Equation (9) is proposed for the initial portion of the reaction, which is kinetically controlled, while equation (10) is proposed for the entire reaction, encompassing the initial reaction as well as the latter part, when diffusion becomes significant as the reaction progresses. These equations were also applied to the weathering of marble in Chicago for an exposure period from 1912 to 1978 (0.1 ppm SO_2 and 60–80% RH), where a gypsum crust of nearly 90 mm was found (Gauri et al., 1989).

An expression for the crust thickness (δ) in terms of deposition velocity and SO_2 concentration is proposed in Coburn et al. (1993):

$$\frac{d}{dt}(\delta) = \frac{1}{\gamma} \frac{M}{\rho} v_d C_{so_2} \tag{11}$$

where γ represents the stoichiometric coefficient of SO_2 in the reaction ($\gamma = 1$ for marble and $\gamma = 2$ for dolomite), M is the mass of solid product formed, ρ is product density, v_d the deposition velocity, and C_{SO_2} the concentration of SO_2 in mol cm^{-3}. The authors employ the equation for calculating the average rate of crust growth for surfaces unexposed to rain washings, assuming 0.01 ppm SO_2 concentration and deposition velocities of 0.08 cm s^{-1} for marble and 0.05 cm s^{-1} for dolomite: crust growth rates of 1.10 μm y^{-1} for marble and 0.79 μm y^{-1} for dolomite are obtained.

4. Field Exposure Tests

Material loss from stone may occur through two main mechanisms. Physical mechanisms, leading to loss through abrasion, erosion, flaking, spalling or cracking, are important in areas of buildings sheltered from rain washout; such mechanisms are difficult to model and do not represent the main portion of material removed from stones with low porosity, such as marble and limestone. Chemical mechanisms, basically due to the interaction of calcium carbonate and rainwater, take place predominantly in the unsheltered, rain-washed areas.

The dissolution of carbonate stone occurs in a number of ways. The first is the karst effect, named after the geological phenomenon where limestone dissolves in rainwater in the absence of air pollution. This process, which is specifically dealt with in Chapter 2, is represented by the following equation:

$$CaCO_3 + H_2O + CO_2 \Leftrightarrow Ca^{2+} + 2HCO_3^- \tag{12}$$

The second mechanism is the acid rain effect. Thus, a chemical reaction of the following type is envisaged for limestone:

$$CaCO_3 \text{ (solid)} + 2H^+ \Rightarrow Ca^{2+} + H_2O + CO_2 \text{ (gas)} \tag{13}$$

In other words, an acid-based reaction proceeds, with the acid rain solution being neutralised through reaction with the calcium carbonate minerals from the limestones.

Several studies have dealt with the major processes affecting the dissolution of rain-washed limestones in the environment by field exposure tests. The approach involves exposing samples of materials in the environment at a number of different sites and measuring the accumulation damage: the advantage is that the samples are exposed to real ambient conditions, but only a limited number of sites are tested and experiments do not generally last long enough, at least ten years being necessary to obtain meaningful results. Three alternative types of measurement are proposed in the literature: surface recession, mass loss and the measurement of solutes dissolved from the surface in rainfall runoff.

Honeyborne and Price (1977) performed exposure trials over a ten-year period to determine the relative effects of pollutants in urban and semi-rural atmospheres, indicating the absence of a direct correlation between environmental pollutant levels and rates of limestone decay. Weber (1985) studied the weathering resistance of different building stones for two years in a closely monitored urban atmosphere at several sites in the city of Vienna. He determined the flux rate of the sulphurous compounds in the atmosphere, taking into account the different bulk densities of the stone under investigation (quartz-sandstones, marble, compact and porous limestones). Theoretical surface erosion for a time period of 100 years was calculated from mass loss and values up to 6.2 mm/100y were obtained, which were considered reasonably fitting with *in situ* observations. Jaynes and Cooke (1987) exposed stone samples on carousels in rain-exposed and rain-sheltered positions under monitored environmental conditions at several sites in southeast England. The weight loss of exposed samples was attributed to both solution and sulphation, the latter being estimated to account for 39% and 44% of loss for Portland and Monk's Park stone, respectively.

In work carried out as part of the United States National Acid Precipitation Assessment Programme (NAPAP), Youngdahl and Doe (1986) and Reddy *et al.* (1986) attempted to quantify acid rain damage on carbonate rocks (Shelburne marble from Vermont and Salem limestone from Indiana), discussing the form of a preliminary equation to relate environmental damage factors. They also showed how much of the environmental damage on stone results in surface material loss, usually by reaction and dissolution processes, but also by the accumulation of surface and sub-surface reaction products which subsequently become detached together with parts of the unreacted stone. Their experiments included quantitative measurements to evaluate erosion of the stone material using gravimetric and interferometric laser profiling techniques, as well as chemical analyses of runoff solution from the stone surface. Marble recession was correlated with hydrogen ion deposition to the rock. They conclude that the contribution of acid rain to the chemical damage ranges between 5–20% for dry deposition of SO_2, between 7–26% for dry deposition of HNO_3 and around 10% for wet deposition.

Working in the same programme, McGee and Mossotti (1992) investigated gypsum accumulation on carbonate stones by examining crusts on limestone and marble samples from monitored sites and also from ancient buildings in the same materials. Gypsum layers on a limestone sample exposed for two years range from 100–250 μm, approximately ten times thicker than the 10–30 μm on the identically exposed marble samples, a discrepancy owing to the different water-accessible porosity, which is 17 and 0.3% for limestone and marble, respectively. Furthermore, gypsum crystal formation was mostly superficial: no crystals were observed below a depth of 0.2 mm, although elevated levels of sulphate (varying from about 10.000–40.000 ppm) were detected within the interior of stones. Particles were observed by SEM to be readily trapped by the bladed network of gypsum crystals that cover the exposed stone surface.

Again within the NAPAP, Baedecker et al. (1992) quantified the effect of the wet and dry deposition of H^+, SO_2 and HNO_3 on stone erosion. The physical measurements of the surface recession at five material exposure sites ranged from 15–30 μm y^{-1} for marble and from 25–45 μm y^{-1} for limestone; these values are double those based on the observed calcium content of runoff solution. The manifest differences between physical and chemical recession are basically attributed to the loss of mineral grains from the stone surface, not measured in the runoff experiments. They conclude from their chemical analyses of runoff solutions that as much as 30% of erosion by dissolution could be due to the wet deposition of H^+ and dry deposition of SO_2 and HNO_3 between rain events. The remaining 70% of erosion by dissolution is attributed to the solubility of calcium carbonate in rain, which is in equilibrium with atmospheric carbon dioxide ("clean rain").

Girardet and Furlan (1991) used pyrolysis and infra-red analyses in an effort to distinguish between the respective contributions of gas and aerosol in the dry deposition of sulphur on field-exposed stone samples, finding that deposits from gaseous sulphur predominates over those from aerosol sulphur. In Milan, a typical example of a large industrialised city, the contribution of aerosol sulphur represents only 2–5% of the total sulphur taken up from sandstone samples, while, for the same stone exposed in a low polluted site, such as a rural area, its contribution rises to 5–20% of the total.

Butlin et al. (1992) reported the results obtained in the National Materials Exposure Programme (NMEP), covering 29 sites located throughout the United Kingdom, where samples of stones (Portland limestone, White Mansfield dolomitic sandstone and Monk's Park limestone) were exposed mounted on freely rotating carousels in rain-exposed and rain-sheltered positions. For the unsheltered samples, the weight loss ranged from 0.16% (Portland limestone) to 1.22% (Monks Park limestone) per year, and a significant correlation was found between weight changes and SO_2 concentration, rainfall volume and hydrogen ion loading (mg H^+ m^{-2}). The sheltered samples, as expected, gained weight as a result of dry deposition (negative correlation). Chemical analyses of soluble ions (Cl^-, NO_3^-, SO_4^{2-}, Na^+, K^+, NH_4^+, Mg^{2+}, Ca^{2+}) by ion chromatography were also performed at different depths in the specimens, and were expressed as variations in concentration (μg g^{-1}) with respect to blank samples. As expected, the data showed an increase in acid species in the sheltered samples evidencing two main effects: the marine effect, with a correlation of Cl^-, Na^+, K^+ Mg^{2+}, Ca^{2+} concentrations with each other and weight change, and a pollution effect, where SO_2 and NO_2 correlate with SO_4^{2-} and NO_3^-, respectively, and with weight change. The SO_2 deposition velocity was also calculated to range from 6 to 15 mm s^{-1}.

A linear equation containing the annual average SO_2 and NO_2, rainfall and rainfall acidity (accounting for at least > 60% of the measured variations in weight change) for unsheltered Portland and White Mansfield stone is proposed:

$$\text{Wt loss} = 0.08 + 0.010 \ [SO_2] - 0.00012 \ [NO_2] \\ + 0.00016 \ R + 0.0026 \ [H^+] \qquad (14)$$

$$\text{Wt loss} = 0.37 + 0.017 \ [SO_2] - 0.0090 \ [NO_2] \\ + 0.00004 \ R + 0.0039 \ [H^+] \qquad (15)$$

where weight change is in %, SO_2 and NO_2 in μg m^{-3}, rain (R) in mm and H^+ in mgH^+ m^{-2}.

Field trials in the United Kingdom (Webb *et al.*, 1992) on Portland limestone exposed at nine sites for periods of up to three years show that the average weight loss over all samples is 0.14 g $m^{-2}d^{-1}$, equivalent to a 24 μm y^{-1} surface retreat, a value within the range reported by Honeyborne and Price (1977) and Jaynes and Cooke

(1987). No residual effect on stone loss due to nitrogen oxide concentrations was found. A model for the chemical dissolution of rain-washed limestone was derived considering the ion and mass balance between the incident rainwater and runoff rainwater and fitting the measured loss rates. The authors attempted to express the most important processes affecting the dissolution of rain-washed limestone, as (1) conversion of the limestone to calcium sulphate by dry deposition of SO_2 and its subsequent dissolution, (2) dissolution of $CaCO_3$ due to normal interaction with H_2O and atmospheric CO_2, and (3) dissolution due to the neutralisation of the excess acidity in rain as a result of air pollution:

$$\text{stone loss} \cong SO_4^{2-} \text{ due to dry deposition}$$
$$+ \text{ "natural" bicarbonate solubility}$$
$$+ \text{ rain acid neutralisation}$$

As a generic case, they propose the following chemical model:

$$\text{stone loss(moles)} \cong ADv_dC_{SO_2}$$
$$+ (K_HK_1P_{CO_2} / 2[H^+]_r) \Sigma (A_iR - E_{vap})$$
$$+ ([H^+]_i / 2) \Sigma A_iR \qquad (16)$$

where C_{SO_2} is the mean SO_2 concentration during the exposure of duration (D), v_d is the deposition velocity, A is the surface area of the stone exposed, A_i is the rainfall interception area, $[H^+]_r$ and $[H^+]_i$ are the volume-weighted mean hydrogen ion concentrations of the runoff and rainfall, respectively, E_{vap} is the volume of rainfall evaporated from the stone sample, K_H and K_1 are the equilibrium constant of carbonate and bicarbonate in equilibrium with atmospheric concentration of 350 ppm CO_2 (Stumm and Morgan, 1970) and R is the amount of rainfall (in mm). They concluded that in determining stone loss, the natural solubility of limestone in water turns out to be the dominant term in the model, accounting for 50–90% of the stone loss. The dry deposition of SO_2 accounts for less than 5–50%, and the neutralisation of rainfall acidity always represents the least significant contribution, being between 0.3 and 3.2%.

In addition, Vleugels *et al.* (1994) investigated the alteration of the surface composition in naked and treated limestones of different origin (French Massangis limestone, Belgian Balegem and Gobertange limestone) after environmental exposure, also comparing the effects of wet versus dry deposition. After a six-month exposure period, the totally exposed surfaces were chemically indistinguishable from the reference interior material, while after one year of exposure, gypsum formed from dry sulphur compound deposition was observed (with a thickness of about 500 μm), together with NH_4^+ on the sheltered surface. However, in this area, the contribution of NO_3^- and Cl^- weathering salts was very small compared to the SO_4^{2-} salts. In particular, the SO_4^{2-}, Cl^- and NO_3^- ranged respectively from 0.01–1.45% in weight, 0.03–1.20% and 0.01–0.56%.

Wittenburg and Dannecker (1994) and Steiger and Danneker (1994) measured the enrichment of soluble salts (particularly nitrate and sulphate) in the surface of typical sandstones (Obernkirchen, Ihrlerstein and Sander sandstone), calculating the deposition velocities of SO_2 and NH_3, and the flux of sulphate, nitrate and ammonium. Furthermore, to identify the contribution of HNO_3, NO_2 and particulate nitrate (p-NO_3^-) to the total input, multiple regression analysis was used, as reported in the following equation:

$$F\text{-}NO_3 = v_{dHNO_3} \cdot c[HNO_3] + v_{dNO_2} \cdot c[NO_2] + v_{dNO_3^-} \cdot [NO_3^-]$$

$$(17)$$

where the deposition velocities for the single species (respectively 0.64 ± 0.06 cm s^{-1} for HNO_3, 0.002 ± 0.004 cm s^{-1} for NO_2 and 0.006 ± 0.03 cm s^{-1} for p-NO_3) appear as coefficients. The product of deposition velocity and average concentration of each component gives the contribution of each species, which turns out to be 88, 8 and 3% for HNO_3, NO_2 and p-NO_3, respectively. It was finally shown that there may well be a significant contribution due to the deposition of calcium-rich particles from the atmosphere to large gypsum accumulations in sandstone.

Henriksen (1994 and 1995) studied dry deposition on stone surfaces by means of field studies conducted in Norway, also calculating the deposition velocities. The penetration depth of SO_2 into the stone was found to be in the range of 1.2 to 5.0 mm, with as much as 90% of the SO_4^{2-} in the upper 0.3 mm.

The deposition velocity of SO_2 and HNO_3 obtained by field exposure tests are summarised in Tables 5 and 6. As can be observed

Table 5. Deposition velocity (v_d) of SO_2 measured in field exposure tests.

Deposition surface	Site	v_d (cm s^{-1})	Reference
Carrara marble vertical	Gettysburg (Pennsylvania)	0.06	Dolske (1995)
Carrara marble	Gettysburg (Pennsylvania)	0.33	Dolske (1995)
Carrara marble	Gettysburg (Pennsylvania)	0.23	Dolske (1995)
Carrara marble	Gettysburg (Pennsylvania)	0.75	Dolske (1995)
Carrara marble	Gettysburg (Pennsylvania)	0.07	Dolske (1995)
Carrara marble	Borregaard (Norway)	0.80	Henriksen (1995)
Carrara marble	Oslo (Norway)	0.31	Henriksen (1995)
Carrara marble	Milan (Italy)	0.38	Pantani et al. (1998)
Carrara marble	Ancona (Italy)	0.12	Pantani et al. (1998)
Pentelic marble	Borregaard (Norway)	0.44	Henriksen (1995)
Pentelic marble	Oslo (Norway)	0.37	Henriksen (1995)
Blue marble	Gettysburg (Pennsylvania)	0.07	Dolske (1995)
Blue marble	Gettysburg (Pennsylvania)	2.20	Dolske (1995)
Blue marble	Gettysburg (Pennsylvania)	0.22	Dolske (1995)
Travertine limestone	Milan (Italy)	0.41	Pantani et al. (1998)
Travertine limestone	Ancona (Italy)	0.18	Pantani et al. (1998)
Vicenza limestone	Borregaard (Norway)	0.80	Henriksen (1995)
Vicenza limestone	Oslo (Norway)	0.60	Henriksen (1995)
Portland limestone	Milan (Italy)	0.89	Pantani et al. (1998)
Portland limestone	Ancona (Italy)	0.45	Pantani et al. (1998)

Table 5. (*Continued*)

Deposition surface	Site	v_d (cm s^{-1})	Reference
Trani limestone	Milan (Italy)	0.40	Pantani *et al.* (1998)
Trani limestone	Ancona (Italy)	0.19	Pantani *et al.* (1998)
Obernkirchen sandstone	Germany	0.41*	Wittenburg and Dannecker (1994)
Obernkirchen sandstone	Germany	0.06†	Wittenburg and Dannecker (1994)
Ihrlestein sandstone	Germany	0.67*	Wittenburg and Dannecker (1994)
Ihrlestein sandstone	Germany	0.42†	Wittenburg and Dannecker (1994)
Sander sandstone	Germany	0.57*	Wittenburg and Dannecker (1994)
Sander sandstone	Germany	0.15†	Wittenburg and Dannecker (1994)

(*) SO_2 concentration lower than 50 μg · m^{-3}.
(†) SO_2 concentration higher than 50 μg · m^{-3}.

Table 6. Deposition velocity (v_d) of HNO_3 measured in field exposure tests.

Deposition surface	Site	v_d (cm s^{-1})	Reference
Carrara marble vertical	Gettysburg (Pennsylvania)	0.06	Dolske (1995)
Carrara marble	Gettysburg (Pennsylvania)	0.17	Dolske (1995)
Carrara marble	Gettysburg (Pennsylvania)	0.18	Dolske (1995)
Carrara marble	Gettysburg (Pennsylvania)	0.16	Dolske (1995)
Carrara marble	Milan (Italy)	0.32	Pantani et al. (1998)
Carrara marble	Ancona (Italy)	0.09	Pantani et al. (1998)
Blue marble	Gettysburg (Pennsylvania)	0.16	Dolske (1995)
Blue marble	Gettysburg (Pennsylvania)	0.17	Dolske (1995)
Travertine limestone	Milan (Italy)	0.32	Pantani et al. (1998)
Travertine limestone	Ancona (Italy)	0.16	Pantani et al. (1998)
Portland limestone	Milan (Italy)	0.54	Pantani et al. (1998)
Portland limestone	Ancona (Italy)	0.62	Pantani et al. (1998)
Trani limestone	Milan (Italy)	0.33	Pantani et al. (1998)
Trani limestone	Ancona (Italy)	0.40	Pantani et al. (1998)
Sandstone	Germany	0.64	Wittenburg and Dannecker (1994)

comparing Tables 3 and 6, the velocities calculated from field studies are much higher than in climate chamber tests, showing the importance of synergistic effects. Surface recession rates on building materials were extensively summarised by Lipfert (1989b), and Table 7 is therefore limited to more recent results.

The United Nation Economic Commission for Europe (UN ECE) Material Exposure Programme measured several parameters of deterioration for two types of stones (Portland limestone and White Mansfield dolomitic sandstone) at 39 test sites in 12 European countries, the United States and Canada under conditions of wet and dry deposition (Kucera and Fitz, 1995). After the eight-year exposure period (Butlin *et al.*, 1998), they concluded that SO_2 is the main contributor to the degradation of calcareous stones in areas of high pollution; wet deposition and natural dissolution are less important and can be considered relevant only in places where pollution concentrations are low. Linear functions for unsheltered Portland and White Mansfield stones have been formulated:

$$WL = -0.0071 \ [SO_2] - 2.4 \ 10^{-4} \ R \tag{18}$$

$$WL = -0.0085 \ [SO_2] - 1.9 \ 10^{-4} \ R \tag{19}$$

where WL is the percentage weight change $((W1-W_0)/W_0 \times 100$, W_1 being the mass after each year of exposure and W_0 the initial mass), $[SO_2]$ the concentration of SO_2 ($\mu g \ m^{-3}$) and R the yearly total rain (mm).

For the sheltered samples, SO_4^{2-} is correlated with weight gain and SO_2 concentration, confirming the predominance of SO_2 as the most reactive pollutant. The high concentrations of Cl^- found in many samples are linked to marine influences and also to industrial pollution (possibly HCl).

Few studies report the modification of the surface structure of exposed samples. From optical and scanning electron microscopy (SEM) observations, the blackening of stones exposed in the NAPAP experiment is reported and attributed to the presence of carbonaceous particles (Ross *et al.*, 1989). Studies on surface modifications using SEM-EDAX were conducted by Smith *et al.* (1994) and Smith and Whalley (1996), revealing the presence of etching at the rain-washed

Table 7. Surface recession rate of marble and limestone from field tests.

Deposition material	Site	Exposure time (y)	Recession rate (mm y^{-1})	Reference
Portland limestone	13 sites (U.K.)	1–3	24	Webb *et al.* (1992)
Shelburne marble (*Vermont*)	5 sites (USA)	10	15–30	Baedecker *et al.* (1992)
Salem limestone (*Indiana*)	5 sites (USA)	10	25–45	Baedecker *et al.* (1992)
Laspra dolomite (*Spain*)	Garston-Westminster	1	41.1–47.6	Grossi *et al.* (1995)
Hontoria limestone (*Spain*)	Garston-Westminster	1	11.8–11.4	Grossi *et al.* (1995)
Portland limestone (*UK*)	Garston-Westminster	1	12–13.9	Grossi *et al.* (1995)
Combe Down limestone (*UK*)	Garston-Westminster	1	40.8–42.7	Grossi *et al.* (1995)

surface, along with crystal gypsum formation and cenospheres deriving from oil and coal combustion containing sulphur and detectable levels of V, Ti and Mn. Colourimetric analyses, performed on stone specimens exposed in a polluted urban area (Milan) for three years showed a relationship between the blackening and exposure time; the deposition of particulate matter rich in carbon particles turns out to incur a significant change in the surface reflectance (Realini *et al.*, 1997).

Brüggerhoff *et al.* (1996) and Simon and Snethlage (1996) have developed exposure programmes which provide a basis for using natural stone samples (sandstone and marbles) as a sensitive material for environmental control under different climatic conditions.

References

Agrawal O.P., Singh T. and Jian K.K. (1986) In *Case Studies in the Conservation of Stone and Wall Paintings*, eds. Bromelle N.S. and Smith P. IIC, London, pp. 165–169.

Alaimo R., Deganello S. and Montana G. (1986–1987) *Mineral. Petrog. Acta*, pp. 271–286.

Alessandrini G. (1992) In *Electron Microscopy*, eds. Lopez-Galindo A. and Rodriguez-Garcia M.I. Publicaciones Universidad Granada, Granada, Vol. 2, pp. 765–769.

Alessandrini G., Bonecchi R., Peruzzi R. and Toniolo L. (1989) In *Proceedings of the 1st International Symposium on The Oxalate Films: Origin and Significance in the Conservation of Works of Art*. Centro Gino Bozza, Milan, pp. 137–150.

Amoroso G.G. and Fassina V. (1983) *Stone Decay and Conservation*. Elsevier, Amsterdam.

Andersson T. (1985) In *Proceedings of the 5th Int. Congress on Deterioration and Conservation of Stone*. Presses Polytechniques Romandes, Lausanne, Vol. 2, pp. 1035–1043.

Arnold A. (1981) In *Proceedings of the Conservation of Stone*. Centro Conservazione Sculture all'Aperto, Bologna, pp. 13–23.

ASTM (American Society for Testing and Materials), (1989) *Soil and Rock; Building Stones; Geotextiles*. Washington D.C., Vol. 04.08, pp. 6–8.

Ausset P., Bannery F. and Lefevre R.A. (1992) In *7th International Congress on Deterioration and Conservation of Stone*, eds. Delgado Rodrigues J., Henriques F. and Telmo Jeremias F. Lab. Nat. Engenharia Civil, Lisbon, pp. 325–334.

Ausset P. *et al.* (1996) *Atmos. Env.* **30**, 3197–3207.

Bacci M., Porcinai S. and Radicati B. (1997) *Appl. Spectro.* **51**, 700–706.

Baedecker P.A., Reddy M.M., Reinmann K.J. and Sciammarella C.A. (1992) *Atmos. Env.* **26B**, 147–158.

Baer N.S. and Berman S. (1983) In *Proceedings of the 76th Annual Meeting of the APCA*. Air Pollution Control Association, Atlanta, No. 5.7.

Behlen A., Wittenburg C., Steiger M. and Dannecker W. (1996) In *Proceedings of the 8th International Congress on Deterioration and Conservation of Stone*, ed. Riederer J. Moller Druck und Verlag, Berlin, Vol. 1, pp. 377–385.

Blaeuer C. (1985) In *Proceedings of the 5th International Congress on Deterioration and Conservation of Stone*. Presses Polytechniques Romandes, Lausanne, Vol. 1, pp. 381–390.

Boke H., Gokturk H. and Caner E.N. (1996) In *Proceedings of the 8th International Congress on Deterioration and Conservation of Stone*, ed. Riederer J. Moller Druck und Verlag, Berlin, Vol. 1, pp. 407–414.

Braun R.C. and Wilson M.J.G. (1970) *Atmos. Env.* **4**, 371–378.

Brocco D., Giovagnoli A., Laurenzi-Tabasso M., Marabelli M., Tappa R. and Polesi R. (1988) *Durab. Build. Mater.* **5**, 393–408.

Bruggerhoff S., Laidig G. and Schneider J. (1996) In *Proceedings of the 8th International Congress on Deterioration and Conservation of Stone*, ed. Riederer J. Moller Druck und Verlag, Berlin, Vol. 2, pp. 861–869.

Butlin R.N., Cooke R.U., Jayne S.M. and Sharp A.S. (1985) In *Proceedings of the 5th International Congress on Deterioration and Conservation of Stone*. Presses Polytechniques Romandes, Lausanne, Vol. 1, pp. 537–551.

Butlin R.N. *et al.* (1992) *Atmos. Env.* **26B**, 189–198.

Butlin R.N., Yates T., Ashall G.J. and Massey S. (1998) *Evaluation of Decay to Stone Tablets. Part 3*. Building Research Establishment, Garston, No. 24.

Camuffo D., Del Monte M., Sabbioni C. and Vittori O. (1982) *Atmos. Env.* **16**, 2253–2259.

Camuffo D. (1998) *Microclimate for Cultural Heritage*. Elsevier, Amsterdam.

Caner E.N., Göktürk, Türkmenoglu A.G. and Eseller G. (1988) In *Air Pollution and Conservation*, ed. Rosvall J. Elsevier, Amsterdam, pp. 279–289.

Carretero M.I. and Galan E. (1996) In *Proceedings of the 8th International Congress on Deterioration and Conservation of Stone*, ed. Riederer J. Moller Druck und Verlag, Berlin, Vol. 1, pp. 311–324.

Chabas A. and Lefevre R. (1996) In *Proceedings of the 8th International Congress on Deterioration and Conservation of Stone*, ed. Riederer J. Moller Druck und Verlag, Berlin, Vol. 1, pp. 415–422.

Cheng R.J., Hwu J.R., Kim J.T. and Leu S.H. (1987) *Anal. Chem.* **59**, 104–106.

Cipriani C. and Franchi L. (1958) *Boll. Serv. Geol. Ital.* **79**, 555–556.

Coburn G.W. *et al.* (1993) *Atmos. Env.* **27B**, 193–201.

Del Monte M., Sabbioni C. and Vittori O. (1981) *Atmos. Env.* **15**, 645–652.

Del Monte M., Sabbioni C., Ventura A. and Zappia G. (1984) *Sci. Total Env.* **36**, 247–254.

Derbez M. and Lefevre R.A. (1996) In *Proceedings of the 8th International Congress on Deterioration and Conservation of Stone*, ed. J. Riederer Moller Druck und Verlag, Berlin, Vol. 1, pp. 359–370.

Dickinson C., Hanef S., Johnson J.B., Thompson G.E. and Wood G.C. (1988) *Europ. Cult. Herit. Newslett. Res.* **2**(4), 13–21.

Dolske D.A. (1995) *Sci. Total Env.* **167**, 15–31.

Esbert R.M., Diaz-Pache F., Alonso F.J., Ordaz J. and Grossi C.M., (1996) In *Proceedings of the 8th International Congree on Deterioration and Conservation of Stone*, ed. J. Riederer. Moller Druck und Verlag, Berlin, Vol. 1, pp. 393–399.

Fassina V. (1988) *Durab. Build. Mater.* **5**, 317–358.

Fobe B., Sweevers H., Vleugels G. and van Grieken R. (1993) *Sci. Total Env.* **132**, 53–70.

Franzini M., Gratziu C. and Wicks E. (1984) *Soc. Ital. Mineral. Petrog.* **39**, 59–70.

Fuzzi S. and Vittori O. (1975) In *Proceedings the Conservation of Stone*. Centro Conservazione Sculture all'Aperto, Bologna, 1975, pp. 651–661.

Garcia-Valles M., Molera J., Vendrell-Saz M. and Veniale F. (1997) In *Proceedings of the 4th International Symposium on the Conservation of Monuments in the Mediterranean*, eds. Moropoulou A., Zezza F., Kollias E. and Papachristodoulou I. Press Line, Athens, Vol. 1, pp. 173–179.

Gauri K.L., Doderer G.C., Lipscomb N.T. and Sarma A.C. (1973) *Studies in Conservation* **18**, 25–35.

Gauri K.L. and Holdren G.C. (1981) *Env. Sci. Technol.* **15**, 386–390.

Gauri K.L. and Gwinn J.A. (1982) *Durab. Build. Mater.* **1**, 217–223.

Gauri K.L., Popli R. and Sarma A.C. (1982) *Durab. Build. Mater.* **1**, 209–216.

Gauri K.L. Chowdhury A.N., Kulshreshtha N.P. and Punuru A.R. (1989) *Studies in Conservation* **34**, 201–206.

Gilardi E.F. (1966) *Absorption of Atmospheric Sulphur Dioxide by Clay, Brick and Other Building Materials*. Ph.D. dissertation, Rutgers State University, University Microfilms 67-8187, Ann Arbor, Michigan, USA.

Girardet F. and Furlan V. (1991) In *Science Technology and European Cultural Heritage*, eds. Baer N.S., Sabbioni C. and Sors A.I. Butterworth-Heinemann Ltd., Oxford, pp. 350–353.

Gobbi G., Zappia G. and Sabbioni C. (1998) *Atmos. Env.* **32**, 783–789.

Grossi C.M., Murray M. and Butlin R.N. (1995) *Water Air Soil Pollut.* **85**, 2713–2718.

Guidobaldi F., Tabasso M.L. and Meucci C. (1985) *Bollettino d'Arte*, 121–134.

Haber J., Haber H., Kozlowski R., Magiera J. and Pluska I.E. (1988) *Durab. Build. Mater.* **5**, 499–548.

Halsey D.P., Dews S.J., Mitchell D.J. and Harris F.C. (1996) In *Processes of Urban Stone Decay*, eds. Smith B.J. and Warke P.A. Donhead Publishing Ltd., London, pp. 53–65.

102   C. Sabbioni

Hanef S.J., Johnson J.B., Dickinson C., Thompson G.E. and Wood G.C. (1992) *Atmos. Env.* **26A**, 2963–2974.
Henley K.J. (1967) In *Proceedings Clean Air Conference.* Blackpool, pp. 55–60.
Henriksen J.F. (1994) In *Proceedings of the 3rd International Symposium on the Conservation of Monuments in the Mediterranean Basin,* eds. Fassina V., Ott H. and Zezza F. Venice, pp. 189–193.
Henriksen J.F. (1995) In *Proceedings of the 5th International Conference on Acidic Deposition Science and Policy.* Goteborg.
Honeyborne D.G. and Price C.A. (1977) *Air Pollution and the Decay of Limestones.* Building Research Establishment, Garston, No. 117/77.
Hutchinson A.J. *et al.* (1992a) *Atmos. Env.* **26A**, 2785–2793.
Hutchinson, A.J. *et al.* (1992b) *Atmos. Env.* **26A**, 2795–2803.
Jaynes S.M. and Cooke R.U. (1987) *Atmos. Env.* **21**, 1601–1622.
Jeannette D. (1980) *Sci. Geol. Bull.* **33**, 111–118.
Jeannette D. (1981) *Sci. Geol. Bull.* **34**, 37–46.
Johansson L.G., Lindqvist O. and Mangio R.E. (1988) *Durab. Build. Mater.* **5**, 439–449.
Jones D. and Wilson M.J. (1985) *Int. Biodeteriorat.* **5**, 99–104.
Johnson J.B. *et al.* (1990) *Atmos. Env.* **24A**, 2585–2592.
Kozlowski R., Magiera J., Weber J. and Haber J. (1990) *Studies in Conservation* **35**, 205–221.
Kozlowski R., Hejda A., Ceckiewicz S. and Haber J. (1992) *Atmos. Env.* **26A**, 3242–3248.
Krumbein W.E. (1988) *Durab. Build. Mater.* **5**, 359–382.
Kucera V. and Fitz S. (1995) *Water Air Soil Pollut.* **85**, 153–163.
La Iglesia A., Garcia del Cura M.A. and Ordonez S. (1994) *Sci. Total Env.* **152**, 179–188.
Lal B.B. (1978) In *Proceedings of the Deterioration and Protection of Stone Monuments.* RILEM, Paris, Vol. 7.8, pp. 1–36.
Lazzarini L. (1993) In *Stone Material in Monuments: Diagnosis and Conservation,* ed. Zezza F. Community of Mediterranean Universities, Bari, pp. 160–168.
Lewin S.Z and Charola A.E. (1978) *Scanning Electron Microscopy* **1**, 695–703.
Lewin S.Z. (1982) In *Conservation of Historic Stone Buildings and Monuments,* ed. Baer N.S. National Academy Press, Washington, D.C. pp. 120–144.
Leysen L., Roekens E., Storms H. and van Grieken R.E. (1987) *Atmos. Env.* **21**, 2425–2433.
Leysen L., Roekens E. and van Grieken R. (1989) *Sci. Total Env.* **78**, 263–287.
Leysen L., Roekens E., van Grieken R. and De Geyter G. (1990) *Sci. Total Env.* **90**, 117–147.
Lipfert F.W. (1989a) *J. Air Pollut. Control Assoc.* **39**, 446–452.
Lipfert F.W. (1989b) *Atmos. Env.* **23**, 415–429.
Livingston, R.A. (1985) In *Proceedings of the 5th International Congress on Deterioration and Conservation of Stone.* Presses Polytechniques Romandes, Lausanne, Vol. 1, pp. 509–516.

Mangio R. and Johansson L.G. (1989) In *Proceedings of the 11th Scandinavian Corrosion Congress.* Stavanger, F45, pp. 1–6.

Mangio R., Langer V. and Johansson L.G. (1991) *Acta Chem. Scand.* **45**, 572–577.

Margolis S.V., Preusser F. and Showers W.J. (1988) In *Materials Issues in Art and Archaeology,* eds. Sayre E.V., Vandiver P., Druzik J. and Stevenson C. Materials Research Society, Pittsburgh, pp. 53–58.

Martinez G.M. and Martinez E.N. (1991) *Studies in Conservation* **36**, 99–110.

Mason B. (1966) *Principles of Geochemistry.* John Wiley and Sons, New York.

McGee E.S. and Mossotti V.G. (1992) *Atmos. Env.* **26B**, 249–253.

Mirwald P.W., Kraus K. and Wolff A. (1988) *Durab. Build. Mater.* **5**, 549–570.

Mirwald P.W. (1991) *Science Technology and European Cultural Heritage,* eds. Baer N.S., Sabbioni C. and Sors A.I. Butterworth-Heinemann Ltd., Oxford, pp. 111–123.

Moropoulou A., Bisbikou K., Torfs K., van Grieken R., Zezza F. and Macri F. (1998) *Atmos. Env.* **32A**, 967–982.

Neill H.L. and Smith B.J. (1996) In *Processes of Urban Stone Decay,* eds. Smith B.J. and Warke P.A. Donhead Publishing Ltd., London, 113–124.

Pantani M., Sabbioni C., Bruzzi L. and Manco D. (1998) In *Proceedings of Air Pollution 98,* pp. 675–684.

Pavia Santamaria S., Cooper T.P. and Caro Calatayud S. (1996) In *Processes of Urban Stone Decay,* eds. Smith B.J. and Warke P.A. Donhead Publishing Ltd., London, pp. 125–132.

Pye K. and Schiavon N. (1989) *Nature* **342**, 663–664.

Realini, M. and Toniolo L. (1996) *The Oxalate Films: Origin and Significance in the Conservation of Works of Art.* EDITEAM, Milan.

Realini M., Negrotti R., Colombo C. and Toniolo L. (1997) In *Proceedings of the 4th Inter Symposium on the Conservation of Monuments in the Mediterranean,* eds. Moropoulou A., Zezza F., Kollias E. and Papachristodoulou I. Press Line, Athens, pp. 253–262.

Reddy M.M., Sherwood S.I. and Doe B.R. (1986) In *Materials Degradation Caused by Acid Rain,* ed. Baboian R. American Chemical Society, Symposium Series 318, Washington, pp. 226–238.

Riederer J. (1973) In *Proceedings of the 1st Coll International Des Pierres en Oeuvre.* La Rochelle, pp. 119.

Rodriguez Navarro C. and Sebastian E. (1994) *Mineral. Mag.* **58**, 781–782.

Rodriguez Navarro C. and Sebastian E. (1996) *Sci. Total Env.* **187**, 79–91.

Roekens E., Bleyen C. and van Grieken R. (1989) *Env. Pollut.* **57**, 289–298.

Roekens E. and van Grieken R. (1989) *Atmos. Env.* **23**, 271–277.

Rogge W.F. *et al.* (1993a) *Env. Sci. Technol.* **27**, 636–651.

Rogge W.F. *et al.* (1993b) *Env. Sci. Technol.* **27**, 1892–1904.

Ross M., McGee E.S. and Ross D.R. (1989) *Am. Mineral.* **74**, 367–383.

Rossi Manaresi R. (1975) *ICOM Committee for Conservation.* 75/5/3/1–24, 4th Meeting, Venice.

Rossi Manaresi R., Alessandrini G., Fuzzi S. and Peruzzi R. (1979) In *Proceedings of the 3rd International Congress on Deterioration and Preservation of Stones*. Venice, pp. 357–374.

Rossi Manaresi R. (1996) In *Proceedings of the 2nd International Symposium on the Oxalate Films: Origin and Significance in the Conservation of Works of Art*, eds. Realini M. and Toniolo L. EDITEAM, Milan, pp. 113–127.

Rosvall J. (1988) *Air Pollution and Conservation*. Elsevier, Amsterdam.

Sabbioni C. and Zappia G. (1991) *Aerobiologia* **7**, 31–37.

Sabbioni C. and Zappia G. (1992a) *Sci. Total Env.* **126**, 35–48.

Sabbioni C. and Zappia G. (1992b) *Water Air Soil Pollut.* **63**, 305–316.

Sabbioni C. (1995) *Sci. Total Env.* **167**, 49–56.

Sabbioni C., Zappia G., Ghedini N. and Gobbi G. (1996a) In *Proceedings of the 8th International Congress on Deterioration and Conservation of Stone*, ed. Riederer J. Moller Druck und Verlag, Berlin, Vol. 1, pp. 1341–1348.

Sabbioni C., Zappia G. and Gobbi G. (1996b) *J. Geophys. Res.* **101**, 19621–19627.

Saiz-Jimenez C. (1993) *Atmos. Env.* **27B**, 77–85.

Saiz-Jimenez C. (1995) *Aerobiologia* **11**, 161–175.

Serra M. and Starace G. (1973) In *Proceedings of the 1st Coll International Des Pierres en Oeuvre*. La Rochelle, pp. 185–188.

Serra M. and Starace G. (1978) In *Proceedings of the Deterioration and Protection of Stone Monuments*. RILEM, Paris, Vol. 3.7, 1–19.

Sharp A., Trudgill S.T., Cooke R.U., Price C., Pickles A. and Smith D. (1982) *Earth Surf. Proc. Landforms* **7**, 387–398.

Simon S. and Snethlage R. (1996) In *Proceedings of the 8th International Congress on Deterioration and Conservation of Stone*, ed. Riederer J. Moller Druck und Verlag, Berlin, Vol. 2, pp. 159–156.

Slinn W.G.N. *et al.* (1978) *Atmos. Env.* **12**, 2055–2087.

Smith B.J., Magee R.W. and Whalley W.B. (1993) In *Alteracion de Granitos y Rocas Afines, Empleados Como Materiales de Construccion*, eds. Hernandez A.V., Ballesteros E.M. and Arnau V.R. Consejo Superior de Investigaciones Cientificas, Madrid pp. 159–162.

Smith B.J., Whalley W.B., Wright J. and Fassina V. (1994) In *Proceedings of the 3rd International Congress on the Conservation of Monuments in the Mediterranean Basin*, eds. Fassina V., Hott H. and Zezza F. Venice, pp. 217–226.

Smith B.J. and Whalley W.B. (1996) In *Proceedings of the 8th International Congress on Deterioration and Conservation of Stone*, ed. Riederer J. Moller Druck und Verlag, Berlin, Vol. 2, pp. 835–848.

Spedding D.J. (1969a) *Atmos. Env.* **3**, 341–346.

Spedding D.J. (1969b) *Atmos. Env.* **3**, 683.

Spiker E.C. *et al.* (1992) *Atmos. Env.* **26A**, 2885–2892.

Spiker E.C., Hosker R.P. Weintraub V.C. and Sherwood S.I. (1995) *Water Air Soil Pollut.* **85**, 2679–2685.

Sramek J. and Eckert V. (1986) In *Case Studies in the Conservation of Stone and Wall Paintings*, eds. Bromelle N.S. and Smith P. IIC, London, pp. 109–111.

Stambolov T. and van Asperen de Boer J.R.J. (1972) In *The Deterioration and Conservation of Porous Building Materials in Monuments: A Literature Review, International Centre for Conservation*. ICCROM, Rome.

Steiger M. and Danneker W. (1994) In *Proceedings of the 3rd International Congress on the Conservation of Monuments in the Mediterranean Basin*, eds. Fassina V., Hott H. and Zezza F. Venice, pp. 179–183.

Stumm W. and Morgan J.J. (1970) *Aquatic Chemistry*. Wiley Interscence, New York.

Subbamaran S. (1985) In *Proceedings of the 5th International Congress on Deterioration and Conservation of Stone*. Presses Polytechniques Romandes, Lausanne, Vol. 2, pp. 1025–1033.

Torfs K. and van Grieken R. (1997) *Atmos. Env.* **31**, 2179–2192.

Trudgill S.T., Viles H.A., Cooke R.U. and Inkpen R.J. (1990) *Atmos. Env.* **24B**, 361–363.

Turpin B.J. and Huntzicker J.J. (1995) *Atmos. Env.* **29A**, 3527–3544.

Twilley J. and Podamy J.C. (1986) In *Case Studies in the Conservation of Stone and Wall Paintings*, eds. Bromelle N.S. and Smith P. IIC, London, pp. 174–179.

Umierski H. (1995) *Untersuchungen zur Schadstoffaufnahme von kalkhaltigen Naturtsteinen unter Berücksichtigung der katalystischen effekte von Übergangsmetallen*. Doktorgrade Dissertation, Universität Münster.

Urquhart D.C.M., Young M.E., MacDonald J., Jones M.S. and Nicholson K.A. (1996) In *Processes of Urban Stone Decay*, eds. Smith B.J. and Warke P.A. Donhead Publishing Ltd., London, pp. 66–77.

Vale J. and Martin A. (1986a) *Durab. Build. Mater.* **3**, 197–212.

Vale J. and Martin A. (1986b) *Durab. Build. Mater.* **3**, 183–196.

Viles H.A., Camuffo D., Fitz S., Fitzner B., Lindqvist O., Lisingston R.A., Maravelaki P.N.V., Sabbioni C. and Warschied T. (1997) In *Saving Our Architectural Heritage*, eds. Baer N.S. and Snethalage R. John Wiley and Sons, Chichester, pp. 95–112.

Vleugels G., Fobe B., Dewolfs R. and van Grieken R. (1994) *Sci. Total Env.* **151**, 59–69.

Warscheid T., Petersen K. and Krumbein W. (1990) *Studies in Conservation* **35**, 137–147.

Webb A.H., Bawden R.J. (1992) Busby A.K. and Hopkins J.N. *Atmos. Env.* **26B**, 165–181.

Weber H. (1985) In *Proceedings of the 5th International Congress on Deterioration and Conservation of Stone*. Presses Polytechniques Romandes, Lausanne, Vol. 2, pp. 1063–1071.

Weber J. (1985) In *Proceedings of the 5th International Congress on Deterioration and Conservation of Stone*. Presses Polytechniques Romandes, Lausanne, Vol. 1, pp. 527–535.

Winkler E.M. (1973) In *Stone: Properties Durability in Man's Environments.* Springer Verlag, New York, p. 230.

Winkler E.M. (1982) In *Conservation of Historic Stone Buildings and Monuments,* ed. Baer N.S. National Academy Press, Washington, D.C., pp. 108–119.

Wittenburg C. and Dannecker W. (1994) In *Proceedings of the 3rd International Congress on the Conservation of Monuments in the Mediterranean Basin,* eds. Fassina V., Hott H. and Zezza F. Venice, pp. 179–183.

Youngdahl C.A. and Doe B.R. (1986) In *Materials Degradation Caused by Acid Rain,* ed. Baboian R. American Chemical Society, Symposium Series 318, Washington, pp. 266–284.

Zappia G., Sabbioni C. and Gobbi G. (1993) *Atmos. Env.* **27A**, 1117–1121.

CHAPTER 4

MECHANISMS OF AIR POLLUTION DAMAGE TO BRICK, CONCRETE AND MORTAR

T. Yates

1. Introduction

The degradation of building materials due to the natural environment is generally relatively slow, and as a result, the service life normally exceeds the time taken for serious damage to occur. Air pollution, however, especially in the form of acidic deposition has accelerated the degradation rate in some materials to such an extent that the costly maintenance is necessary (Building Effects Review Group, 1989; Coote et al., 1989; Cooke and Gibbs, 1993). The focus, in recent years, has been on the increased deterioration rates of architectural structures and monuments, and has in the main been attributed to the effects of sulphur dioxide (SO_2) pollution (House of Commons, 1984; Amorose and Fassina, 1983; Baboian, 1986). The bulk of research work has therefore been concentrated on metals and non-metals such as stone, as a result other building materials such as concrete, brick and mortar have been largely ignored (Yates et al., 1989).

It is known that concrete, brick and mortar are all susceptible to attack by air pollution although the rate and magnitude is usually much lower than for calcareous stones and ferrous metals (Butlin, 1991). However, given that most modern buildings are constructed from concrete, brick and mortar, any degradation can have serious economic consequences.

This chapter aims to review research on the effects of SO_2, oxides of nitrogen (NO_x) and carbon dioxide (CO_2) on concrete and brickwork.

2. Air Pollutants

The air has never been clean in the sense of containing only nitrogen (N_2), oxygen (O_2), CO_2, water (H_2O) and the inert gases because of natural pollutant sources. Man has increased the frequency and intensity of pollutants especially since the industrial revolution. With respect to material degradation, especially brick masonry, the important pollutants to consider are CO_2, SO_2 and NO_x, and "secondary pollutants" formed from these such as sulphuric acid (H_2SO_4), (HNO_3) nitric acid and carbonic acid (H_2CO_3) (Hughes and Bargh, 1982; Wayne, 1985).

In considering air pollutant attack, we must first consider transport processes, concentrations and chemical type of pollutants. The length of time pollutants remain in the atmosphere, the distance they travel, and the atmospheric concentrations they attain will depend on meteorological conditions and deposition processes (Garland, 1978). The processes for transportation from the atmosphere to a surface are usually considered under two main headings — dry and wet deposition.

Dry deposition is defined as the direct collection of gaseous and particulate species on a surface. The processes involved are complex and can be considered in terms of the "deposition velocity" which is a combination of the atmospheric concentration and the resistance of the surface to deposition.

Wet deposition arises from the incorporation of pollutants in cloud droplets ("rainout") their removal by falling precipitation ("washout"), or a combination of the two. Again, this is a complex process, involving the intensity of the rain, its origin and the previous exposure history of the surface (Garland, 1978; Jaynes and Cooke, 1987).

Pollutants which are considered to have an important role in the degradation of building materials are CO_2, SO_2, NO_x, hydrogen chloride (HCl), hydrogen fluoride (HF) and (O_3) along with "secondary pollutants" formed from the above in the atmosphere, such as H_2SO_4 and HNO_3 for example (Wayne, 1985; Franey and Gradel, 1985).

The most abundant minor gas in the atmosphere is CO_2 at an average level of 320 ppm or 0.032% by volume. CO_2 atmospheric concentration is increasing due to industrial smoke and automotive exhausts which can result in localised urban concentrations of up to 2500 ppm (Amorose and Fassina, 1983). The rate of increase of CO_2 concentration has been estimated at 0.1–0.3%/year (Callender, 1958; IPCC, 1990) and has been directly correlated with increasing fossil fuel combustion (Amorose and Fassina, 1983; IPCC, 1990).

The dissolution of CO_2 in the atmosphere will make rain acidic due to the formation of H_2CO_3 (aq). If you assume the background concentration to be 320 ppm, the pH of "natural rain" will be approximately 5.6 by equilibrium calculation (Charola, 1987), which is in agreement with measured values (Lipfert, 1987). The result of increased urban peak concentrations could be rainfall events with a pH as low as 5.2 (Lewry, 1988).

Combustible fossil fuels contain a certain amount of sulphur which results in the formation of SO_2 and sulphur trioxide (SO_3) during burning. The amount of SO_2 produced is approximately between 1 and 10% of the total sulphur oxides (SO_x) and its lifetime in the atmosphere is normally very short due to combination with water vapour to form H_2SO_4 (Amorose and Fassina, 1983; Wayne, 1985). The atmospheric oxidation of SO_2 to H_2SO_4 adds to H_2SO_4 concentrations and it is estimated that 20% of SO_2 emissions are deposited by wet deposition (Lipfert, 1987). Average rural levels of SO_2 have been measured at around 17 ppb rising to a maximum of 105 ppb (Guidabaldi, 1974) in urban areas. Assuming 20% of the total SO_2 is converted to H_2SO_4, a dibasic acid, and a range of 40 to 200 ppb in a highly polluted area, the rainfall could have events with a pH in the range between 2.6 to 2.8 (Lewry, 1988).

Fossil fuel combustion also results in the formation of NO_x, which represents nitric oxide (NO) and NO_2. NO is unstable and is oxidised photochemically to NO_2 in a reaction chain involving O_3 (Amorose and Fassina, 1983). O_3 is also involved in the atmospheric reactions which convert NO_2 to HNO_3, and HNO_3 has been detected in the atmosphere, and nitrates (= acid + aerosol particulates) have been found in the range 0.5 to 1.0 mg cm^{-3} (Guidabaldi, 1974). This would mean an acid equivalent of 0.2 to 0.4 ppm and rainfall events with a pH values between 3.2 to 3.3.

3. Concrete and Cement

3.1. Introduction

Concrete and steel-reinforced concrete are used extensively in the modern building industry (Yates *et al.*, 1989). The weathering and decay of other non-metallic materials have been extensively investigated, for example building stone (Schaffer, 1985). In recent years, the focus has been on the increased deterioration rates of architectural structures and monuments, and this has in the main been attributed to the effects of SO_2 pollution (for example House of Commons, 1984; Amorose and Fassina, 1983; Baboian, 1986). It is known that concrete or reinforced concrete is susceptible to acidic attack (Woods, 1968), but it has not been established that it is significantly affected by acidic deposition from air pollutant sources other than CO_2. In the next section, we will consider the material and its simplified chemistry and properties.

3.2. Cement And Concrete Chemistry

Concrete is made from cement, aggregate and water resulting in a chemically basic material due to the presence of calcium hydroxide (pH ≈ 13).

Three main cement types are commercially available (Shireley, 1975).

(1) Portland cement (PC).
(2) Blended Portland cement (PC blended with other materials such as pulverised fuel ash or ground blast furnace slag).
(3) High alumina cements (also known as calcium aluminate cements).

PCs are the most common cements used in the United Kingdom (Neville, 1981) and this section is focussed on PCs and PC-blended cement. In their unhydrated form, PCs mainly consists of four minerals (Lea, 1970).

(1) Tricalcium silicate, $3CaOSiO_2$ (abbreviated to C_3S).
(2) Dicalcium silicate, $2CaOSiO_2$ (abbreviated to C_2S).
(3) Tricalcium aluminate, $3CaOAl_2O_3$ (abbreviated to C_3A).
(4) Tetracalcium alumino-ferrite, $4CaOAl_2O_3Fe_2O$ (abbreviated to C_4AF).

When mixed with water, PCs undergoes a sequence of hydration reactions which slowly transform the cement paste to a hardened matrix of hydrated products. The most important hydration are those involving C_3S and C_2S which lead to the formation of calcium silicate hydrate (C-S-H) gel (of variable composition and structure) and calcium hydroxide, mainly in the form of Portlandite crystals (Page and Treadaway, 1982). The formation of calcium hydroxide results in a highly alkaline material (pH $\approx 12.5-13.0$) which is sufficient to induce "passivation" of embedded steel in reinforced concrete. The metal is protected from corrosion by a thin surface layer of oxide — γFe_2O_3 (Hausmann, 1967).

The capacity of concrete to act as a physical barrier against the penetration of aggressive environmental components is critical in the degradation processes, especially when reinforcement is being protected. Unfortunately, concrete is not a perfect barrier because of its continuous pore system and tendency to form surface cracks. Water/gas permeability is associated with the water/cement (W/C) ratio (Tonni and Gaudis, 1980), a higher W/C ratio leading to increased porosity and as a result greater permeability.

Due to its high alkalinity, concrete is subject to acidic attack (Woods, 1968) which generally occurs in four ways:

(1) Dissolution of the hydrated and unhydrated cement compounds.
(2) Dissolution of calcareous aggregate.
(3) Physical stresses induced by deposition of soluble salts from the acid/alkali reaction and the subsequent formation of new solid phases within the pore structure.
(4) Salt-induced corrosion of the reinforcing steel.

3.3. The Effect of Carbon Dioxide

Carbonation reactions

The carbonation of concrete normally involves a chemical reaction between atmospheric CO_2 and the products of cement hydration. It is argued that all the hydrates in the cement paste matrix can react with CO_2 at its atmospheric concentration of 0.03% by volume but the important reaction appears to be with calcium silicate hydrate (C-S-H) and calcium hydroxide (Parrott, 1987). This results in the formation of calcium carbonate and silica gel (Brown and Clifton, 1988):

$$Ca(OH)_2 + CO_2 \rightarrow CaCO_3 + H_2O$$

$$10((CaO)_{1.7}SiO_2.4H_2O) + 17CO_2 \rightarrow 17CaCO_3 + 10SiO_2(aq) + 40H_2O$$

Note: $(CaO)_{1.7}SiO_2.4H_2O$ is the approximate composition of C-S-H.

There has been very little work until recently on the mechanisms of C-S-H carbonation and opinions differ as to whether this reaction occurs after or simultaneously with the carbonation of Portlandite. Earlier research has indicated simultaneous carbonation of $Ca(OH)_2$ and C-S-H in a blended cement. Dunster's work (1989) proposed that C-S-H reaction with CO_2 removes calcium-releasing silicate radicals which condense with further silicate anions in C-S-H. Silicate chains of higher molecular weight are initially produced and continued calcium removal causes further cross-linking, eventually producing silica — a 3-D silicate network.

Unhydrated cement compounds can also react with CO_2 at very elevated levels (>10%) but at atmospheric concentrations are regarded as unreactive (Parrott, 1987).

The carbonation reactions are restricted by the amount of moisture present and require the presence of some condensed moisture to become significant. The most favourable conditions are reported as a relative humidity between 50 and 70% in an exposed alternatively wet and dry atmosphere (Beckett, 1984). This maxima can be explained by consideration of the pore water/CO_2 interactions. At low humidities there is insufficient pore water to dissolve the CO_2 resulting in no significant degree of reaction (Hilsdord et al., 1984). Carbonation can occur readily under saturated conditions but

CO_2 has to diffuse through the carbonated surface layer to reach the reaction zone. Gaseous diffusion will be a slow process if the pores are saturated with water resulting in a maximum carbonation rate at intermediate moisture contents (Verbeck, 1958).

Rates of carbonation

The rate of carbonation is thought to be controlled by CO_2 diffusion or by water diffusion and increases with temperature, CO_2, concentration and porosity (Verbeck, 1958; Daimon *et al.*, 1971).

By considering the law of permeability and using Fick's law, several workers have derived expressions of the following form:

$$d = k.\sqrt{t}$$

where $\quad$ d = depth of carbonation
$\quad\quad\quad$ k = coefficient
$\quad\quad\quad$ t = time.

Two such expressions (Ying-Yu and Qui-Dong, 1987) include a term for the partial pressure of CO_2 in their coefficient, k, leading to the expression:

$$D = k_1.\sqrt{P_{CO_2}t}$$

where $\quad\quad P_{CO_2}$ = partial pressure of CO_2.

Expressions of this type appear to hold true for long-term exposure (>10 years) in atmospheric conditions (Fukushima, 1987) but the prediction of carbonation based solely on diffusion relationship is not accurate (Baweja *et al.*, 1987). This is to be expected because these relationships do not account for any properties of the material, such as water/cement ratio which will significantly effect the porosity of the concrete.

The relationship between carbonation depth and CO_2 concentrations of:

$$d = \alpha\sqrt{P_{CO_2}}$$

does not appear to hold for the degree of carbonation observed for all CO_2 concentrations. The carbonation weight gain versus CO_2

concentration would be expected to be a parabolic relationship, but if this data is plotted, the curve obtained does not resemble a parabola. The weight gain observed for up to 1% of CO_2 is very steep and approximately linear, between 1–5% there is a sharp curve, almost a discontinuity, and finally at concentrations >5% CO_2 the weight gain levels out; slowly increasing with CO_2 concentration. This sharp change in gradient between 1–5% of CO_2 could indicate limiting factors in the processes involved or two mechanisms occurring; one below 1% CO_2 and one above 5% CO_2, the area in between being the "change-over" region. Accelerated testing occurs at CO_2 levels of approximately 4% V/V and in light of the above data, the value of using such tests in predicting the carbonation effects on outside structures is brought into question.

The carbonation reaction is known to occur at sites on the pore wall (Lin and Fa, 1987) and the carbonation reaction has been shown to decrease concrete's gas permeability (Zhang, 1985). This indicates that the growth of calcite crystals in the pores is "clogging" them up, probably resulting in a slowing down of transportation processes to the reaction front.

When considering exposure in the natural environment, reaction conditions vary considerably. The conversion of $Ca(OH)_2$ to $CaCO_3$ and any removal of reaction products, in the form of the soluble $(Ca(HCO_3)_2$, are dependent on the excess of dissolved CO_2 (Amorose and Fassina, 1983; Wayne, 1985). This suggests that condensing conditions may dominate the carbonation depth whilst the action of rain may be involved with the leaching of compounds and as a result surface erosion effects.

The effect on material properties

When concrete contains Portland cement, carbonation results in an increase in strength and reduced permeability, neither of which are a problem in practice (Matousek, 1980; British Standard, 1997).

On carbonation, concrete is observed to shrink which appears somewhat paradoxically. This is because the conversion of Portlandite to calcite will result in a molar volume expansion of 11% (Brown and Clifton, 1988). The shrinkage due to drying of the concrete has also been shown not to be a contributory factor (Verbeck, 1958). This leaves the possible shrinkage of C-S-H on carbonation, which appears to be a point of controversy. Sauman (1972) has studied the carbonation of C-S-H on a microstructural level and reports no structural or volume change. Conversely, Brown and Clifton (1988) report a molar volume shrinkage of 80% on carbonation of C-S-H which would certainly explain the phenomena of "carbonation shrinkage" but such a dramatic change would almost certainly lead to shrinkage cracks and a subsequent loss of the improvement in properties. The degree of C-S-H shrinkage also seems very high when the typical values observed in hydrated Portland cement are of the order of 0.1% (Verbeck, 1958) by volume.

The effect of carbonation on Portland cement appears to be one of a slight improvement in material properties. However, when the concrete is steel reinforced, carbonation reduces the pH of the affected material from approximately 12.5 to values as low as 8 which removes the "passivating" effect of the concrete on the embedded steel. Therefore, it is necessary to ensure that the concrete cover is thick enough to prevent corrosion during the lifetime of the structure.

3.4. The Effects of Other Pollutants

Apart from CO_2, the other common pollutants in the atmosphere are SO_2 and NO_x, and their acidic derivatives, H_2SO_4 and HNO_3.

Nitrogen oxides and nitric acid

Very little information is available on the effect of NO_x on concrete. However, it is known that HNO_3 will attack concrete in the following manner (Webster and Kukacka, 1986):

$$Ca(OH)_2 + 2HNO_3 + 2H_2O \rightarrow Ca(NO_3)_2.4H_2O$$

The acid will also cause decomposition of C-S-H into amorphous silica but this reaction is independent of the anion and is dependent on proton attack (Brown and Clifton, 1988). Complete nitric acid attack will potentially cause a large molar weight change (approximately 200% for the above reaction). However, damage does not occur due to salt crystallisation but appears to be due to the high solubility of the calcium nitrate hydrates and their subsequent leaching after formation. If this form of decomposition were to occur throughout the concrete, then the remaining solid volume would be approximately 18% of the original solid volume with a greatly increased porosity. The development of a more porous, weaker cement matrix will expose the sub-surface to further attack and leave it susceptible to other forms of deterioration.

Sulphur dioxide and sulphuric acid

As with the NO_x, very little information is available on the effect of the gaseous pollutant, SO_2, on concrete. It is known, however, that SO_2 and concrete react to yield calcium sulphite.

$$Ca(OH)_2 + SO_2 \rightarrow CaSO_3 + H_2O$$

Under dry conditions, the reaction is limited to the surface but in the presence of moisture and oxygen the reaction becomes more significant. The reaction can now occur in the concrete's sub-surface with oxidation producing some gypsum (Webster and Kukacka, 1986).

Accelerated testing has been used to study the effects of acid precipitation on concrete (Orantie and Ruohomaka, 1988; Kong and Orbison, 1987; Attiogbe and Rizkalla, 1988; Fattuhi and Hughes, 1988) but this had normally taken the form of immersion in H_2SO_4 solutions of pH between 1 and 5 for several months. This type of testing has general drawbacks in its simulation of acidic precipitation. Firstly, in the majority of cases the solution is static, rather than continuously flowing (Fattuhi and Hughes, 1988), which means equilibrium conditions are probably reached, a state which is unlikely in real conditions. The acid solutions used are normally of pHs between 1 and 5, although the most commonly used are between 1 and 2 which is a hydrogen ion concentration of 100 to 10,000 times that observed in rainfall. This high level of acidity appears unrealistic and could lead to different mechanistic processes occurring, especially with respect to the dissolution processes. The use of accelerated testing with H_2SO_4 (aq) solutions are unlikely to model the attack of acidic precipitation but despite this, they are the only indicators available.

The rate of deterioration of concrete has been shown to increase with an increase in cement content, water/cement ratio or acid concentration (Kong and Orbison, 1987; Fattuhi and Hughes, 1988; Sersale, 1998). An accelerated test study (Attiogbe and Rizkalla, 1988) has also shown that acid attack is a surface phenomenon with deterioration starting at the surface of the concrete and progressing inward. A later study conducted at the Abiko Research Laboratory (Abiko, 1995) reached a similar conclusion from a series of tests that exposed samples of concrete to "artificial acid rain". They found that the erosion depth was related to the pH, and which after 4000 mm of rainfall, ranged from 0.2 mm at pH 5.6 to 1.4 mm at pH 2.0. This is in agreement with Lewry and Pettifer (1988) which indicates that acidic precipitation causes surface erosion by leaching calcium ions from the cement matrix of concrete.

Webster and Kukacka (1986) have reported "crust formation" in their review, although no references to validate their statements concerning air pollution as a cause has been quoted. The same case study as above has examined a gypsum crust in a concrete specimen and attributed it to SO_2 deposition but its effects appear insignificant when compared to carbonation.

Other authors (Brown and Clifton, 1988; Figg, 1983) appear to regard the possible reaction of gypsum with other cement phases a problem. Apart from possible salt damage on crystallisation of the gypsum, the expansive reaction with hydrated tricalcium aluminate to produce ettringite ($3CaO.Al_2O_3.3CaSO_4.31H_2O$) is quoted as being a problem. However, the evidence linking such deleterious expansive reactions to acidic deposition is tenuous to say the least.

Experimental investigations have shown that steel reinforcement in an alkaline solution can be considerably damaged by sulphates and nitrates (Hensel, 1985). However, the views of Hensel (1985) and Skoulikidis (1982) that the action of NO and/or SO_2 in small amounts will cause steel reinforcement corrosion in concrete appear unfounded and are not substantiated by experimental evidence.

The view that acidic deposition is not responsible for significant concrete degradation or reinforcement corrosion was supported by the members of an EPA workshop (EPA, 1986), and by a number of other reports since then, for example Harter, 1988, Watts Committee,

1988, and Livingston, 1998. In some cases, they do however suggest further research to substantiate their views. This suggested research should, as a high priority, include any possible synergistic effects between SO_2, NO_x and CO_2. This is a possibility, as SO_2 and NO_x acidic deposition appears to weaken and leach out the cement matrix of concrete. This would result in an easier transportation route to the unreacted Portlandite resulting in an increased carbonation rate.

3.5. Conclusions

The mechanistic paths by which SO_2, NO_x and CO_2 attack cement or concrete appear to depend on the relative atmospheric concentrations. The atmospheric concentration of CO_2 (approximately 320 ppm, Brimblecombe, 1986) is approximately 10,000 times the concentration of SO_2 and NO_x (approximately 20 ppb, Brimblecombe, 1986), resulting in carbonation being a much more significant effect. However, carbonation is a not in practice a problem regarding concrete unless steel reinforcement is being protected. The use of a sufficiently thick layer of good quality concrete should rectify this problem.

Acidic deposition of SO_2 and NO_x is responsible for surface erosion and cracking of the cement matrix. This is only a "cosmetic" problem unless there is a synergistic effect with CO_2 resulting in increased carbonation rates.

4. Brickwork and Mortar

4.1. Introduction

The deterioration of brick masonry is a complex problem because of the two variable components — brick and mortar. Each component can have a large variation in composition and structure resulting in a large number of combinations when they are used together.

Brick and mortar are both susceptible to acidic attack to an extent (Charola and Laezarini, 1986), but little is known about the effects of acid deposition on the bricks of brickwork (Watt Committee, 1988). Brick components can rehydrate, carbonate and adsorb sulphate resulting in possible moisture expansion, cracking and exfoliation.

Mortars will also react with sulphate and other ions, resulting in expansion, erosion and structural deterioration. Transportation of any salts formed by chemical attack can also compound the problem by recrystallising (Hughes and Bargh, 1982).

The composition and firing history of the brick along with the composition and production history of the mortar will complicate the problem still further.

4.2. Brick Manufacture and Composition

Brick manufacture

Bricks are formed by three distinct methods (BIA, 1986):

(1) Soft mud method: soft muds are poured into moulds and dried before firing. This is not a commonly used practice in modern brickmaking.
(2) Stiff mud method: plastic clay is extruded through a die and then the bricks are cut to size.
(3) Semi-dry or dry press method: clay is pressed into a mould.

The second method often causes laminations and these are referred to as weakness planes. Lamination and the associated cracks are often blamed for failures, but Robinson (1984) suggests that this is improper and further investigations are necessary.

During the above "forming" processes, additives are commonly used to colour the brick or inhibit efflorescences.

Apart from the starting materials' composition, the firing temperature, the kiln type and the time kept at the temperature are probably the most important factors in determining the nature and quality of the brick produced (Robinson, 1984).

Three principal types of kiln are used:

(1) Scone kiln: constructed of unfired bricks which are plastered on the exterior. The bricks are stacked so that a channel allows heat to travel by convection. This kiln type is not commonly used.
(2) Periodic kiln: the shape may be beehive, rectangular or square. These kilns are dome-roofed with fire bases around the outside of the wall base.

(3) Tunnel kiln: commonly used because they are efficient and fire uniformly resulting in a cheaper, reproducible product.

The firing process generally controls the physical properties of bricks, differences of 50–100°C can radically alter properties. Firing is carried out by slowly raising the temperature and allowing volatiles to escape and oxidation to occur before raising the temperature further. Generally, water is removed by heating from 100–400°C which results in a material with a porosity of 30–40 vol% (Carlsson, 1988). This is followed by the removal of clay hydroxyl water and oxidation in the range 400–800°C. Above 800°C, the heating results in high temperature phases and fusion products. In the range 800–1400°C, sintering or densification occurs via chemical reactions, grain growth and development of a liquid phase. The porosity decreases as a result of sintering and lead to an increase in strength of the final product.

Generally, the ideal point to stop firing results in a brick which is "steel hard" with a low-water absorption. Overfiring wastes energy and underfiring generally results in poor quality bricks.

Brick composition

Bricks are normally produced from raw materials consisting of at least 50% kaolins and clays which together with water give the necessary plasticity to the materials when forming it into components.

The liquid phase, produced during firing, transforms to glass without crystallisation on cooling. The result is a brick usually composed of a crystalleric silicate phase, mullite $(3Al_2O_3 \cdot 2SiO_2)$, the remaining quartz from the raw materials and some minor phases, which are all bonded together by a glassy matrix. The porosity of the brick could still be as high as 20% and the glassy phase could constitute as much as 60% of the brick. This means that the brick is really a composite with the crystalline phases embedded in the glassy matrix whose typical composition is given in Table 1.

Brick durability

When considering the durability, porosity and permeability are probably the most important physical properties because these

Table 1. The approximate composition of the glass in a ceramic such as brick (after Hermansson and Carlsson, 1978).

Component	Wt.%
SiO_2	71–77
TiO_2	1–4
Al_2O_3	7–16
Fe_2O_3	1–5
MgO	0–2
CaO	0–1
Na_2O	0–3
K_2O	4–8

characterise the accessibility of water to the brick's interior. Thermal and mechanical properties should also be considered in analysing decay. The physical nature of brick should first be considered. Therefore intergranular bonding, cementing matrix, pore shape and size are more important than the properties of individual grains. The durability of brick to weathering is determined by a number of variables, but the critical factors appear to be the porosity and pore system. Large pores and low porosity are widely held to give good durability but there is disagreement over possible critical pore-sizes. The work of Watson, May and Butterworth (Watson *et al.*, 1957; Watson *et al.*, 1962; May and Butterworth, 1962) indicates that pores smaller than 1 μm caused damage whilst those larger than 2 μm, or preferably 10 μm improved the brick durability. Davison (1980) has agreed a critical pore size of 1 μm; <1 μm implies low durability and >1 μm improves durability. Maage (1980 and 1984) investigated the relationship between pore size/volume and frost resistance of brick. He derived an empirical relationship relating to the two which quoted 3 μm as the critical pore size. Robinson (1984) determined the pore size distributions of good and poor durability brick. The critical pores sizes were in agreements with the work of Butterworth *et al.* with pores <1 μm predominant in low quality bricks whilst pores >2 μm were the majority in good bricks. Robinson (1984) also found that certain bricks exhibited a differential pore structure. The surface of brick having a much lower porosity than the centre. This surface "skin" prevents rapid migration of any pressurised water and as a result spalls due to crystallisation processes. Torraca (1985) also indicated that the presence of small pores (<1 μm) increased the susceptibility of brick to freeze-thaw and crystallisation damage.

The pore size distributions were determined in all of this work by mercury intrusion porosity. This method has a number of errors including assuming the mercury contact angle is between 124 and 130° and that the pores are cylindrical in shape (Ross, 1984). Despite this, the technique provides a first approximation of pore size and the pore structure for comparative work.

Robinson (1984) also investigated the effect of firing temperature composition of the raw materials and forming methods on the pore structure and durability. He found that firing did not significantly alter the pore size distribution but did lower the total porosity, thus increasing the final strength. Also insufficient time at the "maturing temperature" at the end of the firing process resulted in the differential pore structure earlier and subsequent failure in use at a later date.

The raw materials appeared to control the pore size distribution of the final brick. Pure clays have small pores and the addition of shale shifts the distribution only slightly. The addition of sand and crushed brick can be used to shift the distribution to larger pore sizes, improving the durability.

Hand moulded and pressed bricks exhibited the best pore size distributions whilst extrusion processes lead to a brick with the majority of pores >1 μm. Although pore size distributions appear to be a good indicator of durability, other factors such as water permeability and mechanical strength must be taken into account.

4.3. Pollutant Attack on Brick

The brick's glassy matrix is similar to acidic silicate glasses, and as a result is probably the most susceptible component to dissolution by neutral, acid or alkaline solutions.

Water attacks glass by leaching alkali ions from the surface and initially is a diffusion controlled reaction (Das and Douglas, 1967). After this period, a second surface controlled reaction becomes dominant which involves dissolution of the alkali ion-depleted layer (Paul, 1987). Dilute acid solutions (pH > 2) attack glass by preferentially dissolving the alkalis and basic oxides (Paul, 1987). Silica loss is less than that observed with water attack because the acid neutralises the leached alkalis, thus keeping the pH below a level where silica dissolution is important. The result of preferential leaching is to produce a layer twice as thick as those observed with water attack (Clark *et al.*, 1979), $1200A^{o}$ as compared to $600A^{o}$ after one hour. The alkali ions are replaced with smaller hydrogen ions inducing stress in the layer which will cause it to crack. The surface layer by this time is several microns thick and further shrinkage occurs if the hydrated silica layer loses water.

The deterioration mechanism above was observed in the glassy matrix of a brick exposed to concentrated H_2SO_4 (Lewin and Charola, 1979). On continued exposure, the leached layer crumbled away exposing the mineral with greater acidic resistance. The only problem is that the environment above is far too aggressive to simulate atmospheric pollutant attack and the rate of attack probably results in surface strain due to shrinkage and as a result loss of material. Robinson (1982) estimated that H_2SO_4 solution formed from pollutants increased the dissolution of the brick ten-fold. However, given that the maximum dissolution of the brick is usually 1%, then the resulting salts, because of their small quantity, only usually damage the appearance, particularly as staining and efflorescence.

Baronio *et al.* (1980) found that a water-saturated, SO_2-containing atmosphere deteriorated bricks by increasing their porosity and pore size. This does not seem like much of a problem as enlarging the pore size leads to greater durability.

El-Shamy's (1975) work on the chemical durability of glasses in acid solution found that HNO_3 attacked more aggressively than H_2SO_4. Although pollutant atmospheres will have some effect on brick, the rate of attack will be very much slower than a modern glass. This is because the glassy matrices' composition is reasonably high in Al^{3+} ions whilst having low levels of Na^+ and K^+ ions. This would lead to a glass of high durability and probably explains why attack on brickwork due to acidic deposition does not appear to be a problem.

A more serious problem can be efflorescence, i.e. the deposition of soluble salts on the brick surface (DOE, 1972) The salts can derive from rehydration of the soluble phase of the brick or from the mortar (Harding and Smith, 1983). This phenomena is normally temporary and only occurs when a new brick first dries out and persistent reccurrence is normally a sign of persistent water penetration (BIA, 1985). However, if salt recrystallisation occurs just below the surface, the result will be surface decay and spalling of the brickwork.

4.4. Mortar and Pollutant Attack

Good quality bricks will have great strength but mortar is just as important a component in brick masonry. Although the nature of mortar has changed considerably over time, modern masonry is usually a mixture of Portland cement (OPC), sand and lime (Charola and Laezarini, 1986).

The effect of acid deposition will depend on the particular mortar used.

Non-hydraulic lime mortars depend on exposure to air to harden, the slaked lime, i.e. $Ca(OH)_2$ reacts with CO_2 in the air (Ashurst and Ashurst, 1988).

$$Ca(OH)_2 + CO_2 \rightarrow CaCO_3 + H_2O$$

On hardening, carbonation and drying will occur simultaneously, resulting in a volume contraction. The setting reaction above results in calcite formation, implying that a fully carbonated lime/sand mortar will behave in a similar manner to a loosely bound sandstone (Hoffman et al., 1976) when considering air pollutant attack. SO_2 deposition can result in gypsum formation, which on recrystallisation could induce mechanical stresses into the mortar matrix (Charola and Laezarini, 1986). Setting plaster, however, is a mixture of Portlandite and calcite which can be further complicated in the case of dolomitic mortars when magnesium phases are also present. Carbonation and attack by SO_2 will occur simultaneously and there is the possibility of interaction. Carbonation, dissolution, re-precipitation and sulphation of the various phases will occur in the pore spaces on absorbed water films (Hoffman *et al.,* 1976).

Hoffman *et al.* (1976) have observed degradation as "flaking off" in the case of lime mortars and as disintegration with dolomitic lime mortars. On the basis of their experimental work, they attributed these phenomena to the sulphate concentrations in the surface area (up to 3 mm deep). This implies salt crystallisation damage but no damage mechanisms have been discussed or investigated.

Hydraulic mortars containing Portland cement are attacked by acidic precipitation which causes softening and disintegration of the surface (Ashurst, 1983). Air pollutant attack on concrete has been described above and will be only briefly summarised here. The main constituents of Portland cement which are susceptible are Portlandite, i.e. $Ca(OH)_2$ and calcium silicate hydrate (C-S-H). SO_2 reacts with $Ca(OH)_2$ to produce calcium sulphite ($CaSO_3$), which in the presence of moisture and oxygen will oxidise to gypsum (Figg, 1983). The resulting sulphate can further degrade the mortar by reacting with calcium aluminate (C_2A) to produce ettringite, i.e. $2CaO.Al_2O_3.3CaSO_4.31H_2O$ (Brown and Clifton, 1988). Gypsum and ettringite formation are expansive reactions and can result in cracking in the mortar or brick depending on water movement.

Martinez-Ramirez *et al.* (1996) used an exposure chamber to examine the effect of different combinations of gases on lime mortar (1:3 lime:aggregate). The results showed that if both water and an oxidant were present, then the formation of salts was greatly increased. Their results also showed that under chamber conditions, SO_2 in combination with water and O_3 was far more reactive than NO or NO_2 under similar conditions.

Longer site exposure studies for mortar are described in Lewry (1992) and Lewry and Butlin (1993). The mortars used included 1:1:6 (cement:lime:sand) and 1:3 (cement:sand) with an air entraining agent. After three years of exposure at a range of urban and rural sites in the UK, the following observations were made:

(1) No visual signs of degradation were observed.
(2) The extent to which soluble slats were deposited was dependent on the pollutant concentrations at the exposure site.

(3) There was an upper limit to the concentration of soluble sulphate. Half the samples reached this limit after one year of exposure, the others took up to three years.

(4) The main product of air pollution was calcium sulphate and no signs of ettringite were found.

This final observation is also reported by Scholl and Knöfel (1991) from their work on cement paste.

A different form of deterioration of mortars is described by Ichitsubo et al. (1997). They describe colour changes in mortars that they attribute to the oxidation of iron in the cement. They link this to the dissolution of calcium hydroxide at the surface of the mortar which results in a concentration of iron.

The carbonation of Portlandite was discussed earlier, however rainwater, acidified with CO_2, leaches $CaCO_3$ from the surface, weakening the cement structure slightly (Charola and Lewin, 1979). In general, it has been observed that the effect of total carbonation doubles the strength of Portland cement mortars (Penkala and Zasum, 1988).

Although sand grading of mortars does not appear to be related to resistance to sulphate attack, the strength and frost resistance has been shown by Harrison (1986) to be reduced with the use of finely graded sands. This appears to be the result of increased porosity and shrinkage. In this context, it is interesting to note that Lewry and Butlin (1993) found that sulphate accumulated more slowly in the mortars made using finer sands.

4.5. Interactions Between Brick and Mortar

Mortar appears to be the only component of brickwork attacked by acidic deposition but the degree of susceptibility to chemical dissolution is fairly low. However, it is possible that salts produced by the reactants of air pollutant attack could cause damage to masonry.

Another source of salts is the bricks themselves which can contain Na_2SO_4. These occur if the fuels used during firing are sulphur-rich or the firing temperature is insufficient to decompose the salt. Na_2SO_4 is very soluble and on evaporation can recrystallise as the anhydrous salt, theriardite, or as the decahydrate, mirabilite, depending on temperature and humidity (Read, 1984). The stresses produced within a material by crystallisation pressure can be great (Winkler, 1973 and 1997) and will be in the case of brick work sufficient to inflict severe damage.

The deterioration of masonry due to salt crystallisation has been proposed by Lewin (1982) to be dependent on the position of where the salt solution evaporates. Salt crystallisation will depend on the capillary migration of the solution and the rate of evaporation.

De La Torre Lopez and Sebastian Pardo (1996) describe some of the longer term weathering effects that they observed on late 19th century church in Granada, Spain. They observed that the original mortar in the joints and the rendering was based on gypsum and is now around 70% gypsum with a very high porosity which was accentuated by dissolution. There was clear evidence that the gypsum had migrated into the bricks and was crystallising there. There was also some calcite in the bricks and this was being attacked by atmospheric SO_2.

Binda and Baronio (1985) have shown that the relative porosities of the brick and mortar are important, with the most porous material being the site of salt-induced deterioration. They proposed that if brick and mortar have similar properties, decay will occur over the masonry unit as a whole and will take a longer time. This appears to be a wide generalisation but warrants further investigations if the durability of masonry could be improved in this way.

The porosity of the brick is also important when considering the setting of the mortar. If the brick contains large pore sizes (> 20 μm), the suction effect is fairly low. Goodwin and West's (1982) work indicates that this produces a stronger mortar/brick bond due to better adherence of the Portland/ettringite layer as it can crystallise in the pores of the brick. However, when you consider the durability of brickwork as a whole, this is in conflict with the work of Harrison (1986) which showed that the mortar has a poorer strength and lower durability when subject to little or no suction from the brick.

4.6. Conclusions

The main mechanism of deterioration in brick masonry appears to be the crystallisation of soluble salts and frost damage. Soluble salts are either inherent to the original materials, introduced during production or possible from air pollutant sources. The resistance of masonry to deterioration is generally related to water penetration as this is a major factor in frost and wetting/drying damage and as the transport medium for salts.

The design and manufacture of the components is also very important, especially composition of the raw materials, production processes and firing conditions. Control of these could lead to low levels of sulphate and control of the pore size distribution, thus improving the durability in the long term.

In conclusion, there appears to be little or no problem concerning the effect of acidic deposition on brick. However acidic deposition on mortar could lead to increased amounts of salts and therefore an increased degree of salt damage on brick masonry as a whole.

References

Abiko (1995) *The Effect of Acid Rain at Concrete Structure (sic)*. Abiko Research Laboratory, Abiko-City, 270-11, Japan.

Amoroso G.G. and Fassina V. (1983) *Stone Decay and Conservation*. Material Science Monographs II, Elsevier.

Ashurst J. (1983) *Mortars, Plasters and Renders in Conservation*. Ecclesiastical Architects' and Surveyors' Association, London, 34–35.

Ashurst J. and Ashurst N. (1988) *Practical Building Conservation; English Heritage Handbook, Vol. 3; Mortars, Plasters and Renders*. Gower Technical Press.

Attiogbe E.K. and Rizkalla S.H. (1988) *ACI Mater. J.* **Nov-Dec**, 481–488.

Baboian R. (1986) *Materials Degradation Caused by Acid Rain, ACS Symposium Series 318*. ACS, USA.

Baronio G. *et al.* (1980) *Proceedings of the 15th ANOIL Congress*, pp. 81–93.

Baweja J. *et al.* (1987) *Proceedings of the 4th International Conference on Durability of Building Materials and Components, Singapore*, pp. 694–701.

Beckett D. (1984) *Corrosion Problems in Medium and High Rise Buildings, 2nd European Seminar on Failure and Repair of Corroded Reinforced Concrete Structures, October 1984, London*. Dye Scientific and Technical Services Ltd, London, 1984.

BIA (1985) *Technical Notes on Brick Construction, 23, Part 1*. BIA, Virginia, USA.

BIA (1986) *Technical Notes on Brick Construction, 9, Part 1*. BIA, Virginia, USA.

Binda L. and Baronio G. (1985) *Proceedings of the 7th International Brick Masonry Conference*, pp. 605–616.

Brimblecombe P. (1986) *Air-Composition and Chemistry.* Cambridge University Press.

British Standard (1997) *BS 5328.*

Brown P.W. and Clifton R.C. (1988) *Durab. Build. Mater.* 5, 409–420.

Building Effects Review Group (1989) *The Effects of Acid Deposition on Buildings and Building Materials in the United Kingdom.* HMSO, London.

Butlin R.N. (1991) *Procs. Roy. Soc. Edinburgh* 97B, 255–272.

Callendar G.S. (1958) *Tellus* 10, 243.

Carlsson R. (1988) *Durab. Build. Mater.* 5, 421–427.

Charola A.E. and Lewin S.Z. (1979) *Scan. Elect. Micro.* 1, 378–386.

Charola A.E. and Laezarini L. (1986) In *Materials Degradation Caused by Acid Rain, ACS Symposium Series 318,* ed. Baboian R. ACS, USA, pp. 250–258.

Charola A.E. (1987) *J. Chem Educat.* 64, 436–437.

Clark D.E. *et al.* (1979) *Corrosion of Glass.* Books for Industry, New York, pp. 22–29.

Cooke R.U. and Gibbs G. (1993) *Crumbling Heritage.* National Power, Swindon, UK.

Coote A.J. *et al.* (1989) In *Acid Rain and the Environment,* ed. Longhurst J.W. London, UK.

Daimon M. *et al.* (1971) *Am. Ceram. Soc.* 54, 423–428.

Das C.R. and Douglas R.W. (1967) *Phys. Chem. Glasses* 8, 178–184.

Davison J.I. (1980) *Proceedings, 2nd Canadian Masonry Symposium, Ottawa, June 1980,* pp. 13–24.

De La Torre Lopez M.J. and Sebastian Pardo E. (1996) *8th International Congress on Deterioration and Conservation of Stone, Berlin,* pp. 325–331.

Department of the Environment (DOE) (1972) *Advisory Leaflet 75.* HMSO, London.

Dunster A.M. (1989) *Adv Cement Res.* 2(7), 99–106.

El-Shamy T.M. (1975) *J. Non-Cryst. Solids* 19, 241–250.

EPA (1986) *Workshop on "Acid Deposition Effects on Portland Cement Concrete and Related Materials".* Final Report, eds. Spence J. and Haynie F. US EPA, Research Triangle Park, North Carolina 27711.

Fattuhi N.I. and Hughes B.P. (1988) *ACI Mater. J.* Nov-Dec, 512–518.

Figg J. (1983) *Chem. Indus.* October, 770–775.

Franey J.P. and Gradel T.E. (1985) *JAPCA* 35, 644–648.

Fukushima T. (1987) *Proceedings of the 4th International Conference on Durability of Building Materials and Components,* pp. 662–669.

Garland J.A. (1978) *Atmos Env.* 12, 349–362.

Goodwin J.F. and West H.W.H. (1982) *Procs. Br. Ceram. Soc.* 30, 23–37.

Guidabaldi F. (1974) *Atti della Sczione 11, del XXIX Congress Nazionale della Associasione Termotecnica Italiana, Florence, September 1974.*

Harding T.D. and Smith R.A. (1983) *Brick Development Assciation (BDA) Design Note 7.* BDA.

Harrison W.M. (1986) *Concrete Res.* **38**, 95–107.

Harter P. (1988) *Acidic Deposition — Materials and Health Effects.* IEA Coal Research.

Hausmann D.D. (1967) *Mater. Prot.* **6**, 19.

Hensel W. (1985) *Bentowerk + Fertigteil-technik* **11**, 714–721.

Hermansson L. and Carlsson R. (1978) *Trans. J. Br. Ceram. Soc.* **77**, 32.

Hilsdord H. *et al.* (1984) *Proceedings of the RILEM Seminar, Hannover,* pp. 202–209.

Hoffman D. *et al.* (1976) *Proceedings of the 2nd International Symposium on Deterioration Building Stone, Athens,* pp. 37–42.

House of Commons Environment Committee (1984) *4th Report, Acid Rain.* HMSO, London.

Hughes R.E. and Bargh B.L. (1982) *The Weathering of Brick: Causes, Assessment and Measurement, Report of the US Geological Survey.* USGS, Illinois USA.

Ichitsubo M. *et al.* (1997) *Semento, Konkurito Ronbunshu* **51**, 450–455 (in Japanese).

Intergovernmental Panel on Climate Change (IPCC) (1990) *Climate Change: The IPCC Scientific Assessment.* Report Prepared for IPCC by Working Group 1, eds. Houghton J.T., Jenkins G.J. and Ephraums J.J. Cambridge University Press, UK.

Jaynes M.S. and Cooke R.U. (1987) *Atmos. Env.* **21**, 1601–1622.

Kong H.L. and Orbison J.G. (1987) *ACI Mater. J.* **March-April**, 110–116.

Lea F.M. (1970) *The Chemistry of Cement and Concrete,* 3rd Ed. E Arnold Ltd, London.

Lewin S.Z. and Charola A.E. (1979) *Proceedings of the Conference on Il Mattone di Venezia,* pp. 189–214.

Lewin S.Z. (1982) *Conservation of Historic Stone Buildings and Monuments.* National Academy Press, Washington D.C., pp. 120–144.

Lewry A.J. (1988) *BRE Note N110/88.* BRE, Watford, UK.

Lewry A.J. and Pettifer K. (1988) *BRE Note N130/88.* BRE, Watford, UK.

Lewry A.J. (1992) *BRE Note N131/92.* BRE, Watford, UK.

Lewry A.J. and Butlin R.N. (1993) *BRE Note N188/93.* BRE, Watford, UK.

Lin X.X. and Fa Y. (1987) *Proceedings of the 4th International Conference on Durability of Building Materials and Components, Singapore,* pp. 686–693.

Liptert F.W. (1987) *Mat. Perform.* **July**, pp. 12–19.

Livingston R.A. (1998) *Acidic Deposition Effects on Portland Cement Concrete.* Forthcoming in the Proceedings of the UN ECE Workshop, Berlin.

Maage M. (1980) *Norwegian Institute of Technical Report No. RM1 80-201.* Trondheim, Norway.

Maage M. (1984) *Mat. Const.* **17**, 345–350.

Martinez-Ramirez S. *et al.* (1996) *8th International Congress on Deterioration and Conservation of Stone, Berlin,* pp. 1557–1563.

Matousek M. (1980) *Proceedings of the International Congress on Chemistry of Cement, Paris*, pp. 764–765.

May J.O. and Butterworth B. (1962) *Science of Ceramics*, ed. Stewart G.H. Academic Press, Vol. 1, pp. 201–221.

Meland I. (1988) *Level of Calcium Hydroxide and Mechanisms of Carbonation in Hardened Blended Cements.* Nordic Concrete Federation Publication No. 6, pp. 169–178.

Neville A.M. (1981) *Properties of Concrete*, 3rd Ed. Pitman Int, London.

Orantie K. and Ruohomaka J. (1988) *Durability of Concrete Structures Protection Against Atmospheric (Acid) Attack with Coating.* Technical Research Centre of Finland, Research Reports 539.

Page C.L. and Treadaway K.W.J. (1982) *Nature* **297**, 109.

Parrott L.J. (1987) *A Review of Carbonation in Reinforced Concrete.* CKCA/BRE Spec. Publn.

Paul A. (1987) *Chemistry of Glasses.* Chapman and Hall, pp. 108–147.

Penkala B. and Zasum H. (1988) *Proceedings of the 6th International Congress on Deterioration and Conservation of Stone, Torun*, pp. 375–380.

Read H.H. (1984) *Ruttley's Elements of Mineralogy*, 26th Ed., p. 233.

Robinson G.C. (1982) *Conservation of Historic Stone Buildings and Monuments.* National Academy Press, Washington D.C., pp. 145–162.

Robinson G.C. (1984) *Ceram. Bull.* **63**, 295–301.

Ross K.D. (1984) *BRE Note 86/84.* BRE, Watford, UK.

Sauman Z. (1972) *Cement Concrete Res.* **2**, 541.

Schaffer R.J. (1985) *The Weathering of Natural Building Stones*, 3rd Ed. BRE Report No. 18, HMSO, London.

Scholl E. and Knöfel D. (1991) *Cement Concrete Res.* **21**, 127–136.

Sersale R. *et al.* (1998) *Cement Concrete Res.* **28**, 19–24.

Shireley D.E. (1975) *Cement and Concrete Association*, Publ. 45, p. 28.

Skoulikidis T. (1982) In *Atmospheric Corrosion*, ed. Ailor W. Wiley, Indiana, USA.

Soroka I. (1979) *Portland Cement and Concrete.* Macmillan Press Ltd, Chap. 6.

Tonni D.E. and Gaudis J.M. (1980) *Corrosion of Reinforcing Steel in Concrete.* ASTM Spec. Tech. Publn. 713, ASTM.

Torraca G. (1985) ICCROM, Rome, pp. 30–32.

Verbeck G. (1958) *ASTM Spec. Publn. N205*, pp. 17–36.

Watson A. *et al.* (1957) *Trans. Br. Ceram. Soc.* **56**, 37–49.

Watson A. *et al.* (1962) *Science of Ceramics*, ed. Stewart G.H. Academic Press, Vol. 1, pp. 187–200.

Watt Report No. 18 (1988) *Air Pollution, Acid Rain and the Environment*, ed. Mellanby K. Elsevier.

Wayne R.P. (1985) *Chemistry of Atmospheres.* Oxford University Press, New York.

Webster R.P. and Kukacka L.E. (1986) *Materials Degradation Caused by Acid Rain.* ACS symposium series 318, ACS, USA, pp. 239–249.

Winkler E.M. (1973) *Stone: Properties, Durability in Man's Environment.* Springer-Verlag, New York.

Winkler E.M. (1997) *Stone in Architecture,* 3rd Ed. Springer, New York.

Woods H. (1968) *Durability of Concrete Construction.* ACI Monograph No. 4.

Yates T.J.S. *et al.* (1989) *Construct. Build. Mater.* **2**, 20–26.

Ying-Yu L. and Qui-Dong W. (1987) *Proceedings of the Construction and Building Material International Conference* ACI, Vol. 2, pp. 1915–1943.

Zhang. Y. (1985) *BRE Note N84/85.* BRE, Watford, UK.

CHAPTER 5

SALTS AND CRUSTS

M. Steiger

1. Introduction

The mechanisms of degradation of inorganic porous building materials such as stone, brick, terracotta, concrete and mortar are commonly classified into chemical and physical mechanisms, including biochemical and biophysical processes. Chemical weathering refers to the dissolution or alteration of the mineral constituents of a material by chemical reactions. Physical weathering includes all processes generating mechanical stress either on a microscopic scale or to whole structures. There is quite a large number of different chemical and physical processes that can cause severe damage, such as granular disaggregation, delamination, spalling and complete structural decay. Unfortunately, the different types of damage can almost never be attributed to a specific damage mechanism. For example, granular disaggregation can be the result of chemical attack, frost damage or many other processes, as well. Moreover, one single process may manifest itself as different types of damage in different materials. Hence in practice, it is generally impossible to deduce the major causes of damage from visual observations. In order to assess the relative importance of different degradation processes and their rates, a detailed understanding of the underlying mechanisms is indispensable.

This chapter is concerned with salt damage which, from its very nature, is a physical damage process. However, salt damage cannot be clearly separated from chemical weathering because salts are also the products of chemical weathering reactions. Hence, the presence of salts in building materials is not only a potential cause of severe damage, but can also be an indicator of the degree of chemical weathering. First, I will briefly review the major sources of salt enrichment in building materials and then discuss the mechanisms of salt damage. Finally, an attempt is made to provide a short review of the most common effects of salts with special emphasis to air pollution.

2. Sources of Salts in Building Materials

2.1. Chemical Weathering

The chemical weathering of building materials involves the attack of water and associated acidity. The H^+ ion attack on the mineral components causes their dissolution and the formation of weathering products. Some examples of chemical weathering reactions relevant to stone deterioration are:

$$CaCO_3 + 2H^+ \rightleftharpoons Ca^{2+} + CO_2 + H_2O \tag{1}$$

$$2KAlSi_3O_8 + 2H^+ + 9H_2O$$
$$\rightleftharpoons 2K^+ + Al_2Si_2O_5(OH)_4 + 4H_4SiO_4^0 \tag{2}$$

$$Mg_5Al_2Si_3O_{10}(OH)_8 + 10H^+$$
$$\rightleftharpoons 5Mg^{2+} + Al_2Si_2O_5(OH)_4 + H_4SiO_4^0 + 5H_2O \tag{3}$$

Acid attack generally causes the mobilisation of metal cations from the parent minerals. The weathering of silicates also leads to the formation of new mineral compounds, e.g. iron oxyhydroxides and clays such as kaolinite, $Al_2Si_2O_5(OH)_4$. Table 1 lists the cations released from the weathering of some common mineral constituents of building materials. It can be seen that chemical weathering reactions mainly lead to the formation of sodium, potassium, magnesium, and calcium salts.

Table 1. Ions released from weathering of some common minerals in building materials.

Mineral name	Idealised formula	Ions released
Carbonate minerals		
Calcite	$CaCO_3$	Ca^{2+}
Dolomite	$CaMg(CO_3)_2$	Ca^{2+}, Mg^{2+}
Feldspars		
Plagioklase feldspar	$Na_xCa_{1-x}Al_{2-x}Si_{2+x}O_8$	Na^+, Ca^{2+}
Microcline (K-feldspar)	$KAlSi_3O_8$	K^+
Clay minerals		
Biotite	$K(Mg,Fe)_3AlSi_3O_{10}(OH)_2$	K^+, Mg^{2+}
Chlorite	$Mg_5Al_2Si_3O_{10}(OH)_8$	Mg^{2+}

Carbon dioxide (CO_2) is a major source of acidity in natural waters and it is the acid most responsible for rock weathering on a geological time scale. The solubility of carbonate materials in water is enhanced in the presence of CO_2 according to the reaction:

$$CaCO_3 + CO_2 + H_2O \rightleftharpoons Ca^{2+} + 2HCO_3^- \qquad (4)$$

The equilibrium solubility of calcite in rain water in equilibrium with atmospheric CO_2 (5.5×10^{-4} moles/L) is about four times higher than in pure water. Thus, even in the absence of other sources of acidity, there is a natural dissolution reaction of carbonate minerals which is commonly referred to as the Karst effect. Although the dissolution rates of silicates are much slower than those of carbonate minerals, silicate minerals are also subject to chemical weathering (Drever, 1994; Lasaga *et al.*, 1994).

2.2. Acid Deposition

Acid deposition is the major source of H^+ in building materials. Here, acid deposition refers to, both, wet deposition, i.e. acid precipitation, and the dry deposition of gaseous and particulate pollutants. Atmospheric acidity is closely related to the atmospheric chemistry of sulphur dioxide (SO_2) and the nitrogen oxides (NO and NO_2) which become oxidised to sulphate and nitrate through gas and aqueous phase processes. The atmospheric chemistry of sulphur and nitrogen compounds is quite complex and can only be briefly summarised, here. More comprehensive reviews on acid deposition (Schwartz, 1989) including its effects on building materials (Fassina, 1988; Butlin, 1991) and the history of air pollution (Brimblecombe and Rodhe, 1988) are available.

The major pathways of acid deposition associated with SO_2 are depicted in Fig. 1. The gas phase oxidation of SO_2, mainly by reaction with OH radicals, ultimately leads to the formation of sulphuric acid (H_2SO_4) which will rapidly nucleate to form aqueous H_2SO_4 droplets, acting themselves as condensation nuclei for cloud droplets. The aqueous phase formation of H_2SO_4 proceeds via absorption of SO_2 in cloud droplets or moist particles and subsequent oxidation reactions. Finally, aerosol sulphate and gaseous SO_2 are subject to below cloud scavenging by rainfall. In effect, a large fraction of the atmospheric SO_2 is removed by acid precipitation. The dry removal of acidic sulphur compounds involves the deposition of gaseous SO_2 and acidic particulate sulphate.

The dry deposition of acidic species on material surfaces involves two different processes (Lipfert, 1989a). First, the transport of pollutants to a surface is controlled by aerodynamic factors, a detailed discussion of which is provided by Hicks (1983). Next, acidic trace

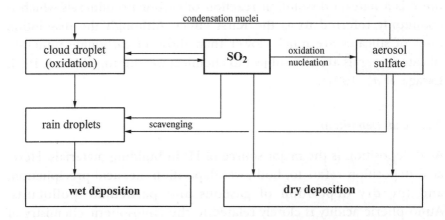

Fig. 1. Major pathways of SO_2 deposition.

species need to be irreversibly deposited to the surface, either by chemical reaction, or by physical processes, e.g. adsorption or dissolution. Thus, the overall rate of dry deposition is closely related to the material properties such as chemical composition and surface wetness, and the reactivity of the atmospheric trace gas of interest. The damage potential of acidic gases is therefore depending on both, their ambient concentrations, and the sensitivity of a particular material to that gas. Graedel and McGill (1986) provide a qualitative review of the sensitivity of various materials to a number of gaseous pollutants. A useful measure for quantitative comparisons of the sensitivity of different materials to dry deposition is the deposition velocity, V_D, defined as the ratio of the surface flux, F, i.e. the deposition per unit area and time, to the ambient concentration of the gas, C:

$$V_D = \frac{F}{C} \qquad (5)$$

Sulphation was early recognised as a cause of decay of carbonate stone (Schaffer, 1932; Kieslinger, 1932) and due to its relatively high concentrations in urban areas, the dry deposition of SO_2 was the subject of many experimental studies. Lipfert (1989a) has presented a literature compilation of SO_2 deposition velocities to various materials. Typical values of V_D reported are largely dependent on the reactivity of the materials indicating that SO_2 dry deposition to many building materials is limited by surface processes rather than aerodynamic factors. The most likely deposition mechanism involves the dissolution of SO_2 in moisture films present on surfaces and the subsequent aqueous phase oxidation of S(IV) resulting in a decrease of the solution pH. The solubility of SO_2 strongly decreases with decreasing pH, thus, further deposition would ultimately stop. However, if the surface film is in contact with sufficiently reactive minerals such as calcite, the solution pH can be effectively buffered by the dissolution reaction (1).

Deposition velocities of SO_2 reported in the literature generally agree with the deposition mechanism described (Lipfert, 1989a). The highest deposition velocities are reported for reactive materials such as cement and limestone. Though much less data are available for sandstones and other less reactive materials, the reported deposition velocities are considerably lower indicating that the reactivity of the materials is the limiting factor. Also, higher deposition velocities are consistently found at high relative humidities (e.g. Spiker et al. 1995), such that aerodynamic factors (wind speed and turbulence) might become the limiting factors for the deposition of SO_2 to the most reactive materials at high relative humidities.

The dry deposition of particles to material surfaces is controlled by physical processes, including diffusion, inertial impaction, and gravitational settling (Hicks, 1983). The deposition velocities are strongly dependent on particle size. Inertial impaction and gravitational settling increase with particle size, whilst diffusion controlled deposition decreases with particle size. This results in a minimum of the deposition velocity for particles in the size range 0.1–1 μm. Most of the aerosol acidity is associated with particles of exactly this size range. Due to the low deposition velocities of acidic particles, their contribution to the total dry deposition of acidity is of minor importance only, which was confirmed in field exposure experiments (Girardet and Furlan, 1991; Wittenburg and Dannecker, 1992).

The total amount of acidity deposited to building materials by wet deposition is given as the product of the H^+ concentration and the total volume of rain deposited on the surface. A number of studies were carried out to determine the effect of acid rain. A useful experimental technique is the collection and analysis of stone run-off solutions (Rönicke and Rönicke, 1972; Reddy et al., 1985; Livingstone, 1986; Cooper, 1986). Any differences in the concentrations between run-off and incident rainfall must be due to interactions with the stone. For carbonate stones, the "excess" calcium concentrations in run-off solutions provides a direct measure of stone dissolution. Here, excess calcium concentration refers to a corrected concentration taking into account contributions from the dry deposition of particles and the calcium concentrations in the incident rainfall itself (Reddy, 1988). Data obtained from run-off water experiments were used to determine the relative contributions of the Karst effect, the neutralisation of acid rain, and the dry deposition of SO_2 (Livingstone, 1992; Baedecker et al., 1992). It was found that in urban areas with significant SO_2 pollution, the dry deposition of local SO_2 was dominant (Roekens and van Grieken, 1989; Livingstone, 1992).

The atmospheric chemistry of the nitrogen oxides is considerably more complex than that of SO_2. The deposition pathways of nitrogen compounds are summarised in Fig. 2. Nitric oxide (NO) is the major oxide formed during the high temperature combustion of fuels. Nitrogen dioxide (NO_2) is formed in the atmosphere by oxidation of NO. The final oxidation product of the nitrogen oxides (NO_x) is nitric acid (HNO_3) which is mainly formed by the reaction of NO_2 with the OH radical. HNO_3 is more volatile than H_2SO_4 and can exist in significant concentrations in the gas phase. However, it is also highly soluble and is readily absorbed by cloud droplets and moist aerosol particles. Another important reaction of HNO_3 is the neutralisation with ammonia (NH_3), which is the most prominent atmospheric alkaline species present in substantial concentrations. This reaction leads to the formation of ammonium nitrate aerosols (NH_4NO_3). Actually, there is a dynamic equilibrium between NH_3, HNO_3 and NH_4NO_3, and both, NH_3 and HNO_3 can revolatilise. Other reactions of NO_X with free radicals, including OH, RO, RO_2 and RCO_3, lead to a number of gaseous nitrogen compounds among which nitrous acid (HONO) and peroxyacetyl nitrate (PAN), $CH_3C(O)OONO_2$, are the

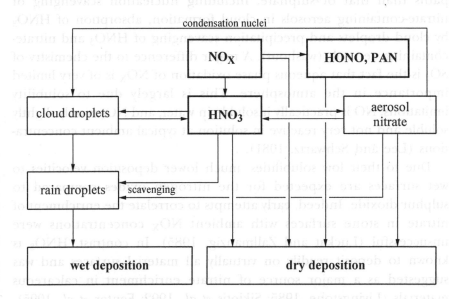

Fig. 2. Major pathways of NO_X deposition.

Table 2. Typical concentrations of some gaseous
nitrogen species in rural and moderately polluted
atmospheres, units are ppb (data from Finlayson-Pitts
and Pitts, 1986).

Pollutant	Rural	Moderately polluted
NO	0.05–20	20–1000
NO_2	1–20	20–200
HNO_3	0.1–4	1–10
HONO	0.03–0.8	0.8–2
PAN	2	2–20
NH_3	1–10	1–10

most prominent species. Thus, the dry deposition of nitrate to
surfaces includes several gaseous precursors and particulate nitrate.
Table 2 lists typical concentration ranges of several gaseous nitrogen
compounds for rural and moderately polluted atmospheres compiled
by Finlayson-Pitts and Pitts (1986).

The incorporation of nitrate into precipitation follows similar
paths than that of sulphate, including nucleation scavenging of
nitrate-containing aerosols in cloud formation, absorption of HNO_3
by cloud droplets and precipitation scavenging of HNO_3 and nitrate-
containing aerosols (washout). A major difference to the chemistry of
SO_2 is the fact that aqueous phase oxidation of NO_X is of very limited
importance in the atmosphere. This is largely due to solubility
limitations, NO is practically insoluble in water, and NO_2 is only slightly
soluble and not very reactive in solution at typical ambient concentrations (Lee and Schwartz, 1981).

Due to their low solubilities, much lower deposition velocities to
wet surfaces are expected for the nitrogen oxides compared to
sulphur dioxide. Indeed, early attempts to correlate the enrichment of
nitrate in stone surfaces with ambient NO_X concentrations were
unsuccessful (Luckat and Zallmanzig, 1985). In contrast, HNO_3 is
known to deposit readily on virtually all material surfaces and was
suggested as a major source of nitrate enrichment in calcareous
materials (Livingstone, 1985; Sikiotis et al., 1992; Fenter et al., 1995).
Due to its reactivity, it is assumed that the deposition velocity of HNO_3

is virtually not depending on surface chemical processes. Thus, turbulent transport to the material surface and the ambient HNO_3 concentration are the limiting factors controlling the delivery rate of HNO_3 to material surfaces. It can be seen from Table 2 that HNO_3 concentrations are considerably lower than those of the precursor gases. Behlen *et al.* (1996) have recently determined deposition velocities of NO, NO_2, HNO_3, HONO and PAN to a calcite cemented sandstone. Considering the ambient concentrations of these gases, they could show that the total deposition of nitrate is mainly due to HNO_3 and NO_2, with only minor contributions of HONO. The contributions of PAN and NO were found to be negligible.

Despite the sensitivity of virtually all materials to HNO_3, the degradation potential as compared to SO_2 might be limited due to the rather low atmospheric concentrations. In fact, there is experimental evidence from exposure experiments that sulphate accumulates more rapidly in calcareous stones than nitrate does (e.g. Jaynes and Cooke, 1987; Baedeker *et al.*, 1992; Sweevers and van Grieken, 1992). However, in view of strongly decreasing SO_2 concentration levels, increasing NO_X emissions, deposition of oxidised nitrogen species may play a more important role in the deterioration of building materials in the future.

The principal role played by NH_3 in the atmosphere is the neutralisation of acidic species such as H_2SO_4 and HNO_3, the main reaction products of which are ammonium sulphate, $(NH_4)_2SO_4$ and NH_4NO_3. Thus, the ammonium ion is an important constituent of both, cloud droplets and the submicron aerosol. It has been already mentioned that NH_3 can revolatilise from aerosols such that NH_3 and HNO_3 can co-exist in the gas phase according to

$$NH_4NO_3 \rightleftharpoons NH_3 + HNO_3 \qquad (6)$$

Interest in the deposition of ammonium (NH_4^+) and NH_3 on building materials and soil mainly arises from the fact that NH_4^+ is biologically oxidised by the process of nitrification (van Breemen *et al.*, 1982; Mansch and Bock, 1998). This two-step process is given by the overall reaction:

$$NH_4^+ + 2O_2 \rightleftharpoons 2H^+ + NO_3^- + H_2O \qquad (7)$$

Thus, the deposition of NH_3 and NH_4^+ to building materials may be regarded as an indirect pathway of acid deposition. Wolters *et al.* (1988) determined rates of nitrification in various sandstones and concrete under idealised laboratory conditions, i.e. optimum temperature and unlimited supply with the substrate. They report very high rates of nitrification, clearly indicating that the limiting factor under ambient conditions will be the delivery of NH_3 or NH_4^+.

Due to its high solubility, NH_3 can deposit to materials through dissolution in moisture films present on surfaces. The ammonia solubility is strongly increasing with decreasing pH, thus in contrast to SO_2, the deposition velocities are considerably higher at low pH (Adema and Heeres, 1995). The dry deposition of NH_3 to material surfaces will therefore depend on the reactivity of the material and its ability to buffer the increase in pH caused by the formation of hydroxyl ions. Carbonate materials are particularly unfavourable in this respect, and it was shown that dry NH_3 deposition to calcareous stone is significantly lower than to silicate stones (Wittenburg and Dannecker, 1994).

Considering that atmospheric concentrations of NH_3 (cf. Table 2) are considerably lower than those of SO_2, it can be concluded that the contribution of NH_3 to the total input of acidity via dry deposition and nitrification is of minor importance, only. Typically, atmospheric concentrations of NH_3 are comparable to those of nitric acid, to which, however, most building materials are much more sensitive. Thus, the degradation potential of gaseous NH_3 appears to be of minor importance except perhaps for buildings directly affected by strong primary NH_3 sources, e.g. stock-farms.

Behlen *et al.* (1997) have determined the wet deposition of NH_4^+ to a brick masonry which was by far exceeding the dry deposition of both, NH_3 and particulate NH_4^+. The wet deposition of NH_4^+ can produce acidity through either of two mechanisms. Firstly, NH_4^+ can be oxidised by nitrification according to reaction (7). Next, NH_4^+ containing solutions can also react acidic if in contact with carbonate minerals, e.g. calcite, according to the following reaction:

$$NH_4^+(aq) + CaCO_3(s)$$
$$\rightleftharpoons Ca^{2+}(aq) + HCO_3^-(aq) + NH_3(g) \qquad (8)$$

However, it appears that this reaction has not yet been systematically studied as a potential source of calcite dissolution. Reaction (8) implies that it might be misleading to characterise wet acid deposition only by the free acidity, expressed as pH or the molar concentration of H^+. Similar arguments apply to bound acidity in the form of undissociated organic acids which might also be present in rainwater (Chebbi and Carlier, 1996).

2.3. Other Sources of Salts

Besides acid deposition, there are a number of other sources of salts in building materials. Sea salt is an important natural source of salts derived from the atmosphere. On a global scale, emissions of sea salt droplets ejected from the oceans are considered as one of the most important primary sources of the atmospheric aerosol (Blanchard and Woodcock, 1980). Sea salt particles are removed from the atmosphere by wet and dry deposition which are the major processes leading to their enrichment in building materials. In contrast to anthropogenic air pollutants, the concentrations of which have dramatically changed during the last centuries (Brimblecombe and Rohde, 1988), the enrichment of sea salt in historic buildings has been continuously progressing over much longer periods of time. Salt accumulations found today are representing an integral effect of sea salt deposition beginning with the time of construction of a building. Experience from many monuments located at coastal sites suggests that deposition and enrichment of sea salt can be a major cause of decay (e.g. Theoulakis and Moropoulou, 1988; Zezza and Macrì, 1995).

Sea salt particles in the atmosphere have a chemical composition very similar to that of bulk sea water. The contribution of the six ions sodium, magnesium, potassium, calcium, chloride and sulphate amounts to > 99% by mass of the total solids dissolved. The major ions, sodium and chloride, account for 85.6% of the bulk sea salt. Oceans have a remarkably uniform composition and it can be assumed, that the relative abundances of the major ions did not significantly change over the milennia. Therefore, the relative abundances of the major constituents can be used as tracers for sea salt deposition (Zappia *et al.*, 1989). It has been shown that the enrichment of sea salt in buildings in coastal environments is considerably variable depending on a number of different influences including environmental parameters, and the geometry and constructional details of a building (Steiger *et al.*, 1997).

Sources of salts in building materials other than atmospheric have been reviewed by Arnold and Zehnder (1991) and are only briefly summarised here. A major source of salts is ground moisture carried into masonry by rising damp in the absence of damp coursing in a building. A more detailed discussion of salts associated with rising damp is provided in a later section. The use of alkaline building and cleaning materials, e.g. portland cement and water glass, can be an important source of salts (Arnold and Zehnder, 1991). Generally, these materials release sodium and potassium hydroxide and carbonate, which can react with salt mixtures already present in a masonry forming new, often more damaging salt mixtures. A simple example is the reaction of sodium carbonate with calcium sulphate according to

$$Na_2CO_3 + CaSO_4 \rightleftharpoons Na_2SO_4 + CaCO_3 \qquad (9)$$

As will be shown below, the damage potential of sodium sulphate is greater than that of calcium sulphate. Other examples can be found in Arnold and Zehnder (1991). Alkaline materials are also excellent substrates for the deposition of SO_2 and CO_2 leading to complex mixtures of carbonates, bicarbonates and sulphates.

Chemicals used for cleaning treatments are sometimes acidic, e.g. containing hydrofluoric acid or formic acid. The chemical reaction of these acids with the mineral constituents of the building materials leads to the formation and enrichment of fluorides or formates (Zehnder and Arnold, 1984). Other sources of salts include de-icing salt, and particularly in historic buildings, the storage of salts in former times.

3. Mobility and Hygroscopicity of Salts

Salts in building materials are typically complex multi-component mixtures. In most practical situations, these salt systems are sufficiently characterised as a mixture of the cations Na^+, K^+, Mg^{2+} and Ca^{2+}, and

Table 3. Salt solubilities relative to gypsum $(CaSO_4 \cdot 2H_2O)$ on a molal basis. Values refer to the thermodynamic stable hydrate at 25°C; the solubility of gypsum at 25°C is 0.0158 moles/kg.

	Cl^-	NO_3^-	SO_4^{2-}	CO_3^{2-}	HCO_3^-
Na^+	389	685	125	178	77
K^+	304	240	44	519	228
Mg^{2+}	367	310	191	0.0795	
Ca^{2+}	461	531	1	0.0097	

the anions Cl^-, NO_3^-, SO_4^{2-}, HCO_3^- and CO_3^{2-}. These salts are commonly referred to as soluble salts. In fact, however, their solubilities differ by orders of magnitude. Table 3 lists the solubilities of the thermodynamic stable forms of these salts at 25°C relative to the solubility of gypsum $(CaSO_4 \cdot 2H_2O)$. It is obvious that gypsum compared to all other salts except the alkaline earth carbonates is a sparingly soluble salt only.

The solubility is an important quantity determining the mobility of salts in porous materials. We might conveniently define mobility as the maximum amount of a salt present in a porous material that can just be dissolved if the pore space is completely filled with water. Salt mobilities can be calculated from the concentration and density of the saturated salt solution, and the porosity and density of the porous material. As an example, Fig. 3 depicts smoothed values of the mobilities of sodium chloride and gypsum as a function of the water accessible pore space for a number of natural stones. Porosities and densities of the stones (48 limestones and 88 sandstones) were taken from Grimm (1990). It is obvious that the mobility of a salt in a particular material is strongly depending on both, the solubility of the salt and the material properties. Comparing the mobilities depicted in Fig. 3 to the typical ranges of sodium and chloride concentrations found in building materials, it can be concluded that sodium chloride in the pore space will be completely dissolved and mobilised in all but the most extreme situations if water penetrates the material. In contrast, gypsum concentrations in building materials typically exceed the mobilities shown in Fig. 3 by several orders of magnitude. Thus,

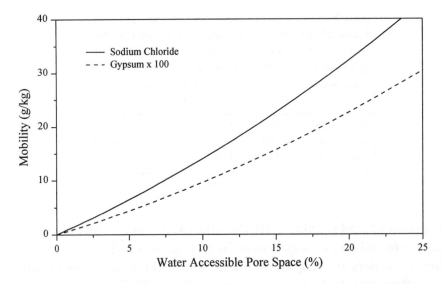

Fig. 3. Mobilities of sodium chloride and gypsum in building stones.

even if the pore space is completely filled with water, only a very small fraction of gypsum can be mobilised.

Apart from gypsum, most other salts are readily dissolved and mobilised in porous materials. It follows that the transport of salts in building materials is largely controlled by the movement of water due to capillary transport. Precipitation, ground moisture and condensation are the major sources of liquid water in building materials. The absorbed water is transported and released later by evaporation. As water evaporates, pore solutions become more concentrated and salts are typically accumulated in evaporation zones.

Arnold and Zehnder (1991) provide a detailed discussion of salt transport and accumulation in walls affected by rising damp based on extensive observations from a number of buildings. Initially, ground moisture penetrating a masonry is a dilute solution which is vertically transported due to capillary rise (cf. Fig. 4). Above ground level moisture evaporates from the wall and the solution becomes more and more concentrated while still being subject to capillary rise. As the solution becomes saturated during transport any further evaporation will cause crystallisation and immobilisation of the salt. In effect,

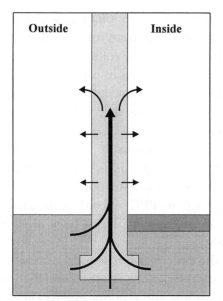 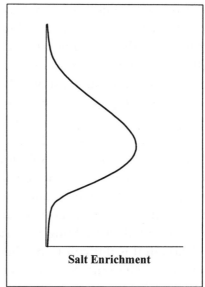

Fig. 4. Moisture transport and vertical profile of salt enrichment in masonry affected by rising damp.

a zone of salt enrichment evolves some distance above ground level. The height of the accumulation zone is dependent on the rate of evaporation and the solubility of the salt. The maximum height is given as the upper limit of capillary rise of moisture, i.e. the height at which the evaporation rate equals the supply of moisture from below.

Pore solutions originating from ground moisture typically contain several ions. During capillary rise and evaporation, less soluble salts will therefore reach saturation earlier than salts of greater solubility, resulting in a fractionation of the salts according to their solubilities. The composition of the pore solution continuously changes during transport and only the very soluble salts are transported as concentrated brine solutions to the upper evaporation zone. Arnold and Zehnder (1991) provide vertical profiles of ion concentrations in the zone affected by rising damp from a number of buildings. Maxima of salt enrichment were observed at heights from 0.5 m to about 3 m above ground level, and the profiles provide evidence for salt fractionation.

It is very difficult, however, to predict the precipitation sequences of minerals from an initially dilute multi-component ionic mixture. In addition to the salt minerals precipitated from a solution of a single salt, a large number of double salts can be formed in mixed solutions. Also, several salts can exist in anhydrous and different hydrated forms, e.g. sodium sulphate can exist as the anhydrous thenardite ($NaSO_4$) and as the decahydrate mirabilite ($Na_2SO_4 \cdot 10H_2O$). Furthermore the solubilities of several salts show considerable variation with temperature. Considering solutions of the major ions found in building materials, about 50 to 60 different evaporite minerals can be precipitated in the temperature range −30 to 40°C. According to Arnold and Zehnder (1991), more than 30 of these minerals have already been detected in building materials.

Though it is difficult to predict the precipitation sequences, due to solubility limitations there are certain restrictions for the composition of brine solutions evolving during transport and evaporation. As Hardie and Eugster (1970) have pointed out, the solubilities of the calcium and magnesium carbonates and of gypsum are so low compared to all other salt minerals of interest, that they provide a chemical divide. Figure 5 depicts the major pathways of fractionation and brine evolution from solutions initially containing Na^+, K^+, Mg^{2+}, Ca^{2+}, Cl^-, NO_3^-, SO_4^{2-}, CO_3^{2-} and HCO_3^-. The evaporation of water from such solutions always leads to the crystallisation of the alkaline earth carbonates, i.e. calcite, dolomite or nesquehonite ($MgCO_3 \cdot 3H_2O$). Thus, concentrated pore solutions cannot contain calcium or magnesium and carbonate at the same time.

If the sum of the calcium and magnesium concentrations exceeds the total carbonate concentration, a carbonate poor solution will evolve due to the precipitation of carbonate minerals. Then, due to its low solubility, gypsum is precipitated next and acts as a second divide. If the initial calcium concentration exceeds that of sulphate a sulphate-poor, type I, solution evolves. Typically, the solubilities are very high in such mixtures, particularly if there are significant relative abundances of calcium or magnesium. Type I solutions are very often found in the upper evaporation zone of masonry affected by rising damp. Several examples can be found in Arnold and Zehnder (1991). If the initial solution contains more sulphate than calcium, a type II,

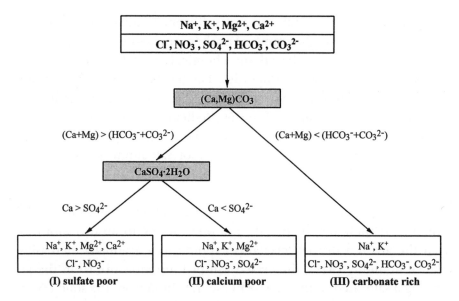

Fig. 5. Evolution of concentrated solutions of major types of hygroscopic salt mixtures.

calcium-poor solution evolves upon evaporation. For example, the evaporation of seawater leads to solutions of the latter type. Only if the carbonate concentration in an evaporating solution exceeds the sum of the calcium and the magnesium concentrations, a carbonate-rich, type III, pore solution can evolve. As was mentioned before, carbonate-rich salt systems are typically found in building materials that have been subjected to inappropriate materials and treatments.

In effect, the alkaline earth carbonates and gypsum are precipitated very early during transport and are thus accumulated at low heights, whilst the further fractionation of the remaining salt systems proves to be quite complicated. Though it is possible to deduce the fractionated crystallisation of complex salt mixtures from experimentally derived solubility diagrams (cf. Mullin, 1993), this approach requires solubility measurements on a huge number of mixture compositions covering the temperature range of interest. For some systems of geochemical importance, e.g. the composition of brines related to the evaporation of seawater, such data exist and were used to describe the formation of evaporite minerals (Braitsch, 1971). However, available data for the more complex salt systems typically found in building materials are not sufficient to adequately describe the required phase diagrams. Another approach to understand evaporite formation is based on chemical modelling of aqueous electrolyte thermodynamic properties. The most frequently used model is the ion interaction approach (Pitzer, 1991 and references therein) which was successfully applied to the prediction of mineral solubilities in natural waters (Weare, 1987). This same model approach has been recently utilised to study the behaviour of salt mixtures relevant to building materials (Price and Brimblecombe, 1994; Steiger and Dannecker, 1995).

Another important property of salts, closely related to the solubility, is their hygroscopicity, i.e. their ability to absorb moisture from the air. If a dry salt crystal is exposed to a gradually increasing relative humidity (RH), the crystal will pick up moisture from the air forming a saturated solution. The phase transition from the solid crystal to the liquid solution only occurs when the RH reaches a certain value, the deliquescence or saturation humidity of the salt. The deliquescence humidity is the RH of air in equilibrium with the saturated salt solution. Deliquescence humidities of several common salts as a function of temperature are shown in Fig. 6. The sodium and potassium salts and all sulphates show deliquescence humidities greater than 70% RH. The deliquescence of the alkaline earth chlorides and nitrates occurs at considerably lower relative humidities (< 60%). In contrast, the deliquescence humidity of gypsum is > 99.9% RH, hence, gypsum is essentially non-hygroscopic.

If a saturated salt solution is subjected to a further increase in the relative humidity, it continues to pick up moisture to achieve equilibrium and the solution becomes progressively more dilute. This is illustrated in Fig. 7 for several salts as the amount of water picked up by 1 g of salt as a function of the relative humidity. It is evident that there is a strong influence of salts on the hygroscopic moisture content of porous materials. Water vapour adsorption and capillary condensation are the major factors controlling the hygroscopic moisture of porous materials in the absence of salts (Camuffo, 1984). Garrecht (1992) studied the hygroscopic moisture uptake of sandstone contaminated with salts. Only considering moisture uptake

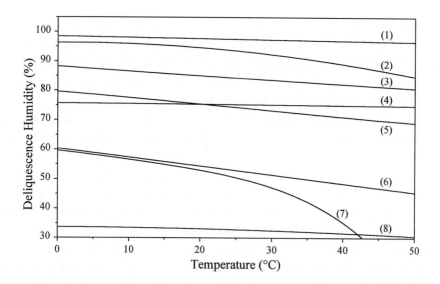

Fig. 6. Deliquescence humidities of several common salts: (1) K_2SO_4; (2) KNO_3; (3) KCl; (4) NaCl; (5) $NaNO_3$; (6) $Mg(NO_3)_2 \cdot 6H_2O$; (7) $Ca(NO_3)_2 \cdot 4H_2O$; and (8) $MgCl_2 \cdot 6H_2O$.

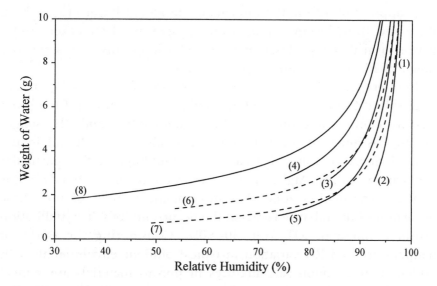

Fig. 7. Water uptake by 1 g of salt as a function of RH at 25°C: (1) K_2SO_4; (2) KNO_3; (3) KCl; (4) NaCl; (5) $NaNO_3$; (6) $Mg(NO_3)_2 \cdot 6H_2O$; (7) $Ca(NO_3)_2 \cdot 4H_2O$; and (8) $MgCl_2 \cdot 6H_2O$.

curves such as those shown in Fig. 7, he could predict to within experimental error the moisture content in the sandstone above the deliquescence humidities of the salts. At humidies below the deliquescence humidities, the hygroscopic moisture content of the stone was not affected by the presence of salts.

The fractionation of salts during capillary rise and evaporation of ground moisture leads to an accumulation of a very hygroscopic salt mixture. In effect, the moisture content in the upper evaporation zone of a masonry affected by rising damp is controlled by the hygroscopicity of the salt mixture. Moreover, as can be seen from Fig. 7, there will be a dynamic equilibrium between the moisture content and the ambient RH. Moisture is picked up as the RH increases and water evaporates from the masonry as the RH decreases. Upon evaporation, the pore solution becomes increasingly concentrated, and eventually, supersaturated with respect to one or several salt minerals which then crystallise out. The situation would be quite simple if the masonry was contaminated with a single salt. Then the salt would crystallise out if the ambient RH drops below its deliquescence humidity. However, the behaviour of salt mixtures is more complex. In their application of the thermodynamic model of Pitzer (1991), both Price and Brimblecombe (1994) and Steiger and Dannecker (1995) have shown that crystallisation in mixed salt systems occurs across a range of RH rather than at specific values.

In effect, the salts in the accumulation zone may either be present in dissolved form or as crystalline solids, only depending, for a given mixture composition, on the ambient temperature and RH. Steiger (1996) has analysed the behaviour of a salt mixture found in the ground moisture zone of a sandstone building. In this particular application, Na^+, K^+, Mg^{2+}, Cl^- and NO_3^- were the most abundant ions in the accumulation zone. Thermodynamic analysis of the phase behaviour of the mixture revealed the occurrence of salt crystallisation within a range of RH from 30–80%. Considering the range of temperature and RH variation in typical outdoor environments, it is evident that common salt mixtures in porous materials are subject to continuous phase changes. In contrast, gypsum, due to its low solubility, behaves different than most other salts. Gypsum is non-hygroscopic and most of the gypsum once deposited in the pore space remains in solid form at all times. Because of its low mobility, gypsum cannot be efficiently transported in porous materials and tends to be accumulated close to where it has been formed. The same arguments apply to other salts of low solubility, e.g. calcium fluoride and the calcium oxalates.

4. Mechanism of Salt Damage

It is generally recognised that crystal growth of salts in porous materials is a major cause of damage. Due to the complex behaviour of salt mixtures, there are a number of phase changes causing crystal growth in building materials including the crystallisation from supersaturated solutions, the change of the state of hydration, and chemical reactions causing the growth of new minerals at the expense of previously deposited phases. Also, the freezing of water and the growth of ice crystals is strongly affected by dissolved salts. Certain very soluble salts can depress the freezing temperature to −50°C and below (Spencer *et al.*, 1990).

The relevant phase changes are reversible processes. For example, a salt crystallises out as water evaporates due to a decrease in the ambient RH, and the salt will re-dissolve as the RH is increasing again. Thus, changes in the environmental parameters, namely temperature and RH, are the driving forces triggering salt damage. Damage does not necessarily occur where the maximum salt concentration in a material is found, but rather, where, for a given climatic situation, the composition of a salt mixture is such that fluctuations of temperature and RH cause damaging cycles of repeated phase changes.

Evans (1970), and more recently, Duttlinger and Knöfel (1993) and Goudie and Viles (1997) provide comprehensive reviews on the available literature on salt weathering, dating back to the early 19th century. Lewin (1989) has pointed out that many studies are qualitative in nature, i.e. they provide detailed descriptions of the manifestation of salt damage for a variety of materials such as delamination, spalling or granular disaggregation. Only few studies, however, propose detailed mechanisms of the processes which can generate the high internal pressures and disruptive stresses necessary to understand the observable weathering phenomena. Moreover, the discussion of the very nature of the processes has often been controversial, and hitherto, agreement among investigators has not been achieved. The question is not *if* crystal growth causes damage, but rather *how* crystal growth can generate disruptive stress.

4.1. Crystallisation Pressure

It is axiomatic that crystallisation can only exert stress if crystals are growing within confined spaces of a porous material. Thus, one necessary condition for salt damage is crystal growth within the pores of a material rather than on top of its surface. The site of crystallisation is determined by both, the rate of evaporation from the exposed surface of the material and the rate of supply of salt solution to the surface (Lewin, 1982). Figure 8 schematically illustrates the evaporation of water from a porous material initially saturated with a dilute salt solution [cf. Fig. 8(a)]. As water evaporates from the material surface, the solution eventually becomes supersaturated resulting in the precipitation of a salt mineral. As long as the rate of capillary transport of fresh solution to the surface equals the rate of escape of water from the surface, the deposition of the salt mineral occurs on top of the surface resulting in the formation of an efflorescence [Fig. 8(b)]. If salt crystals grow exclusively on top of the surface, there will be no damage. However, as more water evaporates the capillary transport of solution to the exposed surface becomes too slow and does not compensate for the escape of water at the surface. As a result, a dry zone evolves beneath the surface and crystal growth now occurs within the pores [Fig. 8(c)]. If the evaporating solution contains a mixture of ions, a second solid might crystallise out in addition to the first one. Salt crystals are then growing within confined spaces of the material and may generate disruptive stress causing damage [Fig. 8(d)].

A second necessary condition for damage to occur is, that a confined crystal continues to grow against a constraint, thus exerting stress. Becker and Day (1916) and Taber (1916) have experimentally shown that crystals can grow against substantial external loads. They called this phenomenon the force of crystallisation, which was later quantitatively measured by Correns and Steinborn (1939) using

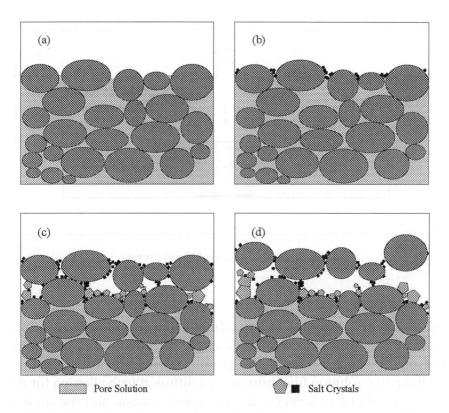

Pore Solution   Salt Crystals

Fig. 8. Evaporation of water and crystallisation of salts in porous materials.

growing crystals of alum ($KAl(SO_4)_2 \cdot 12H_2O$). The experimental setup of these early experiments is illustrated in Fig. 9. It is important to note that the crystals in these experiments continued to grow upon their loaded surfaces despite the fact that the free surfaces were in contact with the salt solution as well. The pioneering experiments of Becker and Day (1916) and Taber (1916) provide evidence that growing crystals can exert stress despite the fact that there is a decrease in the total volume if a salt crystallises from its solution.

In order that a crystal continues to grow upon the loaded surfaces, a solution film must exist seperating at least one of the loaded faces from its constraint, otherwise deposition of mineral matter in the region of contact is impossible. The early observers have been well aware of the importance of such a solution film. Later, Weyl (1959)

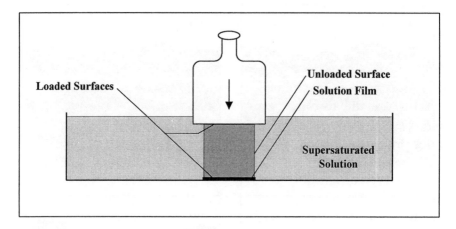

Fig. 9. Growth of a loaded salt crystal.

discussed the properties of the solution film in more detail, which must be of sufficient mechanical strength such that it is not squeezed out of the contact zone of the growing crystal and its support. In addition, the solution film must act as a diffusion path allowing for the exchange of matter between the unstressed solution and the stressed solid. Weyl (1959) gave a derivation of the rather complicated diffusion and mass transport equations and estimated the pressure distribution along the loaded surface of the growing crystal. In accordance with the observations of Becker and Day (1916) and Taber (1916), his model predicts the growth of loaded crystals with a hollow centre.

Another important requirement for the growth of loaded crystals arises from thermodynamic considerations. As a result of the increase of its chemical potential, a solid under stress exhibits a greater solubility than the stress-free solid. It follows that a growing crystal can only generate stress if the external solution is supersaturated. In other words, for a given supersaturation of the solution there is a maximum stress that can be built up. Considering an equilibrium between a solid under stress, σ, and a solution under hydrostatic pressure, p, Correns and Steinborn (1939) derived an expression for the maximum stress as a function of the supersaturation of the solution.

$$\Delta\sigma = \frac{RT}{V} \cdot \ln\frac{C}{C_0} \tag{10}$$

where $\Delta\sigma = \sigma - p$ is the crystallisation pressure, R is the gas constant, T is the absolute temperature, V is the molar volume of the solid, C_0 is the concentration of the saturated solution of the solute under hydrostatic pressure, and C is the actual concentration of the solute. Winkler and Singer (1972) applied equation (10) to calculate crystallisation pressures for a number of common salts. However, to obtain crystallisation pressures exceeding typical tensile strengths of porous materials, they assumed extremely high supersaturation ratios. Lewin (1974) has pointed out that such supersaturations are very unlikely to be achieved in practice.

Apart from the fact that Winkler and Singer (1972) assumed unrealistically large supersaturation ratios, there is yet another reason why crystallisation pressures according to equation (10) are erroneous. A more rigorous analysis of the thermodynamics of solids under anisotropic stress (Paterson, 1973; de Boer, 1979) yields in fact an equation of the same form as equation (10), if the effect of the elastic strain energy is neglected, which was shown to be a reasonable assumption for stresses below about 100 MPa (de Boer, 1979). However, for dissociating solids (i.e. salt minerals), the supersaturation ratio C/C_0 in equation (10) has to be replaced by the respective ratio of the activity products. Thus, to calculate crystallisation pressures the stoichiometry of the dissociation reaction has to be considered. Since activities have to be used instead of concentrations, such calculations are considerably more complicated than implied by equation (10). It appears that Buil (1983) used the correct equation, though he did not provide details of the procedure used to calculate the activities.

Another interesting aspect arises from the fact that there is only a negligible effect of anisotropic stress on the unloaded surfaces of the growing crystal. The supersaturation of the solution with respect to the unloaded surfaces is therefore higher than with respect to the loaded faces. Clearly, the free surfaces are the most favourable faces for growth. The question arises if a sufficient level of supersaturation can be maintained that is required for the growth upon the loaded faces to continue. In principle, this has been confirmed by the experiments of Becker and Day (1916) and Taber (1916). Particularly favourable conditions for growth upon a loaded surface may exist in porous materials. As water evaporates from a porous material, migration of a pore solution towards the surface becomes increasingly slow. A situation might evolve where only the bottom of the growing crystal remains in contact with the supersaturated solution [Fig. 8(c)]. Then the crystal grows upon the loaded bottom surface and the diminishing effect of growth upon the free surfaces is greatly reduced or even absent. Zehnder and Arnold (1989) studied the habits of growing crystals during the drying of a porous material. As the moisture content of the material decreased, they observed columnar and fibrous habits of the growing crystals indicating growth from the bottom.

There is considerable evidence for the existence of the force of crystallisation which can only be explained by the solution film model of Weyl (1959). It is important to note that the condition of crystallisation from a supersaturated solution also applies to other models of the crystallisation pressure. For example, Fitzner and Snethlage (1982) derived an equation for the crystallisation pressure based on the frost damage model, proposed by Everett (1961). According to this model, crystal growth occurs preferentially in large pores due to the differences in the chemical potentials of small crystals compared to large ones. Stress can be generated if large pores are filled and crystal growth continues. However, as was discussed before, a growing crystal can only exert pressure if it is in contact with a solution of sufficient supersaturation. Hence, the maximum stress determined by the supersaturation of the solution also applies to the model of Fitzner and Snethlage (1982). However, the maximum crystallisation pressure of growing salt crystals has not yet been accurately calculated. Therefore, at present, theory is only qualitative and does not permit a complete understanding of field observations and results of laboratory simulation studies.

4.2. Hydration Pressure

Several common salts can exist in different hydrated forms. The phase change from the lower hydrated (or anhydrous) form to the higher

Table 4. Examples for hydration equilibria and associated volume expansion.

Lower hydrate	Hydrated form	Expansion
Na_2SO_4 (thenardite)	$Na_2SO_4 \cdot 10H_2O$ (mirabilite)	314%
$Na_2CO_3 \cdot H_2O$ (thermonatrite)	$Na_2CO_3 \cdot 10H_2O$ (natron)	257%
$MgSO_4 \cdot H_2O$ (kieserite)	$MgSO_4 \cdot 6H_2O$ (hexahydrite)	146%
$MgSO_4 \cdot 6H_2O$ (hexahydrite)	$MgSO_4 \cdot 7H_2O$ (epsomite)	10%

state of hydration causes an increase in volume. Due to this expansion, hydration is considered as an important cause of damage. Table 4 lists examples of hydration–dehydration equilibria relevant to building materials.

The phase diagram of the system $Na_2SO_4 \cdot H_2O$ is depicted in Fig. 10. At high RH, sodium sulphate can only exist as an aqueous solution. At temperatures above the transition temperature (32.4°C), thenardite (Na_2SO_4) crystallises out as the relative humidity drops below the deliquescence humidity (curve 1). At temperatures below 32.4°C, mirabilite ($Na_2SO_4 \cdot 10H_2O$) is the stable solid in equilibrium with a saturated sodium sulphate solution (curve 2). Finally, curve 3 represents the hydration-dehydration equilibrium according to

$$Na_2SO_4 + 10H_2O(g) \rightleftharpoons Na_2SO_4 \cdot 10H_2O \tag{11}$$

Thenardite is the stable solid at low RH, whilst mirabilite can only exist within a limited range of RH at temperatures below 32.4°C. Hydration damage can occur if thenardite is hydrated due to changes in either temperature, or RH, or both (cf. Arnold, 1976). Hence, at constant temperature, hydration damage occurs as the result of an increase in RH. A hydrating crystal can only exert stress if the crystal continues to grow against the confining pore walls. It follows from thermodynamic considerations that, due to the increase of the chemical potential of the stressed crystal, there is a maximum stress that can be exerted. Given the general form of a hydration reaction

$$A \cdot v_1 H_2O + (v_2 - v_1) \, H_2O(g) \rightleftharpoons B \cdot v_2 H_2O \tag{12}$$

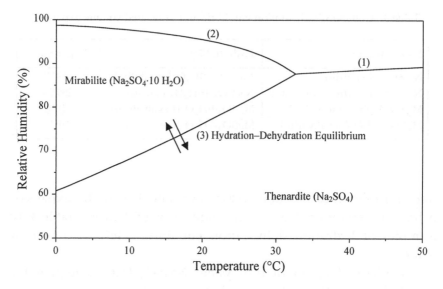

Fig. 10. The Na_2SO_4-H_2O system from 0–50°C.

the maximum hydration pressure is given by the following equation, which was first derived by Mortensen (1933):

$$\Delta\sigma = \frac{(v_2 - v_1)RT}{V_B - V_A} \cdot \ln\frac{P}{P_0} \tag{13}$$

Here, $\Delta\sigma$ is the hydration pressure, V_A and V_B are the molar volumes of the two solids, P is the actual partial pressure of water vapour, and P_0 is the water vapour pressure in equilibrium with the unstressed hydrated solid as shown in Fig. 10 (curve 3) for the thenardite–mirabilite equilibrium. It follows from equation (13) that the maximum hydration pressure is a function of the RH. Winkler and Wilhelm (1970) have calculated the hydration pressures of some common salts using equation (13). At very high RH, they obtained appreciable stresses exceeding the tensile strengths of many materials. However, they did not consider that there is an upper RH limit above which the calculation of hydration pressures is not meaningful. This is illustrated in Fig. 11 for the hydration of thenardite at 25°C. The solid line represents the hydration pressure calculated using equation (13) as a function of RH. The deliquescence of mirabilite occurs at 93.6%

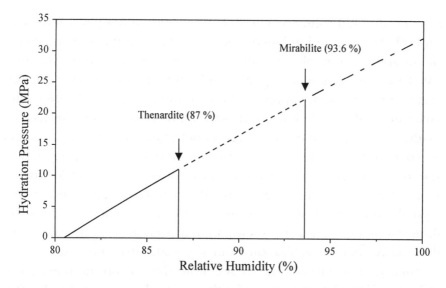

Fig. 11. Hydration pressure associated with the thenardite-mirabilite transition at 25°C.

RH. Thus, at humidities above 93.6% sodium sulphate exists as a solution and no hydration or crystallisation pressure can be exerted. Another important limitation arises from the deliquescence of thenardite occurring at 87% RH. As a thenardite crystal is subjected to a humidity increase, hydration will start as humidity exceeds 80.5% (cf. Fig. 10). Further increasing RH would increase the hydration pressure. As the deliquescence humidity is reached, however, thenardite will pick up water forming a sodium sulphate solution. Hence, the solid state hydration mechanism cannot proceed.

The deliquescence of thenardite represents a metastable equilibrium. The solution formed is saturated with respect to thenardite, but is considerably supersaturated with respect to mirabilite which will be precipitated from that solution. Thus, at RH above the deliquescence humidity of thenardite, the hydration reaction should follow a different reaction mechanism, namely a dissolution-recrystallisation mechanism. Equation (13), however, does not apply to that situation and the stress that might be generated is more realistically described as crystallisation pressure resulting from the growth of mirabilite from a supersaturated solution. The degree of supersaturation will depend on both, the dissolution kinetics of thenardite and the growth kinetics of mirabilite. It is obvious that the maximum hydration pressures calculated according to equation (13) are not meaningful. Similar arguments, of course, apply to the hydration reaction of other salts. Hydration will also follow the dissolution-recrystallisation mechanism as the reaction is initiated by the penetration of water, e.g. rainfall or condensation water, into a porous material containing crystals of the lower hydrated form. Sperling and Cooke (1985) and McMahon *et al.* (1992) carried out experiments to differentiate between the crystallisation and hydration pressures in salt weathering. Using sodium sulphate, they have shown that hydration is significantly less destructive than crystallisation from supersaturated solutions.

Finally, it is interesting to see how the hydration reaction is influenced by the presence of other salts. Using a $NaCl-Na_2SO_4$ mixture as an example, Steiger and Zeunert (1996) have shown that two effects may occur: (1) Depending on the mixing ratio of the salts, the hydration reaction can be entirely suppressed. (2) The solid state hydration is only possible if the respective salt mixture is not deliquescent at the respective RH. In a $NaCl-Na_2SO_4$ mixture, for example, the hydration of thenardite (or dehydration of mirabilite) always occurs in the presence of a solution.

Recent investigations further indicate that the solid state hydration is also kinetically hindered. Charola and Weber (1992) failed to hydrate thenardite even at high RH, which they have attributed to the formation of a mirabilite surface layer which prevents the complete hydration of the thenardite crystal. Using an environmental scanning electron microscope, Doehne (1994) has confirmed the formation of the mirabilite skin on large thenardite crystals. He could also show that the dehydration of mirabilite results in the formation of highly porous aggregates of submicron thenardite crystals which subsequently could be easily rehydrated. However, Doehne (1994) observed that the volume increase of the individual crystals is largely accommodated by the dead space available in the porous aggregates. In both studies, rapid hydration and growth of very large mirabilite crystals in the presence of liquid water were observed.

In conclusion, it appears that the idea of simple volume expansion during the direct hydration of thenardite or other salts in the presence of humid air is not sufficient to understand the deleterious effects of hydration. In particular, the calculation of hydration pressures according to equation (13) is of limited use only. Hydration damage might be more realistically understood as the result of crystal growth from supersaturated solutions if the hydration reaction proceeds via the dissolution-crystallisation mechanism.

5. Atmospheric Pollution and Salt Enrichment

The weathering of building materials causes dramatic changes in the appearance of a building façade, which often follow distinct spatial patterns, clearly dependent on the geometric configuration of the façade and the exposure to the local weather conditions. For example, black and white areas can usually be clearly distinguished on limestone and marble façades (Camuffo *et al.*, 1982; Amoroso and Fassina, 1983) corresponding to rain-sheltered and rain-exposed surfaces (cf. Fig. 12). However, different patterns of macroscopic appearance of sandstone façades are usually observed (Sabbioni and Zappia, 1992; Whalley *et al.*, 1992). Crusts, discolouration, organic growth, soiling, patina etc. are other commonly used terms to describe visual changes of the macroscopic appearance of building materials. However, it often appears that different authors use a different terminology describing the same weathering phenomena and *vice versa*.

Following Zehnder (1982), a crust is generally defined as a compact surface layer, the composition of which is different from the underlying material. The term "black crust" which is often used to describe any form of black deposit on a material surface, however, appears to be ambiguous. There are several different mechanisms that can cause the formation of crusts or deposits. For example, black deposits on sandstone and granite buildings are usually observed on both, sheltered and exposed surfaces (Fig. 12). However, as will be shown below, these deposits represent fundamentally different weathering phenomena. The black colour of such deposits is not necessarily related to a particular mechanism of its formation.

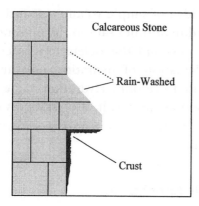 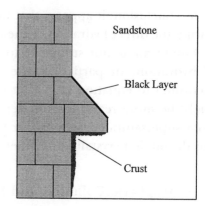

Fig. 12. Typical macroscopic appearance of exposed and sheltered surfaces of calcareous materials (limestone, marble) and sandstone.

Moreover, black discolouration *per se* might not be regarded as damage at all. Since we are more interested in the mechanisms of surface modifications than in their macroscopic appearance, the term "black crust" should be avoided whenever the composition of a crust or deposit is known from microscopic examination or chemical analysis. Even if such information is not available, different types of crust or deposits might be further classified according to their macroscopically visible morphology as deposits either masking or tracing the original material surface (Fitzner *et al,* 1995).

Gypsum crusts are among the most widespread encrustations found on building materials. The formation of sulphation crusts on carbonate stone was previously described by Schaffer (1972) and Kieslinger (1932). According to them, gypsum encrustations are preferentially formed at the rain-sheltered surfaces of carbonate building materials. Also according to Camuffo *et al.* (1982), gypsum crusts occur on surfaces not directly exposed to rainfall which are corresponding to their black areas. In fact, gypsum crusts on buildings in urban environments nearly always appear black. The black colour of gypsum crusts is caused by the embedding of particles originating from a number of different sources including fly ash particles, soot, mineral dust, iron oxides and oxyhydroxides, organic pollutants, and black fungi (e.g. Del Monte *et al.*, 1981; Del Monte and Sabbioni, 1984; Nord and Tronner, 1991;

Urzi *et al.*, 1992; Whalley *et al.*, 1992; Saiz Jimenez, 1993). White gypsum crusts on natural outcrops were observed by Zehnder (1982).

There is no doubt that the formation of gypsum crusts on calcareous stones is the result of the dry deposition of SO_2 to a moisture film on the stone surface followed by the dissolution of calcite and subsequent precipitation of gypsum. Hence, the major damage mechanism is one of chemical weathering. The replacement of calcite by gypsum and the growth mechanism of crusts have been studied in detail, both, for limestone (Schiavon, 1992) and marble (Vergès-Belmin, 1994). It was shown that gypsum crusts on limestone and marble are growing inward which is in contradiction to the growth model of Skoulikidis and Charalambous (1981), who proposed a solid state diffusion model according to which gypsum growth would proceed outward.

Another type of chemical weathering crust on calcareous stones are calcium oxalate crusts. Whewellite ($CaC_2O_4 \cdot H_2O$) and weddellite ($CaC_2O_4 \cdot 2H_2O$) are commonly found on Italian marble and limestone monuments and natural carbonatic outcrops (Del Monte *et al.*, 1987). There has been some controversial discussion on the origin of these crusts in particular with respect to the patina called "scialbatura" which is commonly found on Roman imperial marble monuments (Del Monte and Sabbioni, 1987; Lazzarini and Salvadori, 1989). Biological activity (lichens, algae, etc.) is probably the major source of oxalate crusts (Del Monte *et al.*, 1987), though oxalates may also derive from atmospheric deposition or protective treatments (Lazzarini and Salvadori, 1989; Watchman, 1991).

The formation of gypsum crusts on sandstone is less obvious. However, gypsum crusts are commonly found on calcite-cemented sandstones (e.g. Zehnder, 1982) and nearly calcium-free materials such as quartz sandstone and granite (Whalley *et al.*, 1992; Neumann *et al.*, 1993; Smith *et al.*, 1994). In contrast to calcareous stone, the mechanism of crust formation on sandstone must be different, because both, sulphate and calcium must originate from external sources. One obvious source is the mobilisation of calcium from mortar joints. Another source of calcium is the deposition of particulate matter from the atmosphere, either as gypsum or calcite. Apart from gypsum, the examination of crusts on sandstones and granite by microscopic and chemical analysis shows the presence of a wide range of constituents (e.g. Whalley et al., 1992). Generally, however, the particles embedded in the crusts are similar to those found in gypsum crusts on calcareous stone.

Material loss may occur if crusts are detaching from the stone. Neumann et al. (1997) examined gypsum crusts from several buildings built of quartz sandstones. Based on microstructural changes, they have differentiated crusts detaching with and without adherent stone fragments, respectively. On surfaces perfectly sheltered from rain gypsum growth was found to be essentially restricted to growth on top of the stone surface and no material loss is caused by the detachment of these crusts. On surfaces only partially sheltered from rainfall, as typically found in the transition zones from sheltered to exposed areas, substantial material loss is caused by the detachment of crusts with adherent stone fragments. The infiltration of water in these transition zones is sufficient to permit the migration of gypsum into the underlying stone. Due to the limited supply of water, however, the solution cannot deeply penetrate into the stone resulting in a continuous crystallisation and accumulation of gypsum in the pore space close to the stone surface.

Microscopic examination of crusts confirmed that the available pore space just beneath the stone surface is very often completely filled with gypsum. A characteristic feature of the later stage of the damage process is the complete destruction of the original internal fabric which is replaced by a secondary gypsum supported fabric (Neumann et al., 1997). The resulting loss of cohesion was quantitatively studied by Neumann et al. (1993). They used the number of grain contacts and the distances between quartz grains as a measure of cohesiveness of the fabric. As can be seen from Fig. 13, the number of grain contacts is significantly lower in crusts compared to the original sandstone. Moreover, Neumann et al. (1993) report several examples of crusts in which grain contacts could not be observed at all. At this stage, the quartz grains are embedded in a secondary gypsum fabric. Clearly, stress generated by the crystal growth of gypsum within the confined pore space is the major cause of damage. This damage mechanism is fundamentally different from the chemical weathering process of gypsum growth on calcareous stone.

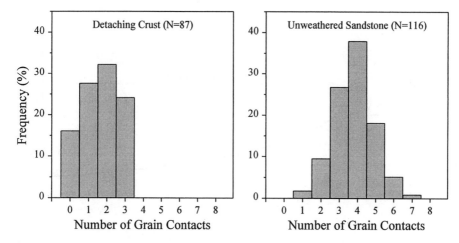

Fig. 13. Frequency distribution of the number of grain contacts per quartz grain in a detaching crust compared to the unweathered sandstone; N is the total number of grains (redrawn from Neumann *et al.*, 1993).

The situation becomes a lot more complicated if a stone surface is exposed to rainfall. If rainfall hits the surface of a porous material, it will be absorbed by capillary transport. As water is freely available at the surface of a porous material, the absorption of water obeys the equation

$$i = A \, t^{1/2} \qquad (14)$$

where i is the cumulative infiltration (mass per unit surface area), and A is the water absorption coefficient. Equation (14) does not strictly apply to the case of driving rain penetrating into a porous material because the supply of water is limited by the rainfall rate. A more rigorous treatment of the problem based on unsaturated flow theory is provided by Hall and Kalimeris (1982). For the simple qualitative approach followed here, equation (14) might be appropriate, however.

It directly follows from equation (14) that the rate of infiltration, $u = di/dt$, decreases as $t^{-1/2}$, resulting in very large rates at small values of t. During the initial phase of absorption, the actual infiltration rate is therefore limited by the rainfall rate, V_0, which is schematically depicted in Fig. 14. As the infiltration rate decreases, a point will be

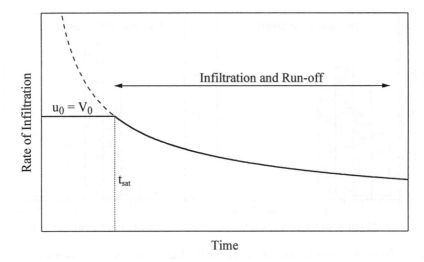

Fig. 14. Supply limited infiltration rate and formation of run-off water.

reached at which the rainfall rate exceeds the infiltration rate. The porous material becomes saturated and the excess water will start to run-off. The time to reach saturation, t_{sat}, depends on the rainfall rate and the water absorption coefficient, A.

Hall and Kalimeris (1982) provide a useful equation for the calculation of t_{sat} for many building materials:

$$t_{sat} = 0.64 \cdot \frac{A^2}{\rho_w^2 V_0^2} \qquad (15)$$

Here, ρ_w is the density of water. Figure 15 depicts the saturation times of typical building materials and their dependence on the rainfall rate, calculated using equation (15). Typically, rates of driving rain are rarely exceeding 2–3 mm/h (Schwartz, 1973; Fazio *et al.*, 1995). It can be seen that saturation times are in the order of minutes to several tenth of hours depending on the water absorption coefficient. Hence, there are different modes of action of precipitation. The acidity of rainfall may lead to direct attack on the mineral compounds of a material. Indirect effects include the dissolution and transport of previously deposited material and the corrosion products of dry deposition. Dissolved matter may be either transported into the

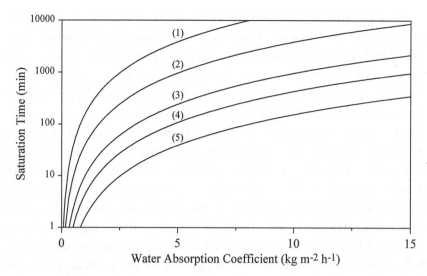

Fig. 15. Dependence of saturation time on water absorption coefficient and rate of driving rain: (1) 0.5 mm/h; (2) 1 mm/h; (3) 2 mm/h; (4) 3 mm/h and (5) 5 mm/h.

porous material, or, may be removed by running water. The significance of these processes is depending on the intensity, duration and frequency of rainfall, its chemical composition, and the properties of the porous material.

The simplest situation that can be considered, is a nearly impermeable material, i.e. a very dense material of low sorptivity such as marble and many limestones. The saturation times of such materials are small and run-off water is the dominant process. This situation corresponds to the surfaces classified as white areas by Camuffo *et al.* (1982). The action of rainfall reduces to the dissolution of calcite and the removal of the reaction products (mainly gypsum) formed due to the dry deposition of acidic species. These processes can cause significant material loss on surfaces exposed to intense rainfall. Lipfert (1989b) provides a review on field exposure studies that were carried out to quantify the material loss of carbonate stones exposed to different environmental conditions. Such data may be used to derive quantitative expressions for the chemical weathering rates of these materials (Lipfert, 1989b; Webb *et al.*, 1992).

The cleansing effect of rainfall is limited due to the relatively low solubility of gypsum. Moreover, the dissolution kinetics might further reduce the maximum amount of gypsum that can be removed with running water. Hence, the cleansing effect depends on both, the frequency and intensity of rainfall and the rate of SO_2 deposition and gypsum formation during dry periods. Gypsum crusts can occur on exposed surfaces if the rate of gypsum formation exceeds the removal rate with running water. In polluted environments, the occurrence of white areas, therefore, is often limited to inclined surfaces, e.g. the top side of cornices. Vertical surfaces, though not sheltered, are less affected by driving rain and the accumulation of airborne particulate matter, and gypsum is also frequently observed on exposed vertical surfaces of buildings in polluted urban areas (Amoroso and Fassina, 1983). Considerable enrichment of gypsum or the formation of crusts on façades exposed to moderate rainfall intensities is particularly often found on calcareous stones of high porosity, such as coarse-grained limestones and calcite-cemented sandstones (Sabbioni and Zappia, 1992; Fitzner and Heinrichs, 1992). Due to the greater saturation times of such materials, rainwater is absorbed and gypsum is transported into the stone. The amount of gypsum in the pore space continuously increases with every wetting and drying cycle, and besides the chemical weathering of calcite, additional damage results from crystal growth of the reaction product (Charola, 1988).

In contrast to marble and limestone, black discolourations often appear more evenly distributed on sandstone and granite surfaces. In particular, their occurrence is not restricted to sheltered areas. Intense black areas are rather found on surfaces frequently wetted by rainfall or run-off. These discolourations often appear as homogeneous black layers firmly attached to the stone surface. In contrast to gypsum crusts, they are tracing the original stone surface and it appears reasonable to adopt the term "thin black layer" as suggested by Nord and Ericsson (1993). Based on microscopic and chemical analysis, black layers might be further classified. In polluted urban environments, black layers mainly consist of insoluble airborne particulates and organic substances (Sabbioni and Zappia, 1992; Nord and Ericsson, 1993; Schiavon, 1993; Schiavon et al., 1995). Microscopic examination confirmed that black layers are homogenous deposits rarely exceeding 200 mm in thickness (Nord and Ericsson, 1993; Begonha and Sequeira Braga, 1996). Black layers, particularly on frequently wetted surfaces, can also originate from biological growth of algae and other microorganisms (e.g. Saiz-Jimenez, 1995; Warscheid and Krumbein, 1996).

Damage mechanisms associated with the presence of thin black layers are still not clear and have been subject to controversial discussion. In many cases, stones covered with black layers show essentially no damage at all. Often, however, the detachment of rather thin scales or flakes is observed. It is very unlikely that damage can be attributed to a single mechanism (Schiavon, 1993). Gypsum may or may not be accumulated in near-surface pores of stones covered with thin black layers. Apparently, gypsum growth and the formation of black layers are different phenomena that can evolve independently. Neumann *et al.* (1997) provide an example for the development of a gypsum crust on a quartz sandstone exposed to rainfall that was originally covered by a thin black layer.

The formation of gypsum crusts on exposed sandstones or granites is only possible if the rate of gypsum formation exceeds the removal of gypsum. Therefore, gypsum crusts on quartz sandstones are usually only observed in polluted areas where the supply with both, calcium and sulphate is sufficient. If gypsum mobilisation by rainfall is predominant, gypsum might be either removed with running water, or, can migrate into the porous material. There are many buildings where the maximum gypsum enrichment is not found at the exposed stone surface, but rather some distance in the interior of the stone. Zehnder (1982) provided several examples from different buildings with gypsum maxima at depths from a few millimetres up to several centimetres. He could further show that the presence of gypsum is correlated with severe damage due to the formation of contour scales.

The transport of gypsum to such great depths requires substantial penetration of water into the stone. In addition, as water evaporates during the drying of the stone, capillary transport back to the surface must not take place. As Zehnder (1982) has pointed out, these conditions are most likely to be met as a result of intense wetting and rapid drying. The evolution of moisture profiles during infiltration

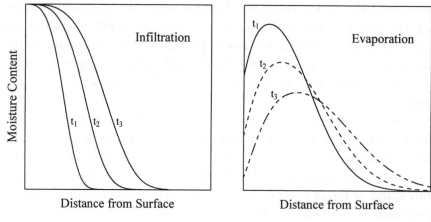

Fig. 16. Moisture profiles at different times, t, during infiltration and evaporation $(t_1 < t_2 < t_3)$.

and drying is schematically illustrated in Fig. 16. Snethlage and Wendler (1997) have numerically simulated the dynamics of wetting-drying cycles using the approach for the parameterisation of moisture transport in porous materials proposed by Künzel (1995). They obtained average moisture distributions for several sandstones with maxima at different distances from the stone surface depending on the moisture transport properties of the stones. Hence, the evaporation of water occurs at some distance from the stone surface corresponding to zones of maximum salt enrichment.

As mentioned before, the amount of gypsum that can be transported into the stone is limited due to its rather low solubility. In effect, gypsum accumulation proceeds slowly. The maximum gypsum concentrations found by Zehnder (1982) and others (Wendler et al., 1991; Steiger et al., 1993) rarely exceed 2–3% by weight. Compared to the enormous enrichment of gypsum in crusts, these concentrations are much lower and are far from filling the available pore space. However, even these moderate concentrations exceed the mobility of gypsum in typical materials by about two orders of magnitude (cf. Fig. 3). Hence, the gypsum once deposited is essentially immobilised. Although every rain event will inevitably add to the further accumulation of gypsum, it should be noted that only a small amount of gypsum crystallises out during each drying cycle. It is not clear whether the growth of such a small amount of gypsum can generate significant stress. This would be possible only if the continuous accumulation of gypsum would preferentially occur in confined intergranular spaces. In a laboratory experiment carried out by Wendler and Rückert-Thümling (1992), however, it was shown that continuous accumulation of gypsum caused significant irreversible expansion of a sandstone. It might be concluded that gypsum is an important cause of contour scaling.

Apart from salts accumulated in the ground moisture zones of buildings, gypsum is very often the most abundant salt in building materials. Atmospheric pollution is by far the most important source of gypsum, and due to its low mobility, gypsum tends to be strongly accumulated in porous materials. The discussion so far reveals that besides the chemical weathering of calcareous materials, the growth of gypsum in the pore space of building materials is a major cause of damage. In the absence of calcite, either as a mineral constituent of the material itself or from external sources, acid deposition can also cause the mobilisation of other cations than calcium. The salts then formed show different behaviour. For example, Fig. 17 depicts a salt profile measured in a sandstone monument. The building is located in a rural environment in Southern Germany where the local air

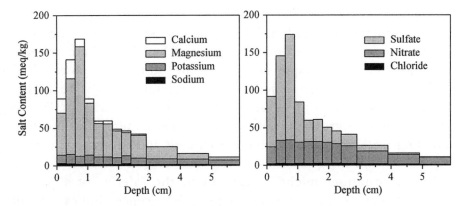

Fig. 17. Salt profile in sandstone from Schloss Weissenstein, Pommersfelden (unpublished results).

pollution level is low and it is expected that there is a significant contribution of acid rain to the total deposition of acidity.

Magnesium, sulphate and nitrate are the major constituents of the salt system, whilst the calcium concentration is of minor importance only. The maximum salt concentration (1.1 wt % total salts) is found at depths of about 6–9 mm. Considering the porosity of the stone and the solubilities in the salt mixture, it can be concluded that apart from gypsum, all other salts will be completely dissolved if rainwater penetrates into the stone. As water evaporates from the stone, salts will crystallise out resulting in a fractionation as discussed before. For the salt mixture given, it can be simply deduced from available solubility data that, apart from gypsum, epsomite ($MgSO_4 \cdot 7H_2O$) will be precipitated first. This is in accordance with the profile in Fig. 17 showing a strong enrichment of magnesium and sulphate at a distance of 6–9 mm from the surface, whilst the remaining ions are more evenly distributed, and are therefore, probably present in dissolved form. Clearly, the profile shown in Fig. 17 only reflects an intermediate state. Depending on the ambient temperature and RH, the drying process may either continue followed by the crystallisation of other solids, or, the system might already be in equilibrium with the surrounding atmosphere. Anyhow, on every wetting-drying cycle, the salts will be completely dissolved and re-distributed. The damage potential of such a hygroscopic salt mixture is obviously much greater than that of an equal amount of gypsum.

It is generally recognised that hygroscopic salts are a major cause of damage in the ground moisture zone of buildings where extreme salt enrichment and rapid destruction are quite often observed. It appears, however, that sometimes their relevance might be under-estimated if they are only present in moderate concentrations. Due to their mobility, hygroscopic salts are usually more evenly distributed in building materials, and extreme enrichment and strong gradients as typically found for gypsum are rarely occurring. Therefore, relevant salt concentrations might be present in the interior of a porous material even if there are no visible efflorescences indicating salt damage. Apart from rainfall and condensation, the water necessary to dissolve these salts can also come from hygroscopic moisture uptake.

In effect, damage is not limited to wetting-drying cycles but can also be continuously initiated during dry periods by relatively minor fluctuations of the ambient RH.

6. Conclusions

Building materials exposed to polluted atmospheres are severely affected by acid deposition. The damage mechanisms associated with acid deposition are reasonably well understood on a qualitative basis. It might be even possible to predict the type of damage for a particular material exposed to a given environment. It is much more difficult, however, to predict rates of damage, i.e. to establish a quantitative relationship between the concentration of air pollutants and the rate of damage. The concept and mathematical form of such damage functions were recently reviewed by Livingstone (1997). However, even for the most simple situation that can be considered — formation of gypsum by dry deposition of SO_2 on a rain-sheltered surface of a calcareous stone — a damage function has not yet been established. Damage functions for other situations, i.e. other materials and different exposures, would be considerably more complex.

The major problem arises from the fact that acid deposition can act in two different ways. First, wet and dry deposition of acidity is the major cause of chemical weathering of materials. Second, acid deposition is a major source of salt accumulation, and therefore, physical weathering. These damage mechanisms follow different rate laws. Chemical weathering rates might be sufficiently well described by the actual environmental conditions (pollutant concentrations, deposition velocities, pH and amount of rainfall). In contrast, the actual rate of salt weathering will largely depend on two influences: (1) the present-day environmental conditions (cyclic variation of meteorological variables), and (2) the concentration and composition of the salt mixture in the material given as the integral salt accumulation from past to present. Not much is known, however, about the influence of the concentration of salts and their relative destructive power on the rate of salt damage in different materials.

In view of these uncertainties, it is also difficult to assess how the overall rates of damage would change in response to clean air legislation. Unquestionably that salt damage would still continue with a rate similar to its present rate even if pollutant concentrations were reduced to zero. However, considering that "every mole of sulphur or nitrogen oxides not emitted is a mole not deposited" (Schwartz, 1989), further accumulation of salts can be significantly retarded by reducing emissions. Materials susceptible to chemical weathering such as calcareous materials would benefit even more from further reduction in acid deposition.

References

Adema E.H. and Heeres P. (1995) *Atmos. Env.* **29**, 1091–1103.
Amoroso G.G. and Fassina V. (1983) *Stone Decay and Conservation.* Elsevier, Amsterdam, pp. 141–147.
Arnold A. (1976) In *2nd International Symposium on the Deterioration of Building Stones.* Athens, pp. 27–36.
Arnold A. and Zehnder K. (1991) In *The Conservation of Wall Paintings*, ed. Cather S. (Getty Conservation Institute, Los Angeles, pp. 103–135.
Baedeker P.A., Reddy M.M., Reimann K.J. and Sciammarella C.A. (1992) *Atmos. Env.* **26B**, 147–158.
Becker G.F. and Day A.L. (1916) *J. Geol.* **24**, 313–333.
Begonha A. and Sequeira Braga M.A. (1996) In *Proceedings of the 8th International Congress on Deterioration and Conservation of Stone*, ed. Riederer J. Möller Druck und Verlag GmbH, Berlin, pp. 371–375.
Behlen A., Wittenburg C., Steiger M. and Dannecker W. (1996) In *Proceedings of the 8th International Congress on Deterioration and Conservation of Stone*, ed. Riederer J. Möller Druck und Verlag GmbH, Berlin, pp. 377–385.
Behlen A., Steiger M. and Dannecker W. (1997) In *4th International Symposium on the Conservation of Monuments in the Mediterranean*, eds. Moropoulou A., Zezza F., Kollias E. and Papachristodoulou I. Technical Chamber of Greece, Athens, Vol. 2, pp. 237–246.
Blanchard D.C. and Woodcock A.H. (1980) *Ann. N.Y. Acad. Sci.* **338**, 330–347.
Boer R.B. de (1979) *Geochim. Cosmochim. Acta* **41**, 249–256.
Braitsch O. (1971) *Salt Deposits, Their Origin and Composition.* Springer, Berlin.
Brimblecombe P. and Rodhe H. (1988) *Durab. Build. Mater.* **5**, 291–308.
Buil M. (1983) In *Werkstoffwissenschaften und Bausanierung*, ed. Wittmann F.H. Ed. Lack und Chemie, Filderstadt, pp. 373–377.
Butlin R.N. (1991) *Procs. Roy. Soc. Edinburgh* **97B**, 255–272.
Camuffo D., Del Monte M., Sabbioni C. and Vittori O. (1982) *Atmos. Env.* **16**, 2253–2259.

Camuffo D. (1984) *Water Air Soil Pollut.* **21**, 151–159.

Charola A.E. (1988) *Durab. Build. Mater.* **5**, 309–316.

Charola A.E. and Weber J. (1992) In *Proceedings of the 7th International Congress on Deterioration and Conservation of Stone*, eds. Delagado Rodrigues J., Henriques F. and Telmo Jeremias F. Lisbon, pp. 581–590.

Chebbi A. and Carlier P. (1996) *Atmos. Env.* **30**, 4233–4249.

Cooper T.W. (1986) *Technol. Ireland*, 32–35.

Correns C.W. and Steinborn W. (1939) *Z. Krist. A* **101**, 117–133.

Del Monte M., Sabbioni C. and Vittori O. (1981) *Atmos. Env.* **15**, 645–652.

Del Monte M. and Sabbioni C. (1984) *Arch. Meteorol. Geophys. Bioclimatol. Series B* **35**, 105–111.

Del Monte M. and Sabbioni C. (1987) *Studies in Conservation* **32**, 114–121.

Del Monte M., Sabbioni C. and Zappia G. (1987) *Sci. Total Env.* **67**, 17–39.

Doehne E. (1994) In *The Conservation of Monuments in the Mediterranean Basin. Proceedings of the 3rd International Symposium*, eds. Fassina V., Ott H. and Zezza F. Venice, pp. 143–150.

Drever J.I. (1994) In *Durability and Change*, eds. Krumbein W.E., Brimblecombe P., Cosgrove D.E. and Staniforth S. John Wiley and Sons, Chichester, pp. 27–39.

Duttlinger W. and Knöfel D. (1993) In *Jahresberichte Steinzerfall–Steinkonservierung Band 3-1991*, ed. Snethlage R. Verlag Ernst and Sohn, Berlin, pp. 197–213.

Evans I.S. (1970) *Rev. Geomorph. Dyn.* **19**, 153–177.

Everett D.H. (1961) *J. Chem. Soc. Faraday Trans.* **57**, 1541–1551.

Fassina V. (1988) *Durab. Build. Mater.* **5**, 317–358.

Fazio P., Mallidi S.R. and Zhu D. (1995) *Build. Env.* **30**, 1–11.

Fenter F.F., Caloz F. and Rossi M.J. (1995) *Atmos. Env.* **29**, 3365–3372.

Finlayson-Pitts B.J. and Pitts J.N. (1986) *Atmospheric Chemistry: Fundamentals and Experimental Techniques.* John Wiley and Sons, New York, pp. 368–369.

Fitzner B. and Snethlage R. (1982) *Bautenschutz Bausanierung* **5**, 97–103.

Fitzner B. and Heinrichs K. (1995) In *Jahresberichte Steinzerfall — Steinkonservierung Band 2-1990*, ed. Snethlage R. Verlag Ernst and Sohn, Berlin, pp. 23–38.

Fitzner B., Heinrichs K. and Kownatzki R. (1995) In *Denkmalpflege und Naturwissenschaft — Natursteinkonservierung I*, ed. Snethlage R. Ernst and Sohn, Berlin, pp. 41–88.

Garrecht H. (1992) *Porenstrukturmodelle für den Feuchtehaushalt von Baustoffen mit und ohne Salzbefrachtung und rechnerische Andwendung auf Mauerwerk* Inst. f. Massivbau u. Baustofftechnologie, Universität Karlsruhe, Karlsruhe, pp. 42–51.

Girardet F. and Furlan V. (1991) In *Science, Technology and European Cultural Heritage*, eds. Baer N.S., Sabbioni C. and Sors A.I. Butterworth-Heinemann, London, pp. 350–353.

Goudie A. and Viles H. (1997) *Salt Weathering Hazards.* John Wiley and Sons, Chichester.

Graedel T.E. and McGill R. (1986) *Env. Sci. Technol.* **20**, 1093–1100.

Grimm W.-D. (1990) *Bildatlas wichtiger Denkmalgesteine der Bundesrepublik Deutschland* Lipp, München.

Hall C. and Kalimeris A.N. (1982) *Build. Env.* **17**, 257–262.

Hardie L.A. and Eugster H.P. (1970) *Mineral. Soc. Am. Spec. Pap.* **3**, 273–290.

Hicks B.B. (1983) In *Conservation of Historic Stone Buildings and Monuments*, ed. Committee on Conservation of Historic Stone Buildings and Monuments. National Academic Press, Washington D.C. pp. 183–196.

Jaynes M.J. and Cooke R.U. (1987) *Atmos. Env.* **21**, 1601–1622.

Kieslinger A. (1932) *Zerstörungen an Steinbauten* Franz Deuticke, Leipzig, pp. 51–82.

Künzel H.M. (1995) *Simultaneous Heat and Moisture Transport in Building Components: One and Two-Dimensional Calculation Using Simple Parameters.* IRB-Verlag, Stuttgart.

Lasaga A.C., Soler J.M., Ganor J., Burch T.E. and Nagy K.L. (1994) *Geochim. Cosmochim. Acta* **58**, 2361–2386.

Lazzarini L. and Salvadori O. (1989) *Studies in Conservation* **34**, 20–26.

Lee Y.N. and Schwartz S.E. (1981) *J. Geophys. Res.* **86**, 11971–11983.

Lewin S.Z., *Studies in Conservation* **19**, 249–252.

Lewin S.Z. (1982) In *Conservation of Historic Stone Buildings and Monuments*, ed. Committee on Conservation of Historic Stone Buildings and Monuments. National Academy Press, Washington, pp. 120–144.

Lewin S. (1989) In *The Conservation of Monuments in the Mediterranean Basin*, ed. Zezza F. Grafo, Brescia, pp. 59–63.

Lipfert F.W. (1989a) *J. Air Pollut. Control Assoc.* **39**, 446–452.

Lipfert F.W. *(1989b) Atmos. Env.* **23**, 415–429.

Livingstone R.A. (1985) In *Proceedings of the 5th International Congress on Deterioration and Conservation of Stone*, ed. Félix G. Presses Polytechniques Romandes, Lausanne, pp. 509–516.

Livingstone, R. (1986) In *Building Performance: Function, Preservation, and Rehabilitation. ASTM STP 901*, ed. Davis G. American Society for Testing and Materials, Philadelphia, pp. 181–188.

Livingstone R.A. (1992) In *Proceedings of the 7th International Congress on Deterioration and Conservation of Stone*, eds. Delgado Rodrigues J., Henriques F. and Jeremias T. Lisbon, pp. 375–386.

Livingstone R.A. (1997) In *Saving Our Architectural Heritage*, eds. Baer N.S. and Snethlage R. John Wiley and Sons, Chichester, pp. 37–62.

Luckat S. and Zallmanzig J. (1985) *Research Report 106 08 006.* Zollern-Institut, Bochum, pp. 58–61.

Mansch R. and Bock E. (1998) *Biodeterioration* **9**, 47–64.

McMahon D.J., Sandberg P., Folliard K. and Mehta P.K. (1992) In *Proceedings of the 7th International Congress on Deterioration and Conservation of Stone*, eds. Delagado Rodrigues J., Henriques F. and Telmo Jeremias F. Lisbon, pp. 705–714.

Mortensen H. (1933) *Petermanns Geograph. Mitt.* **79**, 130–135.

Mullin J.W. (1993) *Crystallisation*. Butterworth-Heinemann, Oxford, pp. 128–170.

Neumann H.-H., Steiger M., Wassmann A. and Dannecker W. (1993) In *Jahresberichte Steinzerfall–Steinkonservierung Band 3–1991*, ed. Snethlage R. Verlag Ernst and Sohn, Berlin, pp. 151–167.

Neumann H.-H., Lork A., Steiger M. and Juling H. (1997) In *Proceedings of the 6th Euroseminar on Microscopy Applied to Building Materials*, ed. Sveinsdóttir E.L. Icelandic Building Research Institute, Reykjavik, pp. 238–249.

Nord A.G. and Tronner K. (1991) *Stone Weathering*. Conservation Institute of National Antiquities, Stockholm, pp. 24–44.

Nord A.G. and Ericsson T. (1993) *Studies in Conservation* **38**, 25–35.

Paterson M.S. (1973) *Rev. Geophys. Space Phys.* **11**, 355–389.

Pitzer K.S. (1991) In *Activity Coefficients in Electrolyte Solutions*, ed. Pitzer K.S. CRC Press Boca Raton, pp. 75–154.

Price C.A. and Brimblecombe P. (1994) In *Preventive Conservation: Practice, Theory and Research*. International Institute for Conservation, London, pp. 90–93.

Reddy M.M., Sherwood S. and Doe B. (1985). In *Proceedings of the 5th International Congress on Deterioration and Conservation of Stone*, ed. Félix G. Presses Polytechniques Romandes, Lausanne, pp. 517–526.

Reddy M.M. (1988) *Earth Surf. Proc. Landforms* **13**, pp. 335–354.

Roekens E. and van Grieken R. (1989) *Atmos. Env.* **23**, 271–277.

Rönicke G. and Rönicke R. (1972) *Deutsche Kunst u. Denkmalpflege* **30**, 57–64.

Sabbioni C. and Zappia G. (1992) *Water Air Soil Pollut.* **63**, 305–316.

Saiz Jimenez C. (1993) *Atmos. Env.* **27B**, 773–785.

Saiz Jimenez C. (1995) *Sci. Total Env.* **167**, 273–286.

Schaffer R.J. (1972) *The Weathering of Natural Building Stones*. Department of Scientific and Industrial Research Special Report No. 18, London, 1932, reprinted 1972, pp. 26–33.

Schiavon N. (1992) In *Stone Cleaning and the Nature, Soiling and Decay Mechanisms of Stone: Proceedings of the International Conference Held in Edinburgh, U.K.*, ed. Webster R.G.M. Donhead, London, pp. 258–267.

Schiavon N. (1993) In *Conservation of Stone and Other Materials*, ed. Thiel M.-J. E & F Spon, London, pp. 271–278.

Schiavon N., Chiavari G., Schiavon G. and Fabbri D. (1995) *Sci. Total Env.* **167**, 273–286.

Schwartz B. (1973) *Berichte aus der Bauforschung* **86**, 3–13.

Schwartz S.E. (1989) *Science* **243**, 753–763.

Sikiotis D., Kirkitsos P. and Delopoulou P. (1992) In *Proceedings of the 7th International Congress on Deterioration and Conservation of Stone*, eds. Delgado Rodrigues J., Henriques F. and Jeremias T. Lisbon, pp. 207–216.

Skoulikidis Th. and Charalambous D. (1981) *Br. Corros. J.* **16**, pp. 70–77.

Smith B.J., Magee R.W. and Whalley W.B. (1994) In *Rock Weathering and Landform Evolution*, eds. Robinson D.A. and Williams R.B.G. John Wiley and Sons, Chichester, pp. 131–150.

Snethlage R. and Wendler E. (1997) In *Saving Our Architectural Heritage*, eds. Baer N.S. and Snethlage R. John Wiley and Sons, Chichester, pp. 7–24.

Spencer R.J., Moller N. and Weare J.H. (1990) *Geochim. Cosmochim. Acta* **54**, 575–590.

Sperling C.H.B. and Cooke R.U. (1985) *Earth Surf. Proc. Landforms* **10**, 541–555.

Spiker E.C., Hosker R.P., Weintraub V.C. and Sherwood S.I. (1995) *WaterAir Soil Pollut.* **85**, 2679–2685.

Steiger M., Neumann H.-H., Ulrich A. and Dannecker W. (1993) In *Jahresberichte Steinzerfall–Steinkonservierung Band 3–1991*, ed. Snethlage R. Verlag Ernst and Sohn, Berlin, pp. 21–33.

Steiger M. and Dannecker W. (1995) In *Jahresberichte Steinzerfall–Steinkonservierung Band 5–1993*, ed. Snethlage R. Verlag Ernst and Sohn, Berlin, pp. 115–128.

Steiger M. (1996) In *Origin, Mechanisms and Effects of Salts on Degradation of Monuments in Marine and Continental Environments*, ed. Zezza F. Protection and Conservation of the European Cultural Heritage Research Report No. 4, Bari, pp. 241–246.

Steiger M. and Zeunert A. (1996) In *Proceedings of the 8th International Congress on Deterioration and Conservation of Stone*, ed. Riederer J. Möller Druck und Verlag GmbH, Berlin, pp. 535–544.

Steiger M., Behlen A., Neumann H.-H., Willers U. and Wittenburg C. (1997) In *4th International Symposium on the Conservation of Monuments in the Mediterranean*, eds. Moropoulou A., Zezza F., Kollias E., Papachristodoulou I. Technical Chamber of Greece, Athens, Vol. 1, pp. 325–335.

Sveevers H. and van Grieken R. (1992) *Atmos. Env.* **26B**, 159–163.

Taber S. (1916) *Am. J. Sci.* **41**, 532–556.

Theoulakis P. and Moropoulou T. (1988) In *Proceedings of the 6th International Congress on Deterioration and Conservation of Stone*, ed. Ciabach J. Nicholas Copernicus University, Torun, Poland, pp. 86–96.

Urzi C., Krumbein W.E. and Warscheid T. (1992) In *The Conservation of Monuments in the Mediterranean Basin. Proceedings of the 2nd International Symposium*, eds. Decrouez D., Chamay J. and Zezza F. Musée d'Histoire Naturelle, Genève, pp. 397–420.

van Breemen N. *et al.* (1982) *Nature* **299**, 548–550.

Vergès-Belmin V. (1994) *Atmos. Env.* **28**, 295–304.

Warscheid T. and Krumbein W.E. (1996) In *Microbially Influenced Corrosion of Materials*, eds. Heitz E., Flemming H.-C. and Sand W. Springer, Berlin, pp. 273–295.

Watchman A.L. (1991) *Studies in Conservation* **36**, 24–32.

Weare J.H. (1987) *Rev. Min.* **17**, 143–176.

Webb A.H., Bawden R.J., Busby A.K. and Hopkins J.N. (1992) *Atmos. Environ.* **26B**, 165–181.

Wendler E., Klemm D.D. and Snethlage R. (1991) In *Materials Issues in Art and Technology II*, eds. Vandiver P.D., Druzik, J. and Wheeler G.S. Materials Research Society, Pittsburgh, pp. 265–271.

Wendler E. and Rückert-Thümling R. (1992) In *Werkstoffwissenschaften und Bausanierung*, ed. Wittmann F.H. Expert Verlag, Ehningen, pp. 1818–1830.

Weyl P.K. (1959) *J. Geophys. Res.* **64**, 2001–2025.

Whalley B., Smith B. and Magee R. (1992) In *Stone Cleaning and the Nature, Soiling and Decay Mechanisms of Stone: Proceedings of the International Conference Held in Edinburgh, U.K.*, ed. Webster R.G.M. Donhead, London, pp. 227–234.

Winkler E.M. and Wilhelm E.J. (1970) *Bull. Geol. Soc. Am.* **81**, 567–572.

Winkler E.M. and Singer P.C. (1972) *Bull. Geol. Soc. Am.* **83**, 3509–3514.

Wittenburg C. and Dannecker W. (1992) *J. Aerosol Sci.* **23**, S869–S872.

Wittenburg C. and Dannecker W. (1994) In *The Conservation of Monuments in the Mediterranean Basin. Proceedings of the 3rd International Symposium*, eds. Fassina V., Ott H. and Zezza F. Venice, pp. 179–183.

Wolters B. *et al.* (1988) In *Proceedings of the 6th International Congress on Deterioration and Conservation of Stone*, ed. Ciabach J. Nicholas Copernicus University, Torun, Poland, pp. 24–31.

Zappia G., Sabbioni C. and Gobbi G. (1989) In *The Conservation of Monuments in the Mediterranean Basin*, ed. Zezza F. Grafo, Brescia, pp. 79–82.

Zehnder, K. (1982) *Verwitterung von Molassesandsteinen an Bauwerken und in Naturaufschlüssen, Beitr. Geol. Schweiz, Geotechn. Ser. 61.* Kümmerly and Frey, Bern, p. 130.

Zehnder K. and Arnold A. (1984) *Studies in Conservation* **29**, 32–34.

Zehnder K. and Arnold A. (1989) *J. Crystal Growth* **97**, 513–521.

Zezza F. and Macrì F. (1995) *Sci. Total Environ.* **167**, 123–143.

CHAPTER 6

ORGANIC POLLUTANTS IN THE BUILT ENVIRONMENT AND THEIR EFFECT ON THE MICROORGANISMS

C. Saiz-Jimenez

1. Introduction

The Northern Hemisphere is experiencing a series of severe ecological problems with forest decline, acidification of lakes, and accelerated corrosion of monuments, buildings and metallic structures.

Stone buildings, monuments, outdoor-exposed sculptures and objects of art have been degraded over the centuries by natural causes. Wind, rain and frosts contribute to a gradual process of ageing and deterioration. The alteration of stone is not therefore a contemporary phenomenon, but was already known in ancient times, and was a preocupation of Greek and Roman writers.

The process of weathering of a rock begins as soon as it is taken from the quarry and comes into contact with atmospheric agents. However, in the last century, industrial and urban activities have modified the composition of the atmosphere, resulting in a more aggressive environment, accelerating the decay of materials.

Thousands of chemicals are emitted directly or indirectly to the atmosphere because of human activities, but most pollutants emitted into the atmosphere are eventually removed through naturally occurring cleansing mechanisms. These removal and deposition processes represent the final stages of a complex sequence of atmospheric phenomena. Once released into the atmosphere, organic and inorganic airborne pollutants may undergo a variety of complex interactions determined by physical, chemical and photochemical processes occurring during their residence. These processes influence the nature of the capture of the pollutants in a sink or reservoir, where they are transformed, immobilised or encapsulated (Schroeder and Lane, 1988).

Buildings and monuments act as repositories of airborne organic pollutants, which accumulate at the surfaces in zones frequently soaked by rainwater but are not washed out. Obviously, this enriches the substratum and anthropogenic compounds may greatly influence the colonisation and growth pattern of microorganisms in stones located in polluted urban environments when compared with the growth of microorganisms in the same stones placed in rural environments. The industrialised society of the 20th century has thus caused a radical change in the conditions of preservation and conservation of stone, and the atmospheric pollution associated with industrialisation is currently modifying the biological spectrum of microorganisms thriving on stones.

Although it has been reported that a great variety of microorganisms colonise stones in urban environments, and their possible role in biodeterioration processes discussed (Saiz-Jimenez, 1995), the studies on different inputs of organic matter to stone, the use of organic pollutants as nutrients, the interactions between anthropogenic compounds and microorganisms, etc. have barely been investigated. In this chapter, recent findings on the effects of environmental conditions upon colonisation and growth of microorganisms, including the deposition of organic compounds on building stones and their possible utilisation by microorganisms are reviewed.

2. Sources of Organic Pollutants in Urban Environments

Organic compounds in the atmosphere arise from biotic (including anthropogenic components), abiotic (volcanic activity and lithospheric erosion) and extraterrestrial origin (meteorites). Only anthropogenic compounds will be considered in this chapter.

The organic species present in the atmosphere comprise gases, lipids and carbonaceous (insoluble) matter. Among the gases, the short chain alkanes and alkenes, aldehydes, ketones and acids are frequent. They originate mainly from spillage of petroleum-derived products as well as from combustion processes. Apart from methane, which is the major biogenic organic gas, a minor amount of volatiles also evolves from microbial activity and odoriferous plants, as for instance, isoprene and monoterpenes.

As stated by Simoneit and Mazurek (1981), urban aerosols generally consist of a mixture of lipid materials emitted locally along the aged material that has been carried into the urban area by winds. Thus, the urban atmospheric environment contains many organic pollutants which are related to incomplete fuel combustion in domestic heating, industrial plants and vehicular exhausts, such as long-chain alkanes and methyl alkanes (Cautreels and van Cauwenberghe, 1978; Boone and Macias, 1987), monocarboxylic and dicarboxylic acids (Kawamura *et al.* 1985; Kawamura and Kaplan, 1987), polycyclic aromatic hydrocarbons (Leuenberger *et al.* 1988), terpenoids (Simoneit, 1984), etc. In addition to organic compounds, carbonaceous matter (mainly graphitic carbon) is common in urban environments. Table 1 shows

Table 1. Classes of compounds identified in aerosols and particulate matter.

Compounds	Range*	Compounds	Range
n-Alkanes	C_7-C_{40}	Alkylnaphthoic acids	$C_{11}-C_{13}$
Isoprenoid hydrocarbons	$C_{10}-C_{20}$	Alkylphenanthroic acids	$C_{15}-C_{17}$
Isoprenoid ketones	$C_{10}-C_{20}$	Alkylcyclohexanes	C_9-C_{29}
Alkan-2-ones	$C_{10}-C_{32}$	Diterpenoid derivatives	$C_{16}-C_{20}$
Alkanols	$C_{11}-C_{28}$	Triterpenoid hydrocarbons	$C_{27}-C_{35}$
n-Fatty acids	C_1-C_{34}	Tricyclic terpane hydrocarbons	$C_{19}-C_{29}$
Hydroxy fatty acids	$C_{10}-C_{26}$	Steranes and diasteranes	$C_{27}-C_{29}$
α, ω-Dicarboxylic acids	C_2-C_{26}	Unresolved hydrocarbons	$C_{14}-C_{31}$
Alkylbenzoic acids	C_7-C_9	Polycyclic aromatic hydrocarbons	$C_{10}-C_{24}$
Alkylbenzenedioic acids	C_8-C_{10}	Oxygen-polycyclic aromatic hydrocarbons	$C_{10}-C_{16}$

* Range denotes number of carbon atoms in the compounds.

the classes of compounds identified in aerosols and particulate matter (Saiz-Jimenez, 1995).

2.1. Hydrocarbons

This class of compounds has been analysed in many urban areas. Typical *n*-alkane distributions range from C_{15} to C_{34} with no carbon number predominance (also termed carbon preference index or CPI). CPI is the sum of the odd carbon number homologues over a specified range divided by the sum of the even carbon number homologues over the same range. This index is an indicator for evaluation of anthropogenic/biogenic contributors in aerosols. Since biogenic *n*-alkanes generally show a strong odd C-numbered predominance, their CPI values are high, whereas *n*-alkanes from petroleum, vehicular exhausts and lubricating oils have a CPI of 1. Thus, the greater the anthropogenic contribution, the more closely the CPI approaches unity. As CPI of *n*-alkanes in urban aerosols is near 1, it was concluded that fuels or partly uncombusted fuels do contribute to these aerosols (Simoneit, 1985).

Aerosols also contain isoprenoid hydrocarbons such as pristane (2, 6, 10, 14-tetramethylpentadecane) and phytane (2, 6, 10, 14-tetramethylhexadecane) which are diagenetic products of phytol and are not primary constituents of most terrestrial biota, thus confirming an origin from petroleum (Simoneit, 1984).

2.2. Acids and Ketones

According to Simoneit and Mazurek (1981), these types of compounds originating from anthropogenic sources are usually only minor components of aerosols. The organic acids (C_1-C_{10}) were detected in many urban aerosols and a similar distribution in motor exhaust from automobiles was found by Kawamura *et al.* (1985). Gasoline and diesel exhaust sampled showed a dicarboxylic acid distribution (C_2-C_{10}) similar to those of air samples, but their concentrations were 28 (gasoline) and 144 (diesel) times higher that the average concentration of atmospheric diacids (Kawamura and Kaplan, 1987). Hence, the data indicate that vehicular emissions are the most important primary source of atmospheric acids. Furthermore, Graedel (1978) reported low molecular weight (up to C_{10}) aldehydes and ketones in diverse anthropogenic emissions.

2.3. Triterpenoid Hydrocarbons

Biological markers are organic compounds present in the geosphere whose structures can be unambiguously linked to the structures of precursor compounds occurring in original source materials (Philp, 1985).

Hopane-type triterpanes are ubiquitous biological markers in fossil fuels and their precursors are widely distributed among organisms. Virtually all crude oils contain the hopane series, which are also found in aerosols. It has been reported (Simoneit, 1985) that the distribution of the $17\alpha(H)$, $21\beta(H)$-hopanes are essentially identical for auto and diesel exhausts, confirming these emissions as the major source of petroleum residues in aerosols.

Gasoline and diesel fuel do not contain these triterpanes, but the same distribution is found in lubricating oils. This indicates that primarily lubricants adsorbed on particulates or as vapour microdroplets impart the molecular indicator signature of petroleum residues to vehicular emissions.

2.4. Polycyclic Aromatic Hydrocarbons

It is generally accepted that airborne polycyclic aromatic hydrocarbons are derived from combustion processes such as the burning of fossil fuel, forest fires and agricultural burning.

Lower polyaromatics (up to benzofluorene) are more abundant in the gas phase while higher polyaromatics are predominantly in the particulate fraction. The higher polyaromatics identified in aerosols are fluoranthene, pyrene, benz(a)anthracene, benzo(a)pyrene, dibenzanthracenes, benzo(ghi)perylene and coronene. The volatile polyaromatics are acenaphthene, biphenyl, fluorene and some heterocyclic species such as dibenzofuran and dibenzothiophene (Simoneit and Mazurek, 1981).

2.5. Carbonaceous Matter

Environmental soot, which is a mixture of various forms of particulate carbon with organic tar and refractory inorganics, is also of importance in aerosols. It has been found that typical aerosols contain 10–30% total carbon. Of this fraction, 20–50% is elemental carbon, less than 5% is carbonate, and the remainder is organic carbon. The colloidal carbon is characterised by a unique morphology defined as aciniform carbon, in which an arrangement of graphitic layers can be observed. This form of carbon is predominant in chimney smoke and engine exhaust, from which the particulate portion of diesel soot is almost exclusively aciniform carbon. About 60% of the carbonaceous soot in the Los Angeles area is graphitic in nature (Simoneit and Mazurek, 1981).

Interestingly, it has been reported that carbonaceous particles consisting of uncombusted coal and oil, coal coke, and intermediates contribute significantly to the interaction with organic compounds. These carbonaceous particles are more sorptive, have high levels of organic matter, and exhibit high specific surface areas. Griest and Tomkins (1984) observed that polycyclic aromatic hydrocarbons were strongly sorbed by and difficult to extract from diesel soot. Carbonaceous particles appear to be responsible for this effect. Chriswell *et al.* (1988) considered that extraction of organic material present in particles is difficult because their matrices contain silica, alumina, metal oxides, and possibly even activated carbon, all of which are excellent adsorbents for organic compounds. These particles appeared to behave not only as carrier of chemicals but also as gypsum nucleating agents (Del Monte *et al.* 1984).

A study of Sagebiel *et al.* (1997) illustrates the importance of diesel vehicles in the production of particulate emissions, as it was reported that 31 diesel vehicles whose age averaged 22 years showed average emissions of 944 mg/km, with one vehicle emitting at a rate of 10,500 mg/km. For comparison, emission rates of total particles were below 6 mg/km for most production catalyst vehicles.

To investigate the possible contribution of diesel soot to organic pollutants and blackening of building stones in urban environments, diesel soot collected from the exhaust of a ten-year-old Sevillian public

Table 2. Main classes of compounds identified in diesel soot.

Compounds	Range*	Compounds	Range
n-Alkanes	$C_{14}-C_{31}$	Alkylnaphthalenes	$C_{10}-C_{17}$
n-Fatty acids	C_6-C_{20}	Alkylfluorenes	$C_{12}-C_{15}$
α, ω-Dicarboxylic acids	C_6-C_{11}	Alkylphenanthrenes	$C_{14}-C_{17}$
Alkylcyclohexanes	$C_{16}-C_{22}$	Alkylbenzoic acids	C_7-C_{10}
Alkylbenzenes	C_{6-21}	Polycyclic aromatic hydrocarbons	$C_{12}-C_{16}$

* Range denotes number of carbon atoms in the compounds.

bus was analysed by Saiz-Jimenez (1994a). The extract was a complex mixture of compounds, the *n*-alkanes being the majority. Some of the identified classes of compounds are listed in Table 2, which agree with those reported in Table 1. In addition, many other compounds were identified up to a total of around 250, among which nitrogen- (e.g. quinolines), oxygen- (benzofurandiones and dibenzofurans), sulphur-heterocyclic compounds (dibenzothiophenes), phenols, alkan-2-ones and alkanols, were also found.

3. Identification of Organic Pollutants in Black Crusts

3.1. *Analytical Methodologies*

Sulphur dioxide has long been recognised as the primary gaseous component of air pollution and a correlation exists between sulphur dioxide concentration and sulphatation of carbonatic stones, because oxidation of sulphur dioxide results in sulphuric acid production. In this process, carbonaceous particles are capable of greatly accelerating the destructive action, and behaves as a catalyst — they contain V, Ti, Fe, Mn, Cu — for the oxidation of sulphur dioxide. Aerosols in the form of sulphuric acid droplets also attack building limestones. Sulphuric acid reaction ultimately results in gypsum formation (Saiz-Jimenez and Bernier, 1981; Del Monte *et al.* 1984).

Wet and dry deposition processes combined with gypsum crystal growth give rise to the formation of a hard, grey-to-black crust, in which airborne organic pollutants, carbonaceous particles, aerosols,

Table 3. Resolving power of some analytical techniques.

Method	Compounds resolved
Thin layer chromatography	<10
High performance liquid chromatography (HPLC)	>20
Gas chromatography (packed column)	<100
Gas chromatography (capillary column)	>200
Pyrolysis-gas chromatography (capillary column)	>500

dust, pollen, spores, and every class of particulate matter are entrapped in the mineral matrix. Therefore, organic species present in weathered surfaces are complex mixtures of many classes of compounds entangled in the gypsum matrix. This prevents direct analysis without prior extraction and separation. In order to eliminate mineral interferences, the samples must be extracted with suitable organic solvents and analysed by means of an adequate technique. Table 3 shows the number of resolved compounds in mixtures from weathered stones when different analytical techniques are applied.

From Table 3, it seems evident that gas chromatography (gc) provides the best means of resolving complex mixtures. Some other techniques, such as for instance, column chromatography with gradient elution or preparative HPLC, help to obtain less complex fractions which permit an easier analysis. Since the state-of-the-art for capillary columns and stationary phases is continuously improving (e.g. the introduction of high temperature capillary columns by Lipsky and Duffy, 1986), the number of compounds resolved on capillary columns possibly could be higher that those stated in Table 3.

Most of the modern analytical techniques currently used require solubility and/or volatility of compounds to be separated and analysed. Thus, much of the work with different types of materials has been concentrated on the analysis of organic molecules with a relatively low molecular weight or those readily volatilised by derivatisation (e.g. methylation). In this respect, samples can be extracted with organic solvent and fractionated by column chromatography, prior to analysis, to separate different classes of compounds and facilitate identification.

There are many fractionation schemes according to the nature of compounds to be studied. However, fractions eluted from the chromatography column with solvent mixtures of increasing polarity often contain hundreds of compounds, which are still difficult to resolve by gc. Furthermore, the final fractions, the most polar compounds, contaminate the analytical instruments, as polar species may be lost on cold or adsorbing sites.

In general, fractionation and separation of organic compounds results in tedious and time-consuming analysis. In addition, contamination of samples with compounds introduced during that preparation (e.g. dialkyl phthalates) is highly probable. Furthermore, a great deal of information is missed when unsuitable methods are applied to the analysis of the insoluble fractions of black crusts. However, solvent extraction, in which relatively large amounts of material were analysed by techniques involving gas chromatography in combination with mass spectrometry (gc/ms) was applied as standard method for studying the black crust composition.

For solvent extraction, 20 g of black crust were ground in an agate mortar and extracted in a Soxhlet apparatus. Two extracts were obtained by successive extractions with 200 ml of toluene (8 hours) and with 200 ml of methanol (8 hours). The use of methanol as solvent is reflected in the appearance of both organic and inorganic materials in the extract. The inorganic materials were eliminated by extraction with bidistilled cold water (three times). Both extracts were mixed and evaporated under vaccuum at low temperature (below 40°C) and redissolved in diethylether. Acidic compounds were derivatised by addition of an excess of diazomethane solution in ether, and consequently, carboxylic groups were converted into methyl esters. Extracts were injected in a gc/ms system. Methods and analytical conditions have been thoroughly described elsewhere (Saiz-Jimenez *et al.* 1991).

3.2. Sites Investigated

The organic compounds present in the crusts collected from monuments located in three cities, Dublin (Ireland), Mechelen (Belgium) and Seville (Spain) were investigated. A sample corresponded to black gypsum crusts from the Custom House, an 18th century building in

Dublin, constructed in 1791 with Portland limestone. In addition to damage to the fabric caused by the aggresive atmosphere of Dublin, and the damage caused by the corrosion of ferrous metals, the Custom House has suffered greatly from the effects of three major fires (Saiz-Jimenez, 1991). Crusts were collected from the balustrade removed from the east front during the restoration works accomplished in 1988. The crusts were situated on the back part of the railing, in a sheltered and rain-protected site. The railing and small columns were eroded and free from the black crust, except on the underside of the railing.

The cathedral of Mechelen (13th–15th century), whose building materials and architecture were characteristic of many large historic buildings throughout Flanders, was also selected. The façades built of Balegem stone, a sandy limestone, are covered by a black gypsum crust, consisting mostly of fine-grained equi-dimensional crystals and a smaller amount of elongated crystals. The cathedral of Mechelen is situated on a straight line between Antwerp and Brussels and the heavily industrialised areas north of Brussels, and especially north of Antwerp, are some 15 and 25 km away, respectively. Therefore, Mechelen lies at the centre of one of the world's most polluted areas and downtown traffic passes around most of the cathedral (Fobe et al. 1995).

From the cathedral of Seville, the Prince Gate, constructed in 1887 was studied. The entrance to this gate is protected by a fence supported by a small limestone wall and pilasters, which are severely deteriorated due to black crust formation. This is related to the fact that one of the main Sevillian bus terminals, operating for more than 20 years, was just in front and less than 10 m far from this wall. The exhaust gases and particulate matter heavily affected the limestone.

3.3. Solvent Extraction of Black Crusts

After extracting the samples with organic solvents, the extracts were methylated and deposited onto a ferromagnetic wire (358°C) inserted in a pyrolysis unit coupled to a gc/ms system. The solvent was removed before introduction in the pyrolysis unit. In this way, very good chromatograms, without interference from solvents, were obtained. The extractable lipid material consisted primarily of hydrocarbons

and fatty acids (as methyl esters), represented by homologous series of
n-alkanes ranging from C_{13} to C_{40}, and n-fatty acids from C_{10} to C_{34}
(Table 4). Furthermore, a few diterpenoids, triterpanes, polycyclic
aromatic hydrocarbons and dialkyl phthalates, were identified as major
peaks.

The range of n-alkanes found in black crusts is determined by the
analytical procedure. This is due to the technical difficulty associated
with the analysis of high molecular weight alkanes (typically > n-C_{35}),
which are less amenable to chromatographic analysis in conventional
capillary gc. It is considered that in the black crusts, in a similar way
to oils and derivatives, the n-alkane range could be greater than those
found in gc studies.

For investigating sources of homologous series of n-alkanes, CPI
was used. CPI of n-alkanes for each sample ranged from 1.0 to 1.7,
denoting an exclusive contribution of petroleum derivatives for the
Seville crust, and progressive inputs of alkanes of biological origin in
the Dublin and Mechelen crusts. A more sensitive method requires
splitting the carbon range into low (C_{15}–C_{24}) and high (C_{25}–C_{34})
ends. In this case, CPI for the Irish extract were 0.8 and 1.9, for the
Belgian extract 1.1 and 2.1, and for the Spanish extract 0.8 and 1.1,
respectively. The lower end is representative of a petrogenic origin, as
the typical CPI of petroleum is 1.0, and petroleum residues are a
major and usually predominant component of the lipids extractable
from aerosols in urban environments (Simoneit, 1986). The high end,
in the case of the Irish and Belgian extracts, demonstrated a plant wax
signature, confirmed by the dominance of n-C_{29}, which indicates a
mixed origin from forest and grassland. Furthermore, burning of
shrubs, litter, trees, etc. produced similar CPI in the high end (Standley

Table 4. Some series of organic compounds present in black crusts.

Location	n-Alkanes	C_{max}	CPI	n-Fatty acids	C_{max}	CPI	Triterpanes
Dublin	C_{13}–C_{35}	C_{29}	1.3	C_{10}–C_{32}	C_{16}	2.9	C_{27}–C_{34}
Mechelen	C_{14}–C_{40}	C_{29}	1.7	C_{12}–C_{34}	C_{16}	5.1	C_{27}–C_{34}
Seville	C_{15}–C_{40}	C_{31}	1.0	C_{10}–C_{34}	C_{22}	1.9	C_{27}–C_{35}

and Simoneit, 1987). On the contrary, the high end of the Spanish extract is characteristic of petroleum.

The source of these hydrocarbons from petroleum is confirmed by the suite of biomarkers found in such materials, as for example the triterpanes, and the hump or unresolved complex mixture (UCM) in the chromatograms of the Irish and Spanish extracts. In fact, these samples exhibited a broad envelope of UCM components ranging approximately from n-C_{16} to n-C_{38} alkanes for the Spanish and n-C_{18} to n-C_{32} for the Irish extract, which have been reported to be composed of highly branched and cyclic hydrocarbons, and ascribed to lubricating oil (Simoneit, 1985) and petroleum (Gough and Rowland, 1990).

According to Simoneit (1986), petroleum contains only minor amounts of long-chain fatty acids, and the homologues $<n$-C_{20} and probably in part $<n$-C_{24} are derived from microbial sources and the homologues $>n$-C_{22} from vascular plant wax. Usually, the major homologue for microbial sources is n-C_{16}, as well as the alkenoic acid n-$C_{18:1}$. However, the distribution of n-fatty acids in diesel engine exhaust and in lubricating oil (Simoneit, 1985) is similar to those reported for the crusts, especially the second one with a bimodal distribution at n-C_{16}/n-C_{18} and n-C_{22}.

Later, Mazurek et al. (1991) split the CPI homologous series of fatty acids (in this case, the sum of the even carbon number homologues over a specified range divided by the sum of the odd carbon number homologues over the same range) into (1) low molecular weight acids (C_6–C_{11}) from non-specific degradation processes (e.g. microbial metabolism, combustion, thermal alteration and photochemical reaction), (2) intermediate molecular weight acids (C_{12}–C_{19}) from recent biogenic sources (e.g. microorganisms and plant waxes; CPI > 2) or from combustion sources (e.g. vehicular emissions; CPI < 2), and (3) high molecular weight acids (C_{20}–C_{33}) derived from epicuticular waxes of vascular plant foliage. CPI for fatty acids from the Irish extract is 2.9, and when broken down into molecular weight categories, the lower CPI in the C_6–C_{11} range (in this and other samples) could not be calculated because this range hardly can be distinguished from background in the chromatogram, CPI for the C_{12}–C_{19} range is 3.4, and CPI for the C_{20}–C_{33} range is 2.4. For the

Belgian extract, CPI in the whole range is 5.1 (5.8 and 4.6, respectively) and for the Spanish extract is 1.9 (1.6 and 2.1, respectively). From all the three extracts, only the CPI for fatty acids in the Spanish sample agree with those of vehicular emissions, whereas it appears that the fatty acid fractions could have a biogenic origin in the Irish and Belgian extracts.

Diterpenoids are regarded as characteristic molecular markers for conifer resins (i.e. methyl dehydroabietate), and retene is an incomplete combustion product of compounds with the abietane skeleton (Standley and Simoneit, 1987). These compounds are frequent in cities in which residential wood combustion is widely used, and therefore could only be observed in Dublin and Mechelen but not in Seville.

Polycyclic aromatic hydrocarbons are the result of combustion-generated airborne particulate matter, and have been identified, among other sources, in smoke particles from plant burning (Standley and Simoneit, 1987) and diesel engine soot (Yu and Hites, 1981). It is well-known that organic compounds adsorbed onto the particulate phase of diesel exhaust possess direct-acting mutagenicity and can be accounted for by polycyclic aromatic hydrocarbons (Bayona *et al.* 1988). Polycyclic aromatic hydrocarbon mixtures encountered in crust extracts are complex because of the presence of alkyl-substituted compounds, as well as the numerous isomeric parent compounds. Generally, compounds from two to six aromatic rings are widely distributed in the three crust extracts, particularly in the Irish one. In addition, ketones, sulphur and nitrogen-substituted compounds were identified.

Petroleum biomarkers are compounds utilized for defining both the fossil origin and the geological source of the petroleum residues (Simoneit *et al.* 1988). In fact, as crude oils usually contain tricyclic terpanes and hopanes, triterpanes have been proposed as sensitive molecular markers of petroleum pollution. The triterpanes of the black crusts are composed of predominantly the $17\alpha(H),21\beta(H)$-hopane series, which is of a petroleum origin. Tricyclic terpanes make them further possible indicators of petroleum. In addition, series of steranes were also present.

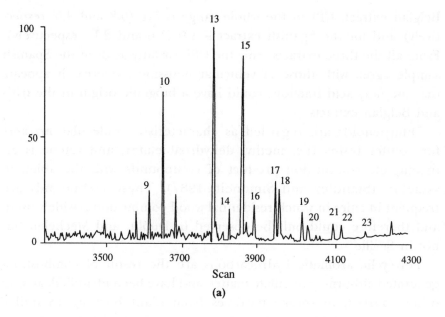

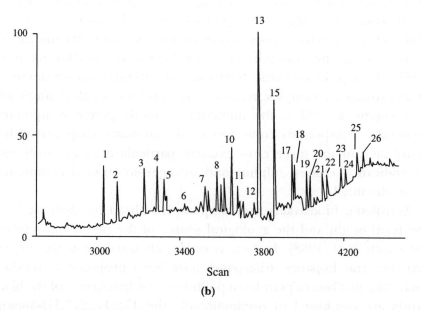

Fig. 1. (a) Mass fragmentogram (m/z 191) of the extract of black crust, Custom House, Dublin. (b) Mass fragmentogram (m/z 191) of the pyrolysate of the black crust from the cathedral of Seville. Peak numbers refer to Table 5.

The hopanes are relatively easy to detect by using gc/ms and ion monitoring since two major fragment ions, m/z 191 and m/z 148 + R, are formed from the parent ion in the ion source of the mass spectrometer. An example of the distribution of the hopanes in the Irish black crust extract is shown in Fig. 1(a) and the identity of the compounds is referred to in Table 5. Interestingly, the pattern of the biomarkers found in black crusts is similar to those reported

Table 5. Triterpanes and steranes identified in black crusts.

Peak	Compound
1	C_{23} tricyclic terpane
2	C_{24} tricyclic terpane
3	C_{25} tricyclic terpane
4	C_{25} tetracyclic terpane
5	C_{26} tricyclic terpane
6	C_{27} tricyclic terpane
7	C_{28} tricyclic terpane
8	C_{29} tricyclic terpane + $5\alpha(H)$, $14\beta(H)$-cholestane
9	$18\alpha(H)$-22,29, 30-trisnorhopane
10	$17\alpha(H)$-22, 29, 30-trisnorhopane
11	24-methyl-$5\alpha(H)$, $14\beta(H)$, $17\beta(H)$-cholestane
12	24-ethyl-$5\alpha(H)$, $14\beta(H)$, $17\beta(H)$-cholestane
13	$17\alpha(H)$, $21\beta(H)$-30-norhopane
14	$17\beta(H)$, $21\alpha(H)$-30-normoretane
15	$17\alpha(H)$, $21\beta(H)$-hopane
16	$17\beta(H)$, $21\alpha(H)$-moretane
17	$17\alpha(H)$, $21\beta(H)$-homohopane 22S
18	$17\alpha(H)$, $21\beta(H)$-homohopane 22R
19	$17\alpha(H)$, $21\beta(H)$-bishomohopane 22S
20	$17\alpha(H)$, $21\beta(H)$-bishomohopane 22R
21	$17\alpha(H)$, $21\beta(H)$-trishomohopane 22S
22	$17\alpha(H)$, $21\beta(H)$-trishomohopane 22R
23	$17\alpha(H)$, $21\beta(H)$-tetraquishomohopane 22S
24	$17\alpha(H)$, $21\beta(H)$-tetraquishomohopane 22R
25	$17\alpha(H)$, $21\beta(H)$-pentaquishomohopane 22S
26	$17\alpha(H)$, $21\beta(H)$-pentaquishomohopane 22R

for crude oil (Philp, 1985), automobile and diesel engine exhausts (Simoneit, 1985).

Studies on the organic composition of black crusts have been carried out by other authors with similar results. Nord and Ericsson (1993) found as main compounds of the black crusts from a church in Stockholm n-C_9 to n-C_{30} alkanes, PAH from three to seven aromatic rings, and some nitrogen- and sulphur-containing polycyclic aromatic species. In the black crusts from historic monuments in Dresden, Machill *et al.* (1997) found n-C_{14} to n-C_{29} alkanes, n-C_{10} to n-C_{25} alkanols, n-fatty acids from C_3 to C_{24}, α,ω-dicarboxylic acids from C_2 to C_9, some hydroxyacids and aromatic acids, PAH from two to five aromatic rings, sulphur- and oxygen-containing polycyclic aromatic species, and a few diterpenoids and carbohydrate derivatives.

3.4. Analytical Pyrolysis of Black Crusts

Sometimes, the required amounts of samples needed for solvent extraction are not available, and an analytical approach feasible at microscale level and able to supply fairly general information on a wide range of classes of compounds is needed. This approach is analytical pyrolysis, a flash thermal desorption and degradation method, also in combination with gc/ms, which allows the collection of fairly general composition data from microgram quantities of samples. This method was applied to a variety of samples obtained from cathedrals, churches, historic buildings and sculptures in Belgium, Ireland, Italy, Spain and the Netherlands, with the best results (Saiz-Jimenez, 1991, 1993, 1994a, b and 1995; Saiz-Jimenez *et al.* 1991).

Analytical pyrolysis was accomplished as described by Saiz-Jimenez (1991). In this case, only a few micrograms of sample were needed for analysis. The sample was applied to a wire and pyrolysed for 10 seconds using an Fe wire with a Curie temperature of 770°C. This temperature has been shown to be adequate for evaporation/pyrolysis of organic materials within mineral matrices (Saiz-Jimenez *et al.* 1991). However, some changes in the thermal breakdown of molecules can be expected, as the crust contains high percentages of gypsum and other salts, and it has been demonstrated that cations influence the thermal behaviour of organic materials (van der Kaaden *et al.* 1983).

Furthermore, some decarboxylation of organic acids was observed (Saiz-Jimenez, 1991).

This method was applied to the study of the black crusts obtained from different building materials in the same monument, the Pardon Gate of the cathedral of Seville, and justified by the scarce amount of sample available. The Pardon Gate is the main entrance to the Orangery Court, remains of the Moorish mosque begun in 1172. This arch is ornamented with arabesques of plaster, all inserted in a limestone wall, which constitutes the cathedral north front. Four terracotta statues representing Saint Peter, Saint Paul and the Annunciation, and a high relief representing the Expulsion from the Temple were added in the 16th century. The four statues are sustained by an alabaster pedestal. Therefore, four types of building materials exist in the Pardon Gate: limestone, plaster, alabaster and terracotta.

Crusts from all these materials, at a height of about 10 m from the ground, were sampled and analysed by analytical pyrolysis. The pyrogram obtained from the black crust formed on terracotta was identical to those of crusts removed from alabaster, limestone or plaster, thus confirming that the same compounds are present in crusts of different materials exposed to the same polluted environment.

The pyrogram was dominated by homologous series of aliphatic hydrocarbons and a few polycyclic aromatic hydrocarbons. The aliphatic hydrocarbons most probably have a double origin, as they may arise either from uncombusted or partially combusted fuels (*n*-alkanes) or from pyrolytic decarboxylation of fatty acids (*n*-alk-1-enes). The polycyclic aromatic hydrocarbons, less abundant than the aliphatic hydrocarbons, were characterised by the presence of indene, naphthalene, fluorene, phenanthrene and pyrene. These compounds have an environmental significance because although present in oil, they also represent the product of incomplete combustion of gasoline and diesel fuel in vehicles (Boyer and Laitinen, 1975).

As observed by SEM and EDX, the black crusts present in the building materials from the Pardon Gate are composed of gypsum crystals originating from oxidation of sulphur dioxide and reaction of the resulting sulphuric acid with the exposed materials. During the process, aerosols and particulates present in the polluted urban atmosphere are entrapped. The carbonaceous particles contain S, Al,

Ca, Fe, Cu, Mn and V as mineral elements, and aliphatic hydrocarbons and polycyclic aromatic hydrocarbons as main organic compounds.

In addition, the black crust obtained from the Prince Gate was investigated. Interestingly, the pyrogram for this crust (taken only 1.7 m from the ground) was different to those available from the Pardon Gate, especially at the end of the analysis programme, characterised by a complex mixture of branched and cyclic hydrocarbons (hump or UCM) which can be seen in the range $C_{20}-C_{34}$ of n-alkanes. This was already observed in the chromatogram from solvent extract. The huge hump indicates a severe contamination by fossil fuels. In the pyrogram, a homologous series of n-alkenes in the range C_6-C_{35} predominates together with the same series of n-alkanes. However, whereas alkenes predominate over alkanes up to C_{20}, the opposite is true for alkanes from C_{21} with a considerable reduction in alkenes as the series progresses. Polycyclic aromatic hydrocarbons up to five rings and alkylbenzenes up to C_{21} were other significant compounds.

No fatty acids were detected in the pyrolysate. Several facts could contribute to this absence, such as the relatively high pyrolysis temperature and the presence of considerable amount of salts. It has been stated that inorganic additives result in the production of unwanted thermal reactions, namely cyclisation and aromatisation reactions of unsaturated fatty acids (Saiz-Jimenez, 1994b and 1995). A decarboxylation process in fatty acids could be expected in the pyrolysis of black crusts, resulting in the formation of alkenyl compounds with one C atom less than the original fatty acid, together with the formation of some other artifacts.

The identification of triterpanes (hopanes and hopenes) and steranes (cholestanes) is of interest in the pyrogram of the black crust from the Prince Gate. They are distributed through the UCM of compounds. Table 5 lists some of the identified compounds from the m/z 191 mass chromatogram [Fig. 1(b)], whereas for instance, in the Pardon Gate crust removed from alabaster only $17\alpha(H),21\beta(H)$-30-norhopane was clearly identified. As all these compounds are found in vehicular emissions, diesel fuel and lubricating oil (Simoneit, 1985), the origin of those encountered in the Prince Gate black crust can directly be ascribed to the vehicular exhausts originated by idle engines in the bus stop.

Interestingly, triterpanes pattern in both solvent extracts [Fig. 1(a)] and pyrolysates [Fig. 1(b)] were similar, indicating that no differences were obtained irrespective of the analytical approach used.

3.5. Simultaneous Pyrolysis/Methylation

In an attempt to solve the bias of data in pyrolysis, mainly the missing of polar compounds, a novel procedure, simultaneous pyrolysis/ methylation, was tested. Solvent extraction versus pyrolysis data was briefly discussed in a previous paper (Saiz-Jimenez, 1991). The major classes of compounds identified in solvent extraction and subsequent gc/ms analysis versus conventional pyrolysis are shown in Table 6. From these data, it was concluded that analytical pyrolysis is a fast screening procedure for identification of organic compounds. The method provides basic information about most classes of compounds present in surfaces of building stones, but for a detailed study and specific search for polar compounds, solvent extraction and concentration should be employed. The disadvantage of pyrolysis with regard to solvent extraction is both the loss of information on polar compounds and the formation of thermal degradation products.

Pyrolysis/methylation consists of the derivatisation of samples containing carboxyl and/or hydroxyl groups with an alkylating reagent. Tetramethylammonium salts of organic acids can be converted to methyl esters and the corresponding by-products in the pyrolysis unit, thus the functional groups are directly protected (Challinor, 1989). The application of pyrolysis/methylation to the study of black crusts seems to be very promising as several series of polar compounds, not previously found in conventional pyrolysis, were identified in the pyrolysates. In fact, Table 6 also shows the major classes of compounds identified from a total of 250 in pyrolytic methylation. The carboxylic acids are recovered as the corresponding methyl esters and the hydroxyls as methoxyls. Throughout this chapter, they are referred to as acids and hydroxyls, their original forms, rather than as derivatised methyl esters and methoxyls.

Table 6. Main classes of compounds identified in the black crust of the Custom House, Dublin, according to the analytical procedure.

Class of compounds	Solvent extraction range[*]	Pyrolysis range	Pyrolysis/methylation range
n-Alkanes	$C_{13}-C_{35}$	C_5-C_{32}	C_9-C_{29}
n-Fatty acids	$C_{10}-C_{32}$	–	C_6-C_{26}
α,ω-Dicarboxylic acids	+	–	C_7-C_{17}
Alkylbenzenes	+	C_6-C_{17}	C_6-C_{26}
Alkylnaphthalenes	$C_{10}-C_{13}$	$C_{10}-C_{14}$	$C_{10}-C_{13}$
Alkylphenanthrenes	$C_{14}-C_{16}$	$C_{14}-C_{17}$	$C_{14}-C_{16}$
Diterpenoids	$C_{18}-C_{20}$	C_{18}	$C_{18}-C_{20}$
Triterpenoids	$C_{27}-C_{34}$	$C_{27}-C_{35}$	+
Polycyclic aromatic hydrocarbons	$C_{10}-C_{22}$	$C_{10}-C_{24}$	$C_{10}-C_{18}$

[*] Range denotes number of carbon atoms.
+ Present, but range not determined; – not found.

The fatty acid series in pyrolysis/methylation is only limited by the resolution of the chromatographic column. This fatty acid series is absent in conventional pyrolysis. The series of dicarboxylic acids was also absent in conventional pyrolysis, and evidenced in pyrolysis/methylation up to a range of C_{17}.

The identification of aromatic acids is to be noted. These included benzoic acid, methylbenzoic acids, benzenedicarboxylic acids, methylbenzenedicarboxilic acids, benzenetricarboxylic acids and naphthalenecarboxylic acid. Cautreels and van Cauwenberghe (1978) identified alkanes, polycyclic aromatic hydrocarbons (PAH) and its alkylated derivatives, dialkyl phthalates, fatty acids and aromatic acids in the gas phase of urban air. Aromatic acidic compounds included phenol, cresols, xylenols, benzoic acid, methylbenzoic acids, hydroxybenzoic acids, benzenedicarboxylic acids and naphthalenecarboxylic acid. These compounds were also encountered in the pyrolytic methylation of the black crusts. Kawamura and Kaplan (1987) suggested that benzoic acids and phenols originate from non-biogenic sources, such as fossil fuel combustion.

Some phenols and quinones were encountered in the pyrolysate mixture. They encompass phenol (as methoxybenzene), cresol (as methoxymethylbenzene) and benzenediol (as dimethoxybenzene), ethoxybenzoic acid ethyl ester, probably as such in the sample, and the methylated derivatives of 3-methoxy-4-hydroxybenzoic acid (vanillic acid), phenylprop-2-enoic acid and 4-hydroxyphenylprop-2-enoic acid (*p*-coumaric acid). In addition, fluorenone and anthracenedione were identified. The 4-hydroxyphenylprop-2-enoic acid could be considered as a tracer of conifer pollen, as *Pinus* pollen contain sporopollenin which comprises *p*-coumaric acid (Mulder *et al.* 1992), whilst the phenylprop-2-enoic acid is probably a pyrolysis product of the original acid, in which a dehydroxylation was produced. Vanillic acid can be traced as a product of residential wood stoves (Hawthorne *et al.* 1988).

The PAH mixtures encountered in black crusts are complex because of the presence of their alkylated derivatives. These compounds derive from combustion-generated airborne particulate matter and have been identified, amongst other sources, in smoke from plant burning (Standley and Simoneit, 1987) and carbonaceous particles from fossil fuel combustion (Simoneit, 1985 and 1986). Basically, species from two to four aromatic rings are distributed in the pyrolysate, the same compounds being previously identified in the solvent extracts (Saiz-Jimenez, 1993). In urban environments, lower PAHs up to benzofluorenes are abundant in the gas phase, whereas higher PAHs are found predominantly in the particulate fraction.

Sulphur- and oxygen-containing polycyclic aromatic species, including polycyclic aromatic ketones, were found in the pyrolysate, thiophene, benzothiophene, dibenzothiophene, benzonaphthothiophene, benzofuran, dibenzofuran, benzonaphthofuran and some of their alkylated derivatives being the identified compounds. Methylthiobenzothiazole and some alkylated pyrrols and indoles were representative of nitrogen-containing compounds.

Lee *et al.* (1977) identified similar PAHs in the combustion of three common fuels. The sulphur-containing species were related to coal combustion products. However, Williams *et al.* (1986) identified these compounds in diesel fuels. Sicre *et al.* (1987) reported that dibenzothiophenes are common in crude oils and coal emissions, therefore their presence is not indicative of a specific origin. Benzofurans are probably related to wood or coal combustion (Lee *et al.* 1977). Bayona *et al.* (1987) found indoles in coal tar fractions, and Ramdahl (1983) identified polycyclic aromatic ketones in diesel exhaust, and wood and coal combustion samples.

Diterpenoids were observed in all three analytical procedures described in Table 6. However, in conventional pyrolysis only retene, an incomplete combustion product with the abietane skeleton, was identified. This compound is used as an indicator of residential coniferous wood combustion (Standley and Simoneit, 1987). Dehydroabietic acid derivatives were identified in both solvent extraction and pyrolytic methylation, which further support a coniferous wood origin.

Triterpenoids were also present in pyrolytic methylation, but due to the absence of molecular ions in the mass spectra, they could not be individualised. However, the series of hopanes were identified in both solvent extraction and conventional pyrolysis studies with a powerful mass spectrometry instrument (Saiz-Jimenez, 1991).

All classes of compounds shown in Table 6 have been previously identified in gas phase, aerosols and particulate matter in urban atmospheres. This is of interest as Grimalt *et al.* (1991) reported the close similarity between the organic composition of black crusts from the Holy Family church (Barcelona, Spain), and airborne particulates, collected by glass fibre filtration, and gas-phase organic compounds, obtained by polyurethane foam adsorption. These facts and the finding of carbonaceous particles entrapped in the voids of gypsum crystals demonstrated that the organic compounds present in the black crusts, covering the building stones in urban environments, are the result of a direct input of air pollutants, the buildings acting as non-selective surfaces passively entrapping all deposited aerosols and particulate matter, from whose analysis a source can be traced.

The series of aliphatic and aromatic acids found in the pyrolysis/methylation mixture clearly indicate that this method protects carboxylic groups through the formation of methyl esters, and hydroxyl groups through the formation of methoxyls. This reveals that conventional pyrolysis is a strongly biased method for the identification of compounds containing oxygenated functional groups. A previous report demonstrated that pyrolysis/methylation is an analytical procedure of great sensitivity for investigating organic compounds in inorganic matrices (Saiz-Jimenez *et al.* 1994). No time-consuming solvent extractions, concentrations, derivatisations and other manipulations which imply the possibility of cross-contamination are required to obtain similar data to those obtained by solvent extraction and subsequent gc/ms analysis.

Finally, Table 7 shows a comprehensive list of the major classes of compounds identified in pyrolysis/methylation of black crusts from the Custom House (Dublin), the cathedral of Seville (Spain), and two new monuments, the balustrade of the stairs next to the Count Belliard

Table 7. Classes of identified compounds in pyrolysis/methylation of black crusts and diesel soot.*

Class of compounds	Dublin	Brussels	Rome	Seville	Diesel soot
n-Alkanes	C_8-C_{31}	C_8-C_{30}	C_8-C_{32}	C_8-C_{30}	C_9-C_{32}
n-Alkenes	C_8-C_{27}	C_9-C_{26}	C_8-C_{24}	C_9-C_{30}	C_8-C_{25}
Isoprenoid hydrocarbons	C_{19}, C_{20}	C_{19}, C_{20}	C_{19}, C_{20}	C_{19}, C_{20}	$C_{18}-C_{20}$
n-Fatty acids[†]	C_6-C_{26}	C_5-C_{30}	C_5-C_{27}	C_4-C_{28}	C_5-C_{25}
α, ω-Dicarboxylic acids[‡]	C_7-C_{17}	C_5-C_{17}	C_7-C_{13}	C_5-C_9	–
Alkylbenzenes	C_6-C_{25}	C_6-C_{22}	C_6-C_{23}	C_6-C_{26}	C_6-C_{27}
Alkylcyclohexanes	–	–	–	–	$C_{15}-C_{24}$
Benzoic acids	+	+	+	+	+
Benzenedicarboxylic acids	+	+	+	+	+
Benzenetricarboxylic acids	+	+	+	+	+
Benzenetetracarboxylic acids	–	–	–	+	–
Polycyclic aromatic hydrocarbons	$C_{10}-C_{16}$	$C_{10}-C_{16}$	$C_{10}-C_{18}$	$C_{10}-C_{16}$	$C_{10}-C_{17}$
Thiophene derivatives	+	+	+	+	+
Furan derivatives	+	+	+	+	+
Pyridine derivatives	–	–	–	+	–
Benzonitriles	+	+	+	+	+
Quinoline derivatives	–	–	–	+	–

* C_8 denotes number of carbon atoms; + detected compound; – not detected compound.
[†] Detected as methyl ester.
[‡] Detected as dimethyl ester.

statue (Brussels) and the pillars of the Barberini Palace fence (Rome). For comparison, the compounds present in diesel soot from a Sevillian public bus is shown (Hermosín, 1995). From the data, it was concluded that the black crusts from different European monuments and buildings contain molecular markers that are characteristic of petroleum derivatives. The overprint of some biogenic components of aerosols over petroleum components from anthropogenic emissions (mainly vehicular) can be illustrated by the dominance of hydrocarbons around n-C$_{29}$ (plant waxes) and fatty acids in the range n-C$_{12}$–C$_{19}$ (microorganisms and plant waxes). Accordingly, the black crusts coating the surfaces of building materials located in urban environments are constituted by a suite of all kind of organic compounds present in aerosols and particulate matter, which are transferred by dry and/or wet deposition. The composition of each crust is governed by the composition of the particular airborne pollutants in the area, but is mainly derived from vehicular emissions, diesel engines having a strong influence. Finally, it is clear that the exposed building materials act as a non-selective surface, passively entrapping all deposited airborne particulate matter and organic compounds which obviously modifies the composition of the materials present in the stone surface.

4. Microbiology of Black Crusts

The study of microbial communities is usually accomplished by using standard culture methods. Microorganisms traditionally are characterised by phenotype (morphology, biochemical tests, lipid composition, etc.). However, it is believed that fewer than 20% of the extant microorganisms have been discovered, and that culture methods are inadequate for studying microbial community composition. Pace (1996) considered that only a small portion, typically far less than 1% of organisms in the environment can be cultivated by standard techniques. There are many reasons for the routine failure of the usual cultivation strategies, the most common is that selective enrichment cultures fail to reproduce the conditions that particular microorganisms require for proliferation in their natural habitat.

Although in the literature many reports on the microbiology of weathered stones in monuments can be found, attention to the microflora of black crusts has barely begun, and only a few studies on the growth of epilithic microorganisms on black crusts (Ortega Calvo *et al.* 1993; Ortega-Calvo and Saiz-Jimenez, 1996) and endolithic filamentous cyanobacterium (*Phormidium* sp.) under black crusts (Saiz-Jimenez, 1991; Saiz-Jimenez *et al.* 1991) can be found. Standard microbiological studies on monuments showed that crust-free stones contained from three to 15 times more CFU than black crusts (Table 8) and that some of the isolated bacteria can degrade pollutants (Table 9). However, when relying on cultural methods to identify species, there is a problem of selectivity and thus the inevitable underestimation of community diversity. Direct exploration

Table 8. Site, description and microbial numbers of stone samples from cathedrals.

Site	Sample code	Description	CFU/g ($\times 10^5$)*
Seville cathedral			
	S1	Pilaster 1, crust-free	8.25
	S2	Pilaster 1, crust-free	7.00
	S3	Pilaster 1, black crust	2.81
	S4	Pilaster 2, black crust	1.29
	S5	Pilaster 3, black crust	0.68
	S6	Pilaster 4, black crust	2.28
Quarry			
	Q1	Ground quarry stone, calcarenite	N.D.[†]
	Q2	Ground quarry stone, limestone	N.D.
Mechelen cathedral			
	M1	Crust-free. East, 1.5 m height	1.40
	M2	Black crust, same as M1	0.09
	M3	Crust-free. North, 1.5 m height	2.54
	M4	Black crust, same as M3	0.58
	M5	Crust-free. Northeast, 2 m height	1.46
	M6	Black crust, same as M5	0.19

* CFU/g presented are for total heterotrophs.
† Not detected.

Table 9. Characterisation and identification of isolates from phenanthrene-degrading enrichments of cathedral samples.

Isolate	Origin[*]	Description	Identification
S4A	Sample S4	Gram-negative rods, motile, oxidase-positive, catalase-positive, nitrate-positive, forms PHB, growth with phenanthrene	*Pseudomonas* sp.
S4B	Sample S4	Gram-negative rods, motile, oxidase-positive, catalase-positive, nitrate-positive, growth with phenanthrene	*Pseudomonas* sp.
S4C	Sample S4	Gram-positive rods, arranged in short chains, slightly motile, oxidase-negative, catalase-positive, forms PHB, growth with phenanthrene	*Nocardia* sp.?
S4D	Sample S4	Gram-positive rods, spore-forming, motile, orange pigment, oxidase-positive, catalase-positive, forms PHB, growth with phenanthrene	*Bacillus* sp.
M4A	Sample M4	Gram-negative rods, motile, oxidase-positive, catalase-positive, nitrate-positive, growth with phenanthrene	*Pseudomonas* sp.
M4B	Sample M4	Gram-positive cocci, motile, oxidase-positive, catalase-positive, nitrate-positive	*Planococcus* sp.

[*] See Table 8.

of the inhabitants of natural microbial communities in the black crusts can be expected to have a significant influence on our understanding of microbial phylogeny and physiology and to ascertain how representative are cultivated species of the phylogenetic diversity within the black crust.

It has been suggested that the presence of an organism on decayed material does not necessarily imply that it has caused the damage observed (Saiz-Jimenez, 1995). The activity of microorganisms in promoting stone deterioration is largely dependent upon the production of corrosive metabolites which can solubilise minerals in a manner similar to chemical agents. Probably, low-frequency isolation cannot be directly correlated with metabolic activity as, for instance, the fungi isolated in culture media may be dormant (spores) and are not necessarily the ones which are functioning in the ecosystem. Because isolation in culture media rich in organic carbon can mask the real fungal distribution in weathered stones and isolate selectively airborne propagules instead of active microorganisms, an approach involving direct study of gypsum crusts was carried out by Saiz-Jimenez (1993). The incubation of the crusts in petri dishes with sterile water revealed a different fungal population to those previously reported by Petersen *et al.* (1988) and De la Torre *et al.* (1991) in weathered stones. In fact, from the Seville cathedral a *Papulaspora*-like fungus with dark chlamydospore balls was the only isolate, with high presence in the tested samples (82%). From Mechelen cathedral the species isolated were *Engyodontium album* (15%), *Botriotrichum piluliferum* (4%), *Ulocladium atrum* (4%) and *Mucor circinelloides* (4%). These fungi most probably use the organic compounds present in the black crusts, which include, among others, aliphatic and aromatic hydrocarbons, fatty acids, triterpanes, etc. (Saiz-Jimenez, 1991). Therefore, it appears that two types of fungal populations can be isolated from stones: those utilising readily available carbon (carbohydrates), which are similar to ubiquitous saprophytic airborne fungi, evidenced by using conventional culture media, and those using petroleum-derivatives originating from deposition of pollutants, capability which probably is present in a relatively minor part of the population utilising carbohydrates.

Although ubiquitous in terrestrial and aquatic environments, the fraction of the total heterotrophic community represented by the hydrocarbon-utilising bacteria and fungi is highly variable, with reported frequencies ranging from 6 to 82% for soil fungi and 0.13 to 50% for soil bacteria (Leahy and Colwell, 1990). Nyns *et al.* (1968) tested a wide range of fungi for their ability to assimilate aliphatic and aromatic hydrocarbons and petroleum fractions. Species of the genera *Fusarium, Penicillium* and *Aspergillus* are particularly active in the assimilation of hydrocarbons. Bossert and Bartha (1984) listed 22 genera of bacteria and 31 genera of fungi able to degrade hydrocarbons.

The most remarkable fact is that the list of genera of fungi growing on hydrocarbons reported by Nyns *et al.* (1968) or the most frequent isolates quoted by Leahy and Colwell (1990) agree well with those isolated from weathered building stones and frescoes (Saiz-Jimenez and Samson, 1981; Petersen *et al.* 1988; Karpovich-Tate and Rebrikova, 1990; De la Torre *et al.* 1991; Saiz-Jimenez, 1993). This finding indicates a possible relationship between fungi and organic pollutants derived from oil combustion.

Advances in molecular biology are now providing the means for solving long-standing problems in microbiology. These techniques have been applied to determining the genetic diversity of microbial communities and to identifying several uncultured microorganisms. Muyzer *et al.* (1993) presented a new approach for directly determining the genetic diversity of complex microbial populations. The procedure is based on electrophoresis of PCR-amplified 16S rDNA fragments in polyacrylamide gels containing a linearly increasing gradient of denaturants (DGGE). In this electrophoresis, DNA fragments of the same length but with different base-pair sequences can be separated. This demonstrated the presence of up to ten distinguishable bands in the separation pattern, which were most likely derived from as many different species constituting these populations, and thereby generated a DGGE profile of the total population. Muyzer and Ramsing (1995) reviewed the potentials and limitations of different molecular techniques which are nowadays used to determine the species composition of microbial communities. A specific limitation of the DGGE approach is, that separation of PCR products obtained from very complex mixtures of bacteria such as those which are found in soils might not be possible. Furthermore, only limited sequence information is obtained with the DGGE approach, because separation is reduced for fragments longer than 500 bp. Felske *et al.* (1997) isolated ribosomes from soils and 16S rRNA was partially amplified via RT-PCR using conserved primers for members of the domain Bacteria. Subsequent sequence-specific separation by temperature-gradient gel electrophoresis (TGGE) in combination with taxon-specific probing, leads to the identification of the metabolically dominant portion of the community. TGGE profilings from black crust from different monuments and locations would permit to know whether there is a specific microbial community for this particular niche or the communities diverge due to environmental effects.

5. Biodegradation of Black Crusts

5.1. Aliphatic Hydrocarbon Biodegradation

Although biodeterioration processes have been discussed by some authors (Petersen *et al.* 1988; Saiz-Jimenez *et al.* 1990; De la Torre *et al.* 1991; Ortega-Calvo *et al.* 1991; Warscheid *et al.* 1991), biodegradation of organic compounds in monuments has rarely been investigated. Lewis *et al.* (1988) suggested that bacteria on stone can be extremely versatile and could maintain their activity during nutrient perturbations, operating at low nutrient levels and utilising what the environment has to offer. As a consequence, bacterial populations may be able to maintain their involvement in the process of stone deterioration during periods of nutrient flux. Recently, both biodegradation of polycyclic aromatic hydrocarbons by heterotrophic bacteria, and gypsum crusts by cyanobacteria have been reported (Ariño *et al.* 1995; Ortega-Calvo and Saiz-Jimenez, 1996) and present new insights on the microbial ecology (and physiology) of bacteria in black crusts.

Warscheid *et al.* (1991) studied chemoorganotrophic bacteria from the uppermost layers of sandstones of German monuments. It was shown that most of the isolated bacteria used a wide range of different carbohydrates, amino acids, fatty acids and hydrocarbons. About 40% of the strains were shown to be potential acid producers, whereas the capability of manganese and iron oxidation was only sporadically found. Kerosene, as a representative mixture of different hydrocarbons detectable in polluted atmospheres, was well metabolised by 70% of the bacteria.

Heath *et al.* (1997) considered that a consequence of the biodegradation of petroleum is that the lower molecular weight compounds are removed preferentially to higher molecular weight compounds greater than n-C_{30} (triacontane). They found that a *Pseudomonas fluorescens* strain was able to totally degrade alkanes with carbon numbers n-C_{20}–C_{25} from a waxy Indonesian oil after 14 days. At the end of the trial, no biodegradation was observed for compounds greater than n-C_{45}, which suggested a carbon number cut off for biological degradation between n-C_{40} and n-C_{45}. However, when the bacteria was acclimated (subjected to a previous growth in high molecular weight alkanes), the utilisation of n-C_{60} as a sole carbon source was observed. This was claimed to be the first tentative evidence of an organism able to utilise such high molecular weight hydrocarbons.

In addition, it has been stressed that the degradation of petroleum by bacteria often results in the progressive depletion of chromatographically resolved hydrocarbons (e.g. *n*-alkanes, acyclic isoprenoid alkanes, alkylbenzenes, alkylnaphthalenes and alkylphenanthrenes) relative to the UCM. Hence, the UCM is thought to comprise compounds which are relatively inert to microbial degradation (Gough *et al.* 1992). Biodegradation of UCM compounds by *Pseudomonas aeruginosa* was proven under laboratory conditions (Robson and Rowland, 1988). This bacterium is common in petroleum products and oil emulsions. The degradation rates were *n*-alkanes > *n*-alkenes > highly branched alkenes > highly branched alkanes + regular and tail-tail isoprenoid alkanes. A further study with *Pseudomonas fluorescens* demonstrated that the UCM rate and extent of degradation was influenced by the molecular structure (Gough *et al.* 1992). However, the ability for oil hydrocarbon biodegradation is not extended. Chosson *et al.* (1991) showed that of 73 aerobic bacteria (*Nocardia, Mycobacterium, Corynebacterium, Arthrobacter, Protoaminobacter, Pseudomonas*, etc.), assessed for their ability to degrade steranes and cyclic triterpanes, only seven Gram-positive strains, belonging to the *Nocardia, Mycobacterium* and *Arthrobacter* genera, were able to produce noticeable effects. A *Nocardia* sp. produced the most extensive biodegradation, the preference observed being $C_{27} > C_{28} > C_{29}$.

Refractory petroleum hydrocarbons (e.g. UCM compounds, triterpanes, steranes, etc.) are usually identified in the surface of building stones (Saiz-Jimenez, 1993) and their degradation probably requires specific bacteria belonging to genera which are widespread in soils and well-known hydrocarbon degraders.

5.2. *Polycyclic Aromatic Hydrocarbon Biodegradation*

Phenanthrene concentrations at contaminated sites, usually associated with wood treatments, coking plants and gas works (Wilson and Jones, 1993) and urban environments (Sturaro *et al.* 1993) are among the highest of individual PAHs; therefore the selection of phenanthrene to estimate the activity of PAH-degrading bacteria and subsequent PAHs degradation in building stones seems an appropriate choice.

Polycyclic aromatic hydrocarbon-degrading bacteria have been used previously as indicators of the microbial activity in polluted environments. For instance, Bogardt and Hemmingsen (1992) detected and enumerated the phenanthrene-degrading bacteria in petroleum-contaminated sites; and phenanthrene-utilising and phenanthrene-cometabolising microorganisms have been evidenced in estuarine sediments (Cerniglia, 1993). For this reason, it was considered that these bacteria could be present on weathered stones in urban environments subjected to heavy air pollution, and a sampling was performed in the cathedrals of Mechelen (Belgium) and Seville (Spain).

Table 10 shows the sites, code samples and extents of phenanthrene mineralisation. Microbial numbers (see Table 8) revealed that most samples had a significant heterotrophic microbial population. Seville and Mechelen samples from crust-free zones contained significantly more heterotrophic microorganisms than samples of black crusts obtained from the same place. Quarry stones had no detectable heterotrophic population (Ortega-Calvo and Saiz-Jimenez, 1997).

Phenanthrene was readily mineralised to CO_2 by the natural microbiota in samples from the cathedrals. Both the rate and extent of mineralisation showed no statistical differences between crust-free or black crust zones for either the Seville or Mechelen sites.

Mineralisation in the crust-free samples S1 and S2 from the cathedral of Seville started rapidly with no apparent lag phase and

Table 10. Mineralisation of phenanthrene in samples from cathedrals.

Sample code*	% Mineralised	Sample code	% Mineralised
Seville		*Mechelen*	
S1	35.0 ± 2.1	M1	23.3 ± 1.4
S2	26.9 ± 2.3	M2	3.2 ± 0.2
S3	15.3 ± 1.7	M3	37.6 ± 9.7
S4	21.4 ± 2.3	M4	28.9 ± 2.7
S5	18.0 ± 1.4	M5	24.5[†]
S6	28.1 ± 1.6	M6	19.5

* See Table 8.
† Values from non-duplicate measurements.

reached a final extent of 35.0 and 26.9% of substrate mineralised after 100 days. An acclimation period of ten days and a lower rate and extent of mineralisation characterised the mineralisation in the black-crust sample S3, where only 15.3% of the compound was converted to CO_2 in 100 days. This sample was obtained from the same place as samples S1 and S2. Mineralisation in samples of black crusts from other pilasters also occurred after an acclimation period but differed in the rates and extents. In these samples, the percentage of phenanthrene mineralised to CO_2 in 100 days ranged from 18.0 to 28.1%. Sample Q1, from a quarry, showed a reduced activity, close to background levels.

Mineralisation of phenanthrene in black crust samples from the cathedral of Mechelen also occurred at lower rates and extents than the corresponding crust-free samples from the same place, and also correlated with microbial numbers. Thus, 37.6% of the compound was converted to CO_2 in crust-free sample M3, which had a total heterotrophic population of 2.54×10^5 CFU/g, whereas the extent of mineralisation in the corresponding black crust sample M4 was 28.9%, with a microbial population of 0.58×10^5 CFU/g. The extent of mineralisation tends to increase as the CFU increases. Mineralisation in a black crust samples, M2, was drastically reduced when compared to other samples, as only 3.2% of phenanthrene was mineralised after 100 days.

Bacteria that were able to grow in media with phenanthrene as the sole source of carbon were isolated from some stone samples with and without black crusts. A short description of each isolate obtained from enrichment cultures is given in Table 9. From sample S4, two Gram-negative rods identified as *Pseudomonas* sp., and two Gram-positive rods, a *Bacillus* sp. and a tentatively identified *Nocardia* sp., were obtained. From sample M3, a Gram-negative rod was identified as *Pseudomonas* sp. and a Gram-positive coccus as *Planococcus* sp. All isolates except *Planococcus* sp. showed capability of growth with phenanthrene.

To investigate the effect of different types of stone materials on the mineralisation of phenanthrene by individual isolates, *Bacillus* sp. S4D, *Pseudomonas* sp. S4B and *Pseudomonas* sp. M3A, which showed the fastest growth, were tested. The bacteria mineralised phenanthrene in liquid culture, at an initial concentration of 0.1 µg/ml, both with and without black crust or crust-free stone samples. The presence of stone samples caused a delay of the phase of maximum mineralisation in the three strains. The final extent of mineralisation was always higher in cultures amended with black crust. However, the influence of stone samples on maximum rates of mineralisation was different depending on the isolate. While *Bacillus* sp. S4D mineralised the substrate at a maximum rate irrespective of whether the medium contained stone or not, the presence of stone induced different maximum rates of mineralisation in the two *Pseudomonas* sp. isolates. Interestingly, the higher rate of mineralisation for each *Pseudomonas* sp. strain occurred according to the origin of the strain (crust-free or black crust sample). Maximum rates of mineralisation by *Pseudomonas* sp. M3A, isolated from crust-free stone, were 0.42 and 0.59 ng/mL/h in the presence of black crust and crust-free stone, while the mineralisation rate in stone-free controls was 0.94 ng/mL/h. *Pseudomonas* sp. S4B, isolated from black crust, mineralised phenanthrene at maximum rates of 0.29, 0.13 and 0.42 ng/mL/h in the presence of black crust, crust-free stone, and stone-free controls, respectively.

The concentration of phenanthrene in solution decreased drastically due to adsorption by the stone, 3.0 and 6.3% of the total phenanthrene being present in the aqueous phase of black crust and crust-free stone suspensions. Those values are in general substantially below the final extents of transformation to CO_2 with black crust and crust-free stone (24.8 and 19.4% for *Bacillus* sp. S4D, 24.6 and 10.6% for *Pseudomonas* sp. M3A, and 21.3 and 6.3% for *Pseudomonas* sp. S4B), indicating that at least part of the adsorbed compound was degraded.

It has been found that black crusts contain toxic compounds such as phenols, benzoic acids, lead, etc. (Fobe *et al.* 1995) which can inhibit bacterial growth. This could explain the decrease of the bacterial population, between three- and ten-fold, with respect to crust-free zones (Table 6). However, these compounds do not inhibit PAH utilisers. On the contrary, a clearly stimulatory effect in ^{14}C-phenanthrene mineralisation was found in cultures of *Pseudomonas* sp. S4B in the presence of black crust suspension, which indicates the presence in the crust of nutrient elements required for an adequate growth and to the adaptation of the bacteria to the polluted environment from which they were isolated.

From literature data, it appears that non-volatile PAHs are degraded by a variety of bacteria, yeasts and fungi (Cerniglia, 1993) which generally belong to autochthonous flora. The PAH-utilising bacteria most frequently reported include several species of *Pseudomonas* and a few of *Arthrobacter, Acinetobacter, Alcaligenes* and *Streptomyces. Pseudomonas* strains are also the most frequent isolates from building stones.

Biodegradation by natural surface and sub-surface soil microorganisms appears to be the process primarily responsible for the removal of PAHs in a multiphase soil system. The biodegradability of two- and three-ring PAHs is extensive, whereas that of four- to six-ring PAHs is considerably less significant (Wilson and Jones, 1993). It has been shown that building stones also contain bacteria able to utilise PAHs. This kind of microflora is not only restricted to fuel-contaminated soils or sediments, PAH-contaminated wastes, etc. but can also be found in urban environments, where the bacteria have adapted to specific site conditions which include high PAH concentrations for long periods. In fact, the Seville public bus service started to operate in the early 1960s, and some of the stops were located just in front of the sampling zone, at about 10 m from it. The city of Mechelen is located on the industrial axis between the major cities of Antwerp and Brussels. Refineries, electrical power plants and non-ferrous industries make this area the most important emitter of industrial SO_2 and

NO_x in Belgium. High concentrations of PAHs are found in the black crust coating the limestone of this cathedral (Saiz-Jimenez, 1993; Fobe *et al.* 1995).

The data indicate that microbial degradation of phenanthrene is common in weathered stones from European cathedrals both in crust-free stones and black crusts. Although the experiments performed in this work required the removal of the samples from the cathedral walls, the rapid and significant phenanthrene mineralisation observed strongly suggests that microbial transformation reactions also occur *in situ.* Mineralisation of organic compounds is characteristic of growth-linked biodegradation and part of the phenanthrene was converted to cell components and degradation products that could remain in the stone. It is also possible that the particular conditions prevailing in the stone niches promote the selection of microorganisms able to transform, either by growth-linked reactions or by co-metabolism, other anthropogenic compounds that have been so far considered as recalcitrant.

Due to its lipophilic nature, phenanthrene is associated to lipids deposited on the stone surface or remains adsorbed to airborne carbonaceous particles. Microorganisms may be able to transform phenanthrene either as such or after spontaneous or microbiologically-induced desorption in the water present in the porous stone. However, it can be expected that phenanthrene entrapment in the weathering black crust during gypsum crystal formation may cause some decrease in bioavailability, and this therefore contributes to its persistence in the stone.

The data herein reported suggest that on the surfaces of building stones, there is an active microflora of PAH-degraders. Therefore, biological activity plays a role in the fate of organic compounds deposited on building stones located in urban (polluted) environments.

5.3. Gypsum Crust Biodegradation

Another biodegradation process in black crusts was due to cyanobacteria. Observations of a wide range of monuments in different enviromental conditions indicate that black crusts covering building stones are colonised by phototrophic microorganisms

(Ortega-Calvo *et al.* 1991 and 1993). In general, the presence of phototrophic microorganisms is more apparent on the north façades of buildings than on the south, because the latter dry out more readily. These organisms are also frequently found on black crusts where they form large patches of green, brown and black cyanobacterial/ algal mats, disfiguring buildings. These patches largely consist of biofilms — cells and other materials immobilised on the substrata and embedded in an organic matrix. Biofilm may also contain significant amounts of adsorbed inorganic materials derived from the substratum (quartz, calcium carbonate and clay) and detritus (dead cells, microbial by-products, etc.). The slimy surface favours the adherence of airborne particles (dust, pollen, spores, oil- and coal-fired carbonaceous particles) giving rise to hard crusts and patinas which are difficult to remove.

The biofilms formed on black crusts are usually characterised by the abundant presence of Chroococcales (*Gloeothece, Gloeocapsa,* etc.). *Gloeothece* is a unicellular cyanobacterium capable of fixing nitrogen aerobically and producing a complex sheath with high amounts of sulphate (Tease *et al.* 1991).

Black crusts deposited in petri dishes with only the addition of sterile water doubled in four weeks the biomass when inoculated with the cyanobacterium *Gloeothece* sp. The result shows that the mineral content of the crusts is enough to support cyanobacterial development. The finding suggest that gypsum might play a role in the cyanobacterial colonisation of blackened monuments in urban environments.

Under laboratory conditions, it was proven that the gypsum present in a black crust from the cathedral of Seville, can be used as a source of sulphur by the cyanobacterium *Gloeothece* sp. (Ortega-Calvo *et al.* 1994). The sulphate released to the medium due to gypsum dissolution was progressively incorporated into the carbohydrate sheath and used for balanced growth.

The sulphate-bonding capacity of the sheath in *Gloeothece* sp. causes a particularly high demand for this anion. This makes sulphur nutrition a relevant aspect in the physiology of this cyanobacterium, as demonstrated by Ariño *et al.* (1995). In fact, when deprived of sulphur, bleached *Gloeothece* sp. cells contain a disintegrated photosynthetic apparatus and accumulations of different kinds of reserve material. Such characteristics are absent in sulphate-sufficient diazotrophic cultures. The inability to obtain from the culture medium an adequate supply of sulphur for protein synthesis, and therefore, for balanced growth induces in *Gloeothece* sp. the immobilisation of fixed carbon in the form of glycogen, polyhydroxybutyrate and cyanophycin. In other words, *Gloeothece* sp. reacts to sulphur starvation by notable changes in its ultrastructure related to alteration in the overall physiological processes, including sheath synthesis, nitrogen metabolism and photosynthesis. The addition of sulphate to a sulphate-deficient medium restored the regular activity of the cyanobacterium.

6. Conclusions

The data herein reported indicate that microorganisms are also able to remove some of the most abundant components of black crusts, such as gypsum and polycyclic aromatic hydrocarbons. Although the use of microorganisms is not foreseen as a method for cleaning façades due to some obvious limitations (dimension of monuments, time needed, wetting of weathered surfaces, economy, etc.), it must be emphasised that nature develops strategies for biodegradation of pollutants in urban environments. In fact, the continuous input of inorganic and organic compounds modifies the chemical composition of building stones, resulting in the selection of microorganisms with specific nutrient requirements or with a defined metabolic capability. Although we are still far from a complete understanding of the physiological diversity of microorganisms and their interactions in the surface of stone monuments, biodegradation of deposited chemicals is envisaged as an important process, in addition to biodeterioration. There is a scope for molecular techniques and particularly for molecular microbial ecology studies which can fill the current gaps in our knowledge on the organisation of microbial communities in monuments and their effect on building materials.

Acknowledgements

This work was supported by the European Commission, contract EVK4-CT-2000-00029.

References

Ariño X., Ortega-Calvo J.J., Hernandez-Marine M. and Saiz-Jimenez C. (1995) Effect of sulphur starvation on the morphology and ultrastructure of the cyanobacterium *Gloeothece* PCC 6909. *Arch. Microbiol.* **163**, 447–453.

Bayona J.M., Tarbet B.J., Chang H.C., Schregenberger C.M., Nishioka M., Markides K.E., Bradshaw J.S. and Lee M.L. (1987) Selective gas chromatographic stationary phases for nitrogen-containing polycyclic aromatic compounds. *Int. J. Env. Anal. Chem.* **28**, 263–278.

Bayona J.M., Markides K.E. and Lee M.L. (1988) Characterization of polar polycyclic aromatic compounds in a heavy-duty diesel exhaust particulate by capillary column gas chromatography and high-resolution mass spectrometry. *Env. Sci. Technol.* **22**, 1440–1447.

Bogardt A.H. and Hemmingsen B.B. (1992) Enumeration of phenanthrene-degrading bacteria by an overlayer technique and its use in evaluation of petroleum-contaminated sites. *Appl. Env. Microbiol.* **58**, 2579–2582.

Boone P.M. and Macias E.S. (1987) Methyl alkanes in atmospheric aerosols. *Env. Sci. Technol.* **21**, 903–909.

Bossert I. and Bartha R. (1984) The fate of petroleum in soil ecosystems. In *Petroleum Microbiology*, ed. Atlas R.M. Macmillan Publishing Co., New York, pp. 434–476.

Boyer K.W. and Laitinen H.A. (1975) Automobile exhaust particulates. Properties of environmental significance. *Env. Sci. Technol.* **9**, 457–469.

Cautreels W. and van Cauwenberghe K. (1978) Experiments on the distribution of organic pollutants between airborne particulate matter and the corresponding gas phase. *Atmos. Env.* **12**, 1133–1141.

Cerniglia C.E. (1993) Biodegradation of polycyclic aromatic hydrocarbons. *Curr. Op. Biotechnol.* **4**, 331–338.

Challinor J.M. (1989) A pyrolysis-derivatisation-gas chromatography technique for the structural elucidationof some synthetic polymers. *J. Anal. Appl. Pyrol.* **16**, 323–333.

Chosson P., Lanau C., Connan J. and Dessort D. (1991) Biodegradation of refractory hydrocarbons biomarkers from petroleum under laboratory conditions. *Nature* **351**, 640–642.

Chriswell C.D., Ogawa I., Tschetter M.J. and Markuszewski R. (1988) Effect of hydrofluoric or hydrochloric acid pretreatment on the ultrasonic extraction of organic materials from fly ash for chromatographic analysis. *Env. Sci. Technol.* **22**, 1506–1508.

De la Torre M.A., Gomez-Alarcon G., Melgarejo P. and Saiz-Jimenez C. (1991) Fungi in weathered sandstone from Salamanca cathedral, Spain. *Sci. Total Env.* **107**, 159–168.

Del Monte M., Sabbioni C. and Vitori O. (1984) Urban stone sulphation and oil-fired carbonaceous particles. *Sci. Total Env.* **36**, 369–376.

Felske A., Rheims H., Wolterink A., Stackebrandt E. and Akkermans A.D.L. (1997) Ribosome analysis reveals prominent activity of an uncultured member of the class Actinobacteria in grassland soils. *Microbiology* **143**, 2983–2989.

Fobe B.O., Vleugels G.J., Roekens E.J., van Grieken R.E., Hermosin B., Ortega-Calvo J.J., Sanchez del Junco A. and Saiz-Jimenez C. (1995) Organic and inorganic compounds in limestone weathering crusts from cathedrals in southern and western Europe. *Env. Sci. Technol.* **29**, 1691–1701.

Gough M.A. and Rowland S.J. (1990) Characterization of unresolved complex mixtures of hydrocarbons in petroleum. *Nature* **344**, 648–650.

Gough M.A., Rhead M.M. and Rowland S.J. (1992) Biodegradation studies of unresolved complex mixtures of hydrocarbons: model UCM hydrocarbons and the aliphatic UCM in petroleum. *Org. Geochem.* **18**, 17–22.

Graedel T.E. (1978) *Chemical Compounds in the Atmosphere.* Academic Press, New York.

Griest W.H. and Tomkins B.A. (1984) Carbonaceous particles in coal combustion stack ash and their interaction with polycyclic aromatic hydrocarbons. *Sci. Total Env.* **36**, 209–214.

Grimalt J.O., Rosell A., Simo R., Saiz-Jimenez C. and Albaiges J. (1991) A source correlation study between the organic components present in the urban atmosphere and in gypsum crusts from old building surfaces. In *Organic Geochemistry. Advances and Applications in the Natural Environment*, ed. Manning D.A.C. Manchester University Press, Manchester, pp. 513–515.

Hawthorne S.B., Miller D.J., Barkley R.M. and Krieger M.S. (1988) Identification of methoxylated phenols as candidate tracers for atmospheric wood smoke pollution. *Env. Sci. Technol.* **22**, 1191–1196.

Heath D.J., Lewis C.A. and Rowland S.J. (1997) The use of high temperature gas chromatography to study the biodegradation of high molecular weight hydrocarbons. *Org. Geochem.* **26**, 769–785.

Hermosín B. (1995) *Efectos de la Contaminación Atmosférica sobre el Patrimonio Histórico. Deposición de Compuestos Orgánicos y Formación de Costras Negras Sulfatadas.* Ph.D. Thesis, University of Seville.

Karpovich-Tate N. and Rebrikova N.L. (1990) Microbial communities on damaged frescoes and building materials in the cathedral of the Nativity of the Virgin in the Pafnutii-Borovskii monasteriy, Russia. *Int. Biodeteriorat.* **27**, 281–296.

Kawamura K., Ng L.L. and Kaplan I.R. (1985) Determination of organic acids (C_1-C_{10}) in the atmosphere, motor exhausts, and engine oils. *Env. Sci. Technol.* **19**, 1082–1086.

Kawamura K. and Kaplan I.R. (1987) Motor exhaust emissions as a primary source of dicarboxylic acids in Los Angeles ambient air. *Env. Sci. Technol.* **21**, 105–110.

222   *C. Saiz-Jimenez*

Leahy J.G. and Colwell R.R. (1990) Microbial degradation of hydrocarbons in the environment. *Microb. Rev.* **54**, 305–315.

Lee M.L., Prado G.P., Howard J.B. and Hites R.A. (1977) Source identification of urban airborne polycyclic aromatic hydrocarbons by gas chromatography mass spectrometry and high resolution mass spectrometry. *Biomed. Mass Spectrom.* **4**, 182–186.

Leuenberger C., Czuczwa J., Heyerdahl E. and Giger W. (1988) Aliphatic and polycyclic aromatic hydrocarbons in urban rain, snow and fog. *Atmos. Env.* **22**, 695–705.

Lewis F.J., May E. and Bravery A.F. (1988) Metabolic activities of bacteria isolated from building stone and their relationship to stone decay. In *Biodeterioration*, eds. Houghton D.R., Smith R.N. and Eggins H.O.W. Elsevier Applied Science London, Vol. 7, pp. 107–112.

Lipsky S.R. and Duffy M.L. (1986) High temperature gas chromatography: the development of new aluminium clad flexible fused silica glass capillary columns coated with thermostable nonpolar phases: Part 1. *HRC-J. High Res. Chrom.* **9**, 376–381.

Machill S., Althaus K., Krumbein W.E. and Steger W.E. (1997) Identification of organic compounds extracted from black weathered surfaces of Saxonean sandstones, correlation with atmospheric input and rock inhabiting microflora. *Org. Geochem.* **27**, 79–97.

Mazurek M.A., Cass G.R. and Simoneit B.R.T. (1991) Biological input to visibility-reducing aerosol particles in the remote arid Southwestern United States. *Env. Sci. Technol.* **25**, 684–694.

Mulder M.M., van der Hage E.R.R. and Boon J.J. (1992) Analytical in source pyrolytic methylation electron impact mass spectrometry of phenolic acids in biological matrices. *Phytochem. Anal.* **3**, 165–172.

Muyzer G., De Waal E.C. and Uitterlinden A.G. (1993) Profiling of complex microbial populations by denaturing gradient gel electrophoresis analysis of polymerase chain reaction-amplified genes coding for 16S rRNA. *Appl. Env. Microbiol.* **59**, 695–700.

Muyzer G. and Ramsing N.B. (1995) Molecular methods to study the organization of microbial communities. *Water Sci. Technol.* **32**, 1–9.

Nord A.G. and Ericsson T. (1993) Chemical analysis of thin black layers on building stone. *Studies in Conservation* **38**, 25–35.

Nyns E.J., Auquiere J.P. and Wiaux A.L. (1968) Taxonomic value of the property of fungi to assimilate hydrocarbons. *Anton. Leeuw.* **34**, 441–457.

Ortega-Calvo J.J., Hernandez-Marine M. and Saiz-Jimenez C. (1991) Biodeterioration of buildings materials by cyanobacteria and algae. *Int. Biodeteriorat.* **28**, 167–187.

Ortega-Calvo J.J., Sanchez-Castillo P.M., Hernandez-Marine M. and Saiz-Jimenez C. (1993) Isolation and characterization of epilithic chlorophyta and cyanobacteria from two Spanish cathedrals (Salamanca and Toledo). *Nova Hedwigia* **57**, 239–253.

Ortega-Calvo J.J., Ariño X., Stal L.J. and Saiz-Jimenez C. (1994) Cyanobacterial sulfate accumulation from black crust of a historic building. *Geomicrobiol. J.* **12**, 15–22.

Ortega-Calvo J.J. and Saiz-Jimenez C. (1996) Polycyclic aromatic hydrocarbon-degrading bacteria in building stones. In *8th International Congress on Deterioration and Conservation of Stone,* ed. Riederer J. Vol. 2, pp. 681–685.

Ortega-Calvo J.J. and Saiz-Jimenez C. (1997) Microbial degradation of phenanthrene in two European cathedrals. *FEMS Microbiol. Ecol.* **22**, 95–101.

Pace N.R. (1996) New perspective on the natural microbial world: molecular microbial ecology. *ASM News* **62**, 463–470.

Petersen K., Kuroczkin J., Strzelczyk A.B. and Krumbein W.E. (1988) Distribution and effects of fungi on and in sandstones. In *Biodeterioration,* eds. Houghton D.R., Smith R.N. and Eggins H.O.W. Elsevier Applied Science London, Vol. 7, pp. 455–460.

Philp R.P. (1985) Biological markers in fossil fuel production. *Mass Spectrom. Rev.* **4**, 1–54.

Ramdahl T. (1983) Polycyclic aromatic ketones in environmental samples. *Env. Sci. Technol.* **17**, 666–670.

Robson J.N. and Rowland S.J. (1988) Biodegradation of highly branched isoprenoid hydrocarbons: a possible explanation of sedimentary abundance. *Org. Geochem.* **13**, 691–695.

Sagebiel J.C., Zielinska B., Walsh P.A., Chow J.C., Cadle S.H., Mulawa P.A., Knapp K.T., Zweidinger R.B., Snow R. (1997) PM-10 exhaust samples collected during IM-240 dynamometer test on in-service vehicles in Nevada. *Env. Sci. Technol.* **31**, 75–83.

Saiz-Jimenez C. and Bernier F. (1981) Gypsum crusts on building stones. A scanning electron microscopy study. In *6th Triennial Meeting ICOM Commitee for Conservation, Ottawa.* Paper 81/10/5, p. 9.

Saiz-Jimenez C. and Samson R.A. (1981) Microorganisms and environmental pollution as deteriorating agents of the frescoes of Santa Maria de la Rabida, Huelva, Spain. In *6th Triennial Meeting ICOM Committee for Conservation, Ottawa.* Paper 81/15/5, p. 14.

Saiz-Jimenez C. (1991) Characterization of organic compounds in weathered stones. In *Science, Technology and European Cultural Heritage,* eds. Baer N.S., Sabbioni C. and Sors A.I. Butterworth-Heinemann, Oxford, pp. 523–526.

Saiz-Jimenez C., Hermosin B., Ortega-Calvo J.J. and Gomez-Alarcon G. (1991) Applications of analytical pyrolysis to the study of cultural properties. *J. Anal. Appl. Pyrol.* **20**, 239–251.

Saiz-Jimenez C. (1993) Deposition of airborne organic pollutants on historic buildings. *Atmos. Env.* **27B**, 77–85.

Saiz-Jimenez C. (1994a) Modern concepts on the origin and structure of terrestrial humic substances: the alkylaromatic network approach. In *Humic Substances in the Global Environment and Implications on Human Health,* eds. Senesi N. and Miano T.M. Elsevier, Amsterdam, pp. 71–90.

Saiz-Jimenez C. (1994b) Production of alkylbenzenes and alkylnaphthalenes upon pyrolysis of unsaturated fatty acids. A model reaction to understand the origin of some pyrolysis products from humic substances? *Naturwissenschaften* **81**, 451–453.

Saiz-Jimenez C., Hermosin B., Ortega-Calvo J.J. (1994) Pyrolysis/methylation: a microanalytical method forinvestigating polar organic compounds in cultural properties. *Int. J. Env. Anal. Chem.* **56**, 63–71.

Saiz-Jimenez C. (1995) Deposition of anthropogenic compounds on monuments and their effect on airborne microorganisms. *Aerobiologia* **11**, 161–175.

Schroeder W.H. and Lane D.A. (1988) The fate of toxic airborne pollutants. *Env. Sci. Technol.* **22**, 240–246.

Sicre M.A., Marty J.C., Saliot A., Aparicio X., Grimalt J. and Albaiges J. (1987) Aliphatic and aromatic hydrocarbons in different sized aerosols over the Mediterranean sea: occurrence and origin. *Atmos. Env.* **21**, 2247–2259.

Simoneit B.R.T. and Mazurek M.A. (1981) Air pollution: the organic components. *CRC Crit. Rev. Env. Control* **11**, 219–276.

Simoneit B.R.T. (1984) Application of molecular marker analysis to reconcile sources of carbonaceous particulates in tropospheric aerosols. *Sci. Total Env.* **36**, 61–72.

Simoneit B.R.T. (1985) Application of molecular marker analysis to vehicular exhaust for source reconciliations. *Int. J. Env. Anal. Chem.* **22**, 203–233.

Simoneit B.R.T. (1986) Characterization of organic constituents in aerosols in relation to their origin and transport: a review. *Int. J. Env. Anal. Chem.* **23**, 207–237.

Simoneit B.R.T., Cox R.E. and Standley L.J. (1988) Organic matter of the troposphere-IV. Lipids in Harmattan aerosols of Nigeria. *Atmos. Env.* **22**, 983–1004.

Standley L.J. and Simoneit B.R.T. (1987) Characterization of extractable plant wax, resin, and thermally matured components in smoke particles from prescribed burns. *Env. Sci. Technol.* **21**, 163–169.

Sturaro A., Parvoli G. and Doretti L. (1993) Plane tree bark as a passive sampler of polycyclic aromatic hydrocarbons in an urban environment. *J. Chromatogr.* **643**, 435–438.

Tease B.E., Jurgens U.J.M., Golecki J.R., Heinrich U.R., Rippka R. and Weckesser J. (1991) Fine structural and chemical analyses on inner and outer sheath of the cyanobacterium *Gloeothece* sp PCC 6909. *Anton. Leeuw. Int. J. Gen. Mol. Microbiol.* **59**, 27–34.

van der Kaaden A., Haverkamp J., Boon J.J. and de Leeuw J.J. (1983) Analytical pyrolysis of carbohydrates. I. Chemical interpretation of matrix influences on pyrolysis-mass spectra of amylose using pyrolysis-gas chromatography-mass spectrometry. *J. Anal. Appl. Pyrol.* **5**, 199–220.

Warscheid T., Oelting M. and Krumbein W.E. (1991) Physico-chemical aspects of biodeterioration processes on rocks with special regard to organic pollutants. *Int. Biodeteriorat.* **28**, 37–48.

Williams P.T., Bartle K.D. and Andrews G.E. (1986) The relation between polycyclic aromatic compounds in diesel fuels and exhaust particulates. *Fuel* **65**, 1150–1158.

Wilson S.C. and Jones K.C. (1993) Bioremediation of soil contaminated with polynuclear aromatic hydrocarbons (PAHs): a review. *Env. Pollut.* **81**, 229–249.

Yu M.-L. and Hites R.A. (1981) Identification of organic compounds on diesel engine soot. *Anal. Chem.* **53**, 951–954.

CHAPTER 7

AIR POLLUTION DAMAGE TO METALS

J. Tidblad and V. Kucera

1. Introduction

It has been known for several centuries that air pollutants emitted by burning of fossil fuels have a serious impact on metals exposed in the built environment. In addition to the loss of mechanical strength and failure of protective coatings, the release of heavy metals to the environment is of concern. Calculations of the economic impact of sulphur pollutants on technical materials have indicated that the costs for corrosion damage to materials including metals are substantial.

There are many parameters that can influence the atmospheric corrosion of metals, the degradation process is a complex interplay between chemical, physical and biological parameters. The present treatment focuses on air pollutants and their effect on the built environment, i.e. the aspects specific to the urban situation. It is, however, important to recognise that atmospheric corrosion is a process that occurs even in the absence of pollutants and that the interplay between natural and anthropogenic factors determine to which extent urban conditions affects and accelerates the "natural" or background atmospheric corrosion.

The present treatment serves as an overview of the effects of acidifying air pollutants to metallic materials. It is divided in two main sections and three additional sections. Section 2 describes the general role of atmospheric pollutants in the corrosion process. Despite the general characteristics of the pollutants, every metal is unique and a selection of metals is described separately in Sec. 3. The quantification of effects, in particular the assessment of life times and calculation of corrosion costs deserves a special treatment given in Sec. 4. Finally, the trends in corrosion, in Europe as well as the developing countries are given in Sec. 5. This chapter is not a review. Instead of a complete list, references have been selected that are suitable for further reading including general references on atmospheric corrosion (Kucera and Mattson, 1987; Kucera and Fitz, 1995; Leygraf and Graedel, 2000) and are cited once at the most appropriate place.

2. Effects of the Environment

The most important pollutants acting as corrosive agents are sulphur and nitrogen compounds, including secondary pollutants and particulates. It has clearly been demonstrated that pollutants enhance the natural corrosion process for several metallic materials even if it has so far been difficult to quantify the direct contribution of individual pollutants other than sulphur dioxide (SO_2) in the weathering process.

Depending on the way pollutants are transported from the atmosphere to the corroding surface, two types of deposition processes are recognised in atmospheric corrosion — dry deposition and wet deposition. Wet deposition refers to precipitation whereas dry deposition refers to the remaining processes, including gas-phase deposition and particle deposition.

2.1. The Multi–Pollutant Situation

In many European countries, the concentration of SO_2 has decreased significantly in the last decades while the concentrations of nitrogen pollutants, ozone (O_3) and particulates remain at high levels. Nitrogen dioxide (NO_2) has also decreased significantly but not as much as SO_2. Traditionally and with right, SO_2 has been regarded as the most important corrosion stimulator. With the reduced levels, SO_2 is no longer the only main pollutant. In addition to the added effects of many individual pollutants, synergistic effects of i.a. SO_2 in combination with NO_2 or O_3 need to be considered. This complex pollution situation with effects of a number of important single-pollutants in combination with several synergistic effects, described in the following, is named the "multi-pollutant situation". The general acceptance of the new multi-pollutant situation was recently manifested in the multi-effect, multi-pollutant Protocol to Abate Acidification, Eutrofication and Ground-level Ozone adapted in 2000 in Gothenburg within the Convention on Long-Range Transboundary Air Pollution of the United Nations Economic Commission for Europe.

2.2. Climate

Climate has both indirect and direct effects on atmospheric corrosion of metals. Indirect effects include for example the less mixing and higher concentration levels of pollutants present when there is a temperature inversion. The present treatment only describes direct effects. The dominating part of the research activities has been performed in Europe and North America and therefore the effect of acidifying pollutants has mainly been investigated in regions situated in temperate climate.

Atmospheric corrosion is an electrochemical process and proceeds only when the surface is sufficiently wet, the degradation process may be considered as discontinuous. This is illustrated in Fig. 1 where it is seen that the instantaneous corrosion rate varies strongly with time, several orders of magnitude. The time of wetness (TOW) is a commonly used concept for metals that refers to the time when corrosion occurs, i.e. when a moisture layer is present. This concept is useful when describing the degradation process and when classifying different climatic regions from a corrosion point of view but is difficult to calculate from readily available meteorological data. Therefore, recently developed dose-response functions include annual averages of temperature and relative humidity, which is more easily available.

Dry deposition of pollutants is greatly influenced by temperature (T) and relative humidity (Rh). T and Rh are important since they are the main factors that determine the thickness of the moisture layer in absence of rainfall. The corrosion rate increases with Rh, starting

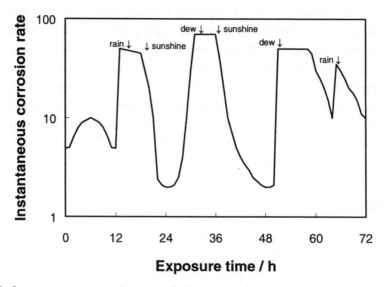

Fig. 1. Instantaneous corrosion rate during a few days of exposure (adapted from Kucera and Mattson, 1987).

from a critical humidity value, where the adsorbed water layer begins to act as an electrolyte. At low temperatures, the corrosion rate increases with increasing annual temperature since the fraction of time above the water freezing point also increases. At higher temperatures, the corrosion in non-marine areas decrease with increasing temperature, which has been attributed to periods with a surface temperature above the ambient temperature, partly related to sun radiation. An elevated surface temperature leads to a faster evaporation of moisture after rain or condensation periods and to a decrease of the thickness of the adsorbed water layer and, consequently, to a decrease of the time when the metal surface is wet. Figure 2 show an example of the temperature effect for zinc based on a dose-response function presented in Sec. 3 (equation 2).

2.3. Gaseous Pollutants

Trace gases of importance for atmospheric corrosion include O_3, H_2O_2, SO_2, H_2S, COS, NO_2, HNO_3, NH_3, HCl, Cl_2, $HCHO$ and $HCOOH$.

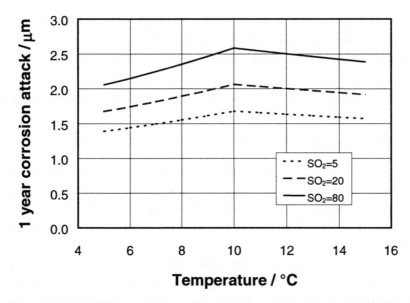

Fig. 2. Corrosion attack of unsheltered zinc vs. temperature calculated from equation (2) with Rh = 75% and a constant wet deposition term, 0.029Rain[H⁺], corresponding to 0.6 μm. Curves are shown for the SO_2 concentrations 5, 20 and 80 μg/m³.

The present treatment focuses on SO_2, NO_2, O_3 and HNO_3 — pollutants that are especially important in urban environments. Recently, carboxylic acids have also earned some attention due to the increased use of methanol and ethanol as alternative fuels. Especially formic acid is a strong acid, which makes the potential effect high. However, data is scarce both on estimated future pollutant levels and atmospheric corrosion effects.

SO_2

In the field of atmospheric corrosion, SO_2 is the single most investigated gaseous pollutant. Its detrimental effect on materials in general has long since been an indisputable fact. Sulphur compounds emitted to the atmosphere were suspected to be the cause of acidification even towards the end of the 19th century. Systematic laboratory exposures in the 1930s demonstrated the corrosive effect of SO_2 on metals. This was also later proven by field exposures and SO_2 was for a long time considered to be the main corrosive pollutant. Even with decreased levels, it is important to remember that SO_2 is still one of the main actors in the new multi-pollutant situation, although not as dominant as in the past.

SO_2 is dissolved in the moisture layer forming sulphite and, after oxidation, sulphate:

$$SO_2(g) + H_2O \rightarrow H^+ + HSO_3^- \rightarrow ... \rightarrow 2H^+ + SO_4^{2-} \qquad (1)$$

This process results in an acidification of the moisture layer, which enhances the corrosion process. Sulphate is also frequently present in corrosion products. The SO_2 deposition rate depends mostly on the material, it is often higher for sensitive materials, and varies between 0.01 and 2 cm/s.

Figure 3 shows a plot of one-year corrosion values of carbon steel, with selected cities in Europe highlighted. Worth noting is that the SO_2 pollution levels and corrosion rates today in most European cities are significantly lower (see Sec. 5). Quantitative relationships, i.e. dose-response functions, expressing the corrosion attack as a function of SO_2 and other parameters exist for several metals (see Sec. 3).

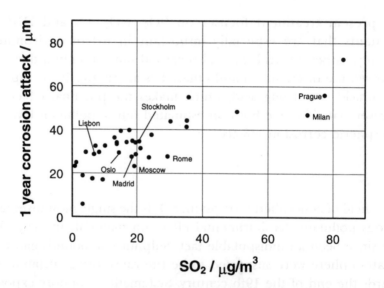

Fig. 3. Corrosion attack of unsheltered carbon steel exposed in the UN ECE exposure programme during the period September 1987 to August 1988 vs. SO_2 concentration.

Synergistic effects of NO₂ and O₃ in combination with SO₂

In contrast to SO_2, the effects of NO_2, and especially O_3, are not well documented (Arroyave and Morcillo, 1995; Tidblad and Kucera, 1996). Field exposure programmes have failed to detect effects attributable to NO_2 or O_3 with the exception of O_3 and copper (see Sec. 3.4). Laboratory exposures of several metals including zinc, copper and nickel have led to the following picture regarding synergistic effects.

The term "synergistic effect" means that the corrosion attack for a material exposed to a mixture of gases is greater than the sum of effects from individual single-gas exposures. Significant synergistic effects have been found in the laboratory for both O_3 and NO_2, however, the relative magnitude of the effects depends on the metal. For both O_3 and NO_2, the synergistic effect is attributed to the promotion of equation (1), i.e. the oxidation of S(IV) to S(VI). The reaction between SO_2 and O_3 is most likely stoichiometric while the reaction between SO_2 and NO_2 needs further investigation. With few exceptions,

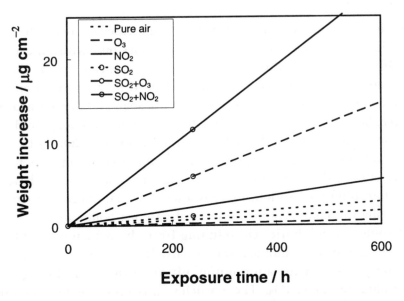

Fig. 4. Weight increase of nickel exposed in air at 70% Rh based on results after 240 hours and assuming a linear time dependence. The influence of additions of O_3 (100 ppb), NO_2 (100 ppb) and SO_2 (90 ppb) including combinations are shown.

there seem to be synergistic effects. The synergistic effects increase with relative humidity for most metals. However, at high relative humidities, aluminium and iron show no SO_2 + NO_2 synergism. When comparison is possible, the synergistic effect of SO_2 + O_3 is stronger than SO_2 + NO_2, except for nickel, which is illustrated in Fig. 4.

HNO_3

Nitric acid (HNO_3) is a secondary pollutant formed by the oxidation of NO_2. It can reach appreciable concentration levels in urban areas — reports are usually in the interval 1–7 $\mu g/m^3$ but in some cases the concentration can be above 10 $\mu g/m^3$. HNO_3 is a strong acid with a high deposition velocity that is almost independent of the humidity, which makes it relatively more harmful for dry and warm climates. The magnitude of its effect relative to other pollutants such as SO_2 is so far not investigated for metals.

2.4. Particles

Particles normally increase but may in some cases decrease the deterioration rates of metals. The effect of particles on corrosion can also be either direct or indirect. Anthropogenic particles can be divided into primary and secondary. The primary particles are directly emitted from combustion, have a relatively short life span and deposit near the source. Secondary particles are smaller, less than 2 μm, long-lived, and are the result of chemical reactions amongst other pollutants i.a. SO_2, NO_2, volatile organic compounds and ammonia. Compared to the many effects, described in the following, that particles can have on the atmospheric corrosion, there is a lack of quantitative assessments especially the comparison to effects of gaseous pollutants.

Droplets of sulphuric acid resulting from the oxidation of SO_2 can be neutralised by ammonia-forming particles of ammonium sulphate. Similarly, particulate ammonium nitrate can form through the neutralisation of HNO_3. The particles thus enhance the corrosion by providing corrosion stimulators, especially sulphate is frequently found in corrosion products. Also, when water is adsorbed and then evaporates, gaseous ammonia is released to the atmosphere resulting in an acidification of the aqueous layer that may increase the corrosion rate.

Particles may in some cases also decrease the corrosion rate if they are basic by neutralising the surface water film formed on the degraded material.

Salt containing particles play an important role in atmospheric corrosion and this is related to their ability to increase the time of wetness. They are hygroscopic and starts to absorb water when the relative humidity exceeds a critical level, determined ideally by the equilibrium properties of a saturated solution of the salt. In practice, however, the particles deposited on the surface are a mixture of different compounds resulting in a difficulty to define a critical relative humidity. The ionic content of the particles prolongs the time of wetness also by reducing the freezing point of the adsorbed water below 0°C.

2.5. Wet Deposition

SO_2 and NO_2 both contribute to the increased acidity of wet deposition, in addition to their direct effects. The effect of wet deposition is relatively well quantified for many metals. Dose-response functions based on field exposures that includes the effect of acid rain exists for several metals. On the other hand, similar to particles, the effect of wet deposition can be either detrimental or beneficial, depending on the conditions. In principle, the wet deposition can have two effects on the corrosion process. First, it transports chemically active compounds present in the rain to the surface, thereby increasing the corrosivity of the moisture layer. Second, it washes away compounds previously deposited on the surface, with the opposite effect. Thus, for a specific material and environment the choosing of a sheltered exposure condition rather than unsheltered may, or may not, increase the corrosion rate as is illustrated in Table 1, where it is shown that aluminium has a higher corrosion rate in sheltered position in contrast to the other metals.

The effect of wet deposition has traditionally been regarded as minor compared to the effect of dry deposition. However, as is illustrated in Fig. 5, the effect depends strongly on pH and amount

Table 1. Corrosion attack in mm after eight years of exposure in the UN ECE exposure programme in unsheltered and sheltered positions. Intervals are shown corresponding to lower and upper quartiles (25- and 75-percentiles).

Metal	Unsheltered	Sheltered
Weathering steel	37–59	27–54
Zinc	9.2–18	2.2–3.4
Copper	3.4–5.3	1.8–3.5
Bronze	2.9–5.4	1.6–3.0
Aluminium	1.0–1.7	1.1–3.1

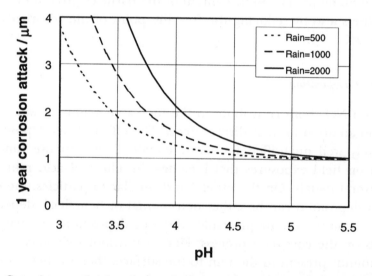

Fig. 5. Corrosion attack of unsheltered copper vs. pH of precipitation calculated from equation (3) and a constant dry deposition term, $0.0027[SO_2]^{0.32}[O_3]^{0.79}Rh \cdot e^{f(T)}$, corresponding to 1.0 μm. Curves are shown for the precipitation (Rain) values of 500, 1000 and 2000 mm.

of precipitation. At about pH 4.5 and lower, the effect cannot be neglected.

In addition to compounds deposited on the surface, corroded metal ions are removed by the rain-wash. This release of metals to the environment is an environmental concern due to the potential effect that the heavy metals can have on ecosystems. However, when assessing the impact it is important to recognise first that the metal release is significantly lower than the total corrosion rate, which also includes metal ions left on the corroded surface in the corrosion products, and second that only part of the released metal is biologically active. Also, the metal runoff depends on the amount of precipitation reaching the surface, which means that the amount of precipitation as well as the inclination of the surface is of outmost importance when assessing run-off values.

3. Effects on Metallic Materials

In its unprotected form, examples of use for carbon steel include mounting and fittings and forged objects. Cast iron has been used in fences and gates. The main use is, however, in its protected form, which makes the economic importance of unprotected steel relative low. Zinc is commonly used as a pigment in zinc-rich coatings and in hot dip galvanising. An alternative use is rolled zinc sheet for roofing and rainwater drains. In buildings, aluminium is mainly used as a roof and facade material. By 1897, the San Gioacchino church in Rome was already covered with aluminium. The main uses of copper include roofs and window ledges. Historic and cultural monuments made of cast bronze are common especially in statues.

3.1. Effects on Metals in General

A typical metal exposed to dry air spontaneously forms a metal oxide of thickness 1 to 5 nm. Thus, it is the nature of the thin metal oxide, whether it is passive or active, stable or unstable, protective or not protective, that governs the atmospheric corrosion behaviour of that metal, rather than the properties of the base metal. The relative humidity when adsorbed water have properties equal to those of bulk water varies from 50% to 90% depending on the metal and this interval coincides with that of the "critical relative humidity" above which corrosion rates starts to be significant (see Sec. 2.2). Atmospheric corrosion of metallic materials is an electrochemical process in which the overall reaction can be divided into an anodic and a cathodic reaction. Under most atmospheric conditions the anode reaction, i.e. metal dissolution, is the rate-limiting step. With time, a film of corrosion products form that usually hinders the transport of ions, which lowers the anodic reaction rate and, hence, the atmospheric corrosion rate. The initially formed corrosion products are oxides, hydroxides, and even oxyhydroxides. During prolonged exposure, however, a variety of corrosion products may form including compounds originating from the atmosphere, i.e. sulphate, nitrate and chloride. The atmospheric corrosion of metals is a complex process and the extent of deterioration as well as the mechanism varies considerably depending on the metal.

For several metals that are frequently used in buildings, including objects of cultural heritage, dose-response functions on damage have been obtained. The functions are of outmost importance for development of systems for classification of corrosivity of environments, for mapping of areas with increased risk of corrosion, and for calculation of cost of damage caused by deterioration of materials. A dose-response function links the dose of pollution, measured in ambient concentration and/or deposition, to the rate of material corrosion.

Metals dealt with specifically in this chapter include ferrous metals (Graedel and Frankenthal, 1990), zinc and galvanised steel (Graedel, 1989a; Odnevall, 1994; He, 2000), copper (He, 2000; Graedel *et al.*, 1987; Strandberg, 1997), bronze (Strandberg, 1997) and aluminium (Graedel, 1989b).

3.2. Ferrous Metals

The corrosion rate of carbon steel is generally high, about 20–60 mm after one year of exposure in non-marine areas, the value to a great extent depending on the SO_2 concentration (see Fig. 3). Weathering steel behaves in a different way than an ordinary carbon steel at prolonged exposure and the corrosion rate is stabilised at fairly low levels since a protective rust layer forms during the first two to four years of exposure. The decrease of the SO_2 influence with time for unsheltered weathering steel is a known fact for this material, which is illustrated in Fig. 6. The practical implication is that weathering steels are a very useful material for unsheltered exposure in SO_2

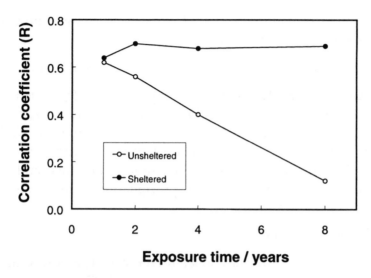

Fig. 6. Correlation coefficient, R, vs. exposure time. R is calculated by comparing SO_2 concentration with corrosion attack of weathering steel exposed in unsheltered or sheltered position in the UN ECE exposure programme.

polluted atmospheres. They have, however, similar corrosion rates as carbon steel in sheltered positions as the creation of a protective rust layer needs regular washing and drying at rain events.

3.3. Zinc and Galvanised Steel

As for many other metals, SO_2 is the most important corrosion stimulator. In addition, wet deposition may result in dissolution of protective corrosion products, which leads to an increase of the deterioration rate. In marine areas, the deposition of chlorides is also a determining factor. The corrosion attack after one year of exposure is about 1–4 µm in non-marine areas.

Zinc is the only one of the common engineering metals for which a carbonate plays an important role in atmospheric corrosion. After an initial formation of zinc hydroxycarbonate, a zinc hydroxysulfate is formed in the urban environment. Deposition of chlorides may eventually also lead to the formation of a zinc chlorohydroxysulfate.

Zinc were exposed in the UN/ECE materials exposure programme and a dose-response function

$$ML = 1.4[SO_2]^{0.22}e^{0.018Rh}e^{f(T)}t^{0.85} + 0.029Rain[H^+]t \qquad (2)$$

was obtained where ML is the mass loss in g/m^2, $[SO_2]$ is the SO_2 concentration in $\mu g/m^3$, Rh is the relative humidity in %, $f(T)$ is a function of temperature in °C, equal to $0.062(T-10)$ when T is lower than 10°C and $-0.021(T-10)$ when T is higher than 10°C, t is the time in years, Rain is the amount of precipitation in mm/year and $[H^+]$ is the hydrogen ion concentration in precipitation in mg/l.

3.4. Copper and Bronze

The corrosion chemistry of copper is complex. The corrosion products consist of a mixture of corrosion products where cuprite (Cu_2O) and basic sulphates, including anthlerite and brochantite constitute the dominating part on surfaces exposed long-term in the urban environment. The formation of cuprite is especially difficult to quantify in terms of environmental parameters and the formation rate can initially be very high especially in rural areas, which have high O_3 concentrations. Copper is the only metal where it has been proven by field exposures that O_3 as well as SO_2 have an important role. After longer exposure times, the relative importance of cuprite decreases in favour of sulphate and the degradation characteristics become similar to those of bronze. The corrosion attack on copper and bronze after one year of exposure is about the same, 1–2 mm, or slightly lower compared to zinc.

Copper is one of the metals that were exposed in the UN ECE materials programme and a dose-response function

$$ML = 0.0027[SO_2]^{0.32}[O_3]^{0.79}Rh \cdot e^{f(T)}t^{0.78} + 0.050Rain[H^+]t^{0.89}$$

$$(3)$$

was obtained where ML is the mass loss in g/m^2, $[SO_2]$ is the SO_2 concentration in $\mu g/m^3$, $[O_3]$ is the O_3 concentration in $\mu g/m^3$, Rh is the relative humidity in %, $f(T)$ is a function of temperature in °C, equal to $0.083(T-10)$ when T is lower than 10°C and $-0.032(T-10)$ when T is higher than 10°C, t is the time in years, Rain is the amount of precipitation in mm/year and [H$^+$] is the hydrogen ion concentration in precipitation in mg/l.

Bronze used for sculptures usually contain copper, 1–10% tin, 1–10% zinc and 0–6% lead. Bronze is also one of the metals that were quantified in the UN/ECE materials programme as:

$$ML = 0.026[SO_2]^{0.44}Rh \cdot e^{f(T)}t^{0.86} + (0.029Rain[H^+]$$
$$+ 0.00043Rain[Cl^-])t^{0.76} \tag{4}$$

where ML is the mass loss in g/m^2, [SO$_2$] is the SO$_2$ concentration in μg/m^3, Rh is the relative humidity in %, f(T) is a function of temperature in °C, equal to 0.060(T–11) when T is lower than 11°C and −0.067(T–11) when T is higher than 11°C, t is the time in years, Rain is the amount of precipitation in mm/year, [H$^+$] is the hydrogen ion concentration in precipitation in mg/l and [Cl$^-$] is the chloride ion concentration in precipitation in mg/l. Figure 7 shows observed versus predicted values based on the dose-response function.

3.5. Aluminium

The corrosion attack on aluminium is generally lower compared to the other metals, about 0.1 to 0.2 μm after one year of exposure in non-marine areas. The low corrosion rate is due to a protective

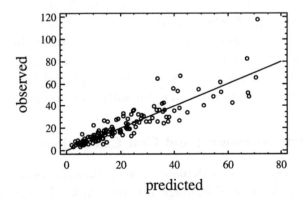

Fig. 7. Observed vs. predicted values for mass loss of unsheltered bronze using equation (4).

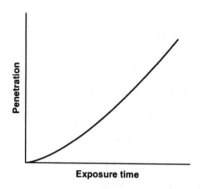

Fig. 8. Sketch of corrosion-time curve for rain-sheltered aluminium in a heavy industrial atmosphere (adapted from Kucera and Mattson, 1987)

aluminium oxide film that forms initially, γ-Al_2O_3. The film may grow and also be transformed into other compounds, such as hydrated aluminium oxides, hydroxides and sulphates, during the atmospheric corrosion process. As for many other metals, the corrosion rate of aluminium is affected by SO_2 to a great extent but aluminium is particularly sensitive to chlorides, which may result in increased pitting formation. Sulphate species dominate in corrosion products formed in urban areas.

Looking at Table 1, aluminium differs from the other metals by having higher corrosion attack values in sheltered position compared to unsheltered. A possible time development in sheltered position is illustrated in Fig. 8 where particles may accelerate corrosion by adsorbing moisture and acidifying compounds from the atmosphere thus producing an acid medium on the surface. Due to the non-existing cleaning by rain, these conditions prevail for long periods and the corrosion rate may increase with time initially.

4. Life Time Assessment and Cost Calculations

In order to be able to calculate costs, a damage function needs to be obtained. A physical damage function links the rate of material corrosion due to the pollution exposure given by the dose-response function to the time of replacement or maintenance of the material.

Performance requirements determine the point at which replacement or maintenance is considered to become necessary. If the performance requirements can be described in terms of a critical degradation level it is possible to transform a dose-response function into a damage function (Kucera *et al.*, 1996; Tidblad *et al.*, 1999).

For assessment of direct costs of corrosion damage caused by air pollutants, a model has been developed and used first in three cities in Europe: Stockholm, Sarpsborg and Prague, and subsequently also for a rough estimation for whole Europe (Kucera *et al.*, 1993; UN ECE Workshop, 1997). The estimated total cost savings were 9.5×10^9 \$/year for SO_2-induced corrosion in Europe resulting from the implementation of the second Sulphur Protocol within UN ECE. The model is shown in Fig. 9. The estimated economic damage can be calculated according to the equation

$$K_a = K \cdot S \cdot (Lp^{-1} - Lc^{-1}) = K \cdot S \cdot f \qquad (5)$$

where K_a is the additional cost for maintenance/replacement, K is the cost per surface area of material, S is the surface area of material, L_p is the maintenance interval (life time) in polluted areas and L_c is the maintenance interval in clean areas. The maintenance frequency, f, is the fraction of material that is subject to actions each year as a result of increased pollution. Equation (5) has to be applied for each material individually. Furthermore, each component in equation (5) can, in principle, have a spatial dependence, i.e., the cost, exposed areas and lifetimes may each depend on the geographical location. In order to make a cost estimation, each block in Fig. 9 needs to be addressed. Depending on the available data the treatment of individual boxes will be more or less comprehensive.

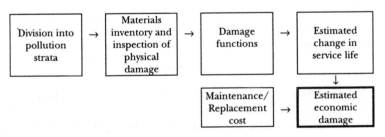

Fig. 9. General approach for assessing cost of corrosion damage (Kucera *et al.*, 1993).

5. Corrosion Trends

Trend effects were assessed as part of a UN ECE exposure programme on materials (ICP Materials) and the following is brief summary of the findings (Tidblad *et al.*, 1999). The aim of the trend exposure was to elucidate the environmental effects of pollutant reductions achieved under the Convention on Long-Range Transboundary Air Pollution and identify extraordinary environmental changes that result in unpredicted materials damage. The trend exposure consisted of re-peated one-year exposure of steel and zinc on the 39 test sites during the period 1987–1995.

Figure 10 shows an overview of the trends during the period 1987 to 1994. There are comparatively few trend exposures of steel and zinc while pollution data are available each year. In order to present the results on a common scale, all values are compared to the initial value,

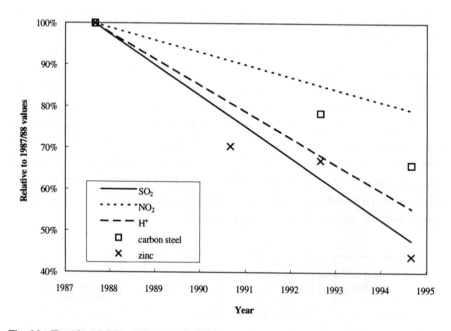

Fig. 10. Trends of SO_2, NO_2, acidity (H^+), and corrosion of unsheltered unalloyed carbon steel and zinc. All values are expressed relative to the initial (1987/1988) value. For environmental data the average trend during the eight-year period is indicated instead of the individual annual averages.

corresponding to the 1987/88 period (100%). Of the environmental parameters measured in the network of urban and rural sites of ICP Materials, only SO_2, NO_2 and H^+ exhibit trends. All of these are decreasing with SO_2 having the strongest trend and NO_2 the weakest. For O_3, no specific trends were observed. The decreasing trend in the concentration of acidifying air pollutants has resulted in decreasing corrosion rates of the exposed materials. Both carbon steel and zinc show decrease of corrosion rate in unsheltered as well as in sheltered positions.

Figure 11 shows a summary of the analysis of unsheltered and sheltered carbon steel and zinc separating the relative contributions of dry and wet deposition assuming that effects of dry deposition is attributed to SO_2 while effects of wet deposition is attributed to H^+ deposition.

SO_2 is the largest single contributing factor to the decreasing corrosion trends. The decreasing H^+ in precipitation is a contributing factor, its effect is, however, much smaller than that of dry deposition. The decrease in corrosivity is generally larger than expected from

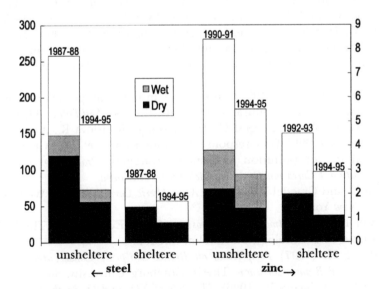

Fig. 11. Average mass loss of unalloyed carbon steel and zinc in g/m² for unsheltered and sheltered positions. Each bar consists of parts corresponding to dry and wet deposition as well as a part not directly related to SO_2 and H^+ load.

the drop of SO_2 and H^+ concentrations. This part cannot be directly related to a specific pollutant and reflects the multi-pollutant character of the process of material degradation.

References

Arroyave C. and Morcillo M. (1995) The effect of nitrogen oxides in atmospheric corrosion of metals. *Corros. Sci.* **37**(2), 293–305.
Graedel T.E., Nassau K. and Franey P. (1987) Copper patinas formed in the atmosphere I. Introduction. *Corros. Sci.* **27**(7), 639–657.
Graedel T.E. (1989a) Corrosion mechanisms for zinc exposed to the atmosphere. *J. Electrochem. Soc.* **136**(4), 193C–203C.
Graedel T.E. (1989b) Corrosion mechanisms for aluminum exposed to the atmosphere. *J. Electrochem. Soc.* **136**(4), 204C–212C.
Graedel T.E. and Frankenthal R.P. (1990) Corrosion mechanisms for iron and low alloy steels exposed to the atmosphere. *J. Electrochem. Soc.* **137**(8), 2385–2394.
He W. (2000) *Corrosion Rates and Runoff Rates of Copper and Zinc as Roofing Materials.* Licentiate Thesis, Royal Institute of Technology, Stockholm, Sweden.
Kucera V. and Mattson E. (1987) Atmospheric corrosion. In *Corrosion Mechanisms.*, ed. Mansfeld F. Marcel Dekker, New York, pp. 211–284.
Kucera V., Henriksen J., Knotkova D. and Sjöström C. (1993) Model for calculations of corrosion cost caused by air pollution and its application in three cities. *10th European Corrosion Congress, Barcelona,* Paper No. 084, p. 12.
Kucera V. and Fitz S. (1995) Direct and indirect air pollution effects on materials including cultural monuments. *Water Air Soil Pollut.* **85**, 153–165.
Kucera V., Tidblad J., Leygraf C., Henriksen J., Kreislova K., Ashall G. and Stöckle B. (1996) Dose-response relations from the UN ECE project as a tool for air pollution abatement strategies. *Proceedings of the 13th International Corrosion Congress, Melbourne, Australia.*
Leygraf C. and Graedel T.E. (2000) *Atmospheric Corrosion.* John Wiley & Sons Inc., New York.
Odnevall I. (1994) *Atmospheric Corrosion of Field Exposed Zinc.* Thesis, Royal Institute of Technology, Stockholm, Sweden.
Strandberg H. (1997) *Perspectives on Bronze Sculpture Conservation. Modeling Copper and Bronze Corrosion.* Thesis, Göteborg University, Sweden.
Tidblad J. and Kucera V. (1996) *The Role of NO_x and O_3 in the Corrosion and Degradation of Materials,* Swedish Corrosion Institute, Stockholm, Sweden.
Tidblad J., Mikhailov A.A. and Kucera V. (1999) Allocation of corrosion trends to dry and wet deposition based on ICP materials data. In *Quantification of*

Effects of Air Pollutants on Materials. Proceedings of the UN ECE Workshop on Quantification of Effects of Air Pollutants on Materials. Berlin, May 25–28, 1998., ed. Fitz S., Texte 24/99. Umweltbundesamt Berlin.

Tidblad J., Kucera V., Mikhailov A.A., Henriksen J., Kreislova K., Yates T., Stöckle B. and Schreiner M. (1999) Final dose-response functions and trend analysis from the UN ECE project on the effects of acid deposition. *Proceedings of the 14th International Corrosion Congress, Cape Town, South Africa.*

UN ECE Workshop (1997) Economic evaluation of air pollution damage to materials. In *Proceedings of the UN ECE Workshop on Economic Evaluation of Air Pollution Abatement and Damage to Buildings including Cultural Heritage,* eds. Kucera V., Pearce D. and Brodin Y.W., Report 4761. Swedish Environmental Protection Agency, Stockholm, Sweden.

CHAPTER 8

THE EFFECT OF AIR POLLUTION ON GLASS

J. Leissner

1. Introduction

Glass products that are used outdoors undergo degradation which is of considerable practical importance. Loss in clarity or transmission of ordinary window glass or of stained glass windows is one example. Lighting fixtures that decline in output as a result of weathering mean lost energy. Solar reflectors and collectors that employ glass components must continue to perform at a high level of efficiency if they are to be practical energy collecting systems, which justify the enormous capital outlay in construction. Glass panels that lose their gloss are no longer aesthetically appealing; worse yet if they fail mechanically because of environmentally induced changes.

Historical stained glass windows are among the most important items of cultural heritage of Europe. However, mediaeval stained glasses and their paintings suffer heavy deterioration and are often completely destroyed by environmental corrosive influences. Many parameters such as humidity, temperature changes, inadequate conservation treatments, microorganisms, gaseous pollutants and synergetic interactions contribute to the deterioration of these objects.

There is no doubt that this is a corrosion process of the glass. The main reason lies in the much lower durability of the chemical composition of historic stained glasses compared to contemporary glasses.

2. The Corrosion Process

Glass can be chemically attacked by environmental agents or airborne contaminants may adhere to the glass surface. The deterioration of stained glass windows for example takes place, since manufacture, both indoors and outdoors although when storing the glasses in a museum environment the corrosion rate can be slowed down — a rather unsatisfactory protection measure because stained glass windows gain their sphere of harmony only in their original architectural setting. Additionally major problems occur with the paint layers which mainly consist of a lead silicate glass. These are affected by weathering in the same manner as the stained glass, but the weathering resistance can be better or worse than the glass substrate. Even when the paint is relatively stable, the underlying corroding glass will loosen the paint layers into flakes. Sometimes stained glass is very durable, but the black stain from enhanced corrosion results in a complete loss of the painting. A negative image of the painting is typically observed.

In general, weathering is the interaction of a glass surface with chemical agents in the atmosphere which can lead to three different effects on the glass:

- loss of transparency and gloss
- loss of paint layers
- loss of glass material

The reasons and basic reactions of corrosion will be explained in the following sections.

2.1. Chemical Composition and Structure of Glass

Glass is normally regarded as a hard, brittle and transparent material. The properties of glass however are mainly due to its chemical composition and manufacturing process. The main component of glass is silica (sand), the network former. In order to make a workable glass various compounds, such as network modifiers (alkali and earth alkali compounds) and colourants (metal oxides) are added and result in a marked effect upon the structure and properties of the resulting material (Scholze, 1988; Vogel, 1979).

Glass which was made during the Middle Ages usually has mainly a low content of silica and a high content of potassium and calcium plus some other minor compounds, often phosphates, magnesia and alumina, etc. (Newton and Davison, 1989). The higher the content of silica, the higher the chemical durability and temperature required to melt the glass. The basic molecular structural units for a glass are the silicon atoms which are surrounded tetrahedrically by oxygen (Scholze, 1988). These SiO_4^- tetrahedrons form a three-dimensional irregular non-crystalline network (Fig. 1).

By adding the network modifiers (calcium and potassium), this network is loosened because oxygen bridges are opened and new ionic bonds are formed (Fig. 2). This results in a higher corrosion

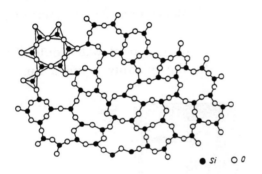

Fig. 1. Two-dimensional network of pure silica glass.

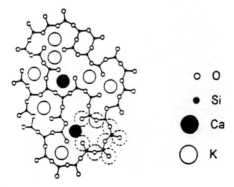

Fig. 2. Two-dimensional network of a mediaeval glass.

sensitivity. Medieval glass is a multi-component material of various oxides and its composition can vary significantly from glass to glass. This can be seen in stained glass windows where relatively good preserved glasses are side-by-side with heavily corroded ones.

2.2. Principle Corrosion Reactions

There are many parameters which contribute to the deterioration of glass, therefore the complex corrosion process is not completely understood even today. An enormous number of publications have dealt with its investigation. The most prevalent corrosive agent is water and the existence of water is in nearly all cases necessary to initiate corrosion. In the literature, three basic reaction paths for corrosion are described: corrosion under neutral, acidic or alkaline conditions (Scholze, 1988; Fitz, 1991; Marschner, 1985; Sanders and Hench, 1973; Müller, 1992).

The first step is the adsorption of water onto the glass surface (as illustrated in Fig. 3), the quantity being a function of relative

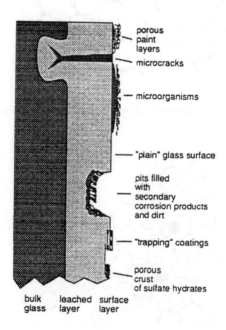

Fig. 3. Schematic representation of the weathering process.

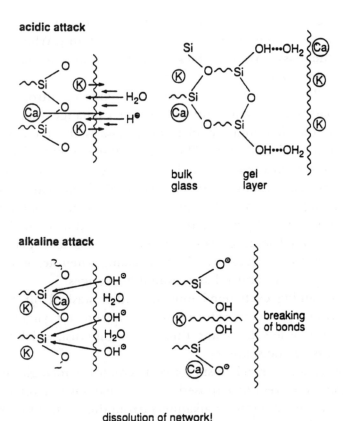

Fig. 4. Scheme for the main glass corrosion reactions.

humidity and glass type. Water then reacts with the glass surface through a sequence of steps involving an exchange of ions (leaching) and etching (first order reaction). In leaching, the network-modifier ions such as potassium and calcium are exchanged against protons deriving from acidic compounds, typically acetic acid or sulfuric acid. This process leads to the formation of a gel-layer which plays an important role for the preservation of stained glass windows. In Fig. 4, the reactions are seen schematically.

The presence of water is necessary for glass to react with compounds from its environment. Under normal conditions, it can be assumed that there is always a certain amount of water at the glass surface. The quantity of water depends mainly on the surface properties, the temperature and the relative humidity. When this water layer on the glass surface is thick enough, it is able to dissolve gases from the atmosphere.

Acid pollutant gases such as sulphur dioxide or nitrogen dissolve on the glass surface and initiate the ion exchange reactions. The reason for the deterioration of stained glass is the increased levels of environmental pollution. The higher the content of alkali and earth alkali ions in the glass, the larger the number of sites which are available for ion exchange reactions. These alkali and earth alkali ions can then migrate towards the surface resulting in a voluminous gel layer of silica. This layer can reach about 100 μm in thickness, for example, in mediaeval glasses. Further reactions take place in this gel layer, which can cause cracks and flaking when accompanied by drastic temperature and relative humidity changes.

The remaining exchanged ions are either washed off by rain or form corrosion products, as for example, gypsum and syngenite resulting in a more or less dense and thick corrosion crust, which will result in a decreasing transparency.

There is another mechanism for corrosion for stained glasses, which have flaws and are not exposed to rain that washes off corrosion products. Under neutral or alkaline surroundings, dissolution of the glass network can occur which means that the strong silicon oxygen bonds are broken; the silica network is totally disrupted and the glass structure is destroyed. In the initial stage, this will be indicated by a cloudy surface that eventually becomes roughened and covered with tenacious reaction products. Hollow glasses of the 17th and 18th century are particularly affected by this type of corrosion. These have a very high potassium content and a very low calcium content, and are called weeping glasses (Fitz, 1991).

$$Si-O^-K^+ + H_2O \rightarrow Si-OH + K^+ + OH^-$$
$$Si-O-Si + OH^- \rightarrow Si-OH + Si-O^-$$

2.3. Different Stages of Surface Corrosion

Surface corrosion can be detected in different stages in stained glass windows. Nearly all medieval glasses are affected by the type of corrosion described above, although sometimes the glass seems to be unaltered. However, if the glass is put under a microscope the changes can be seen clearly. Figures 5(a) to (e) illustrate main alterations of stained glasses and show the progressing surface corrosion as seen under the electron microscope. The glasses are model glasses and exhibit identical corrosion behaviour and similar chemical composition to medieval stained glasses.

Fig. 5(a). Firepolished new glass surface.

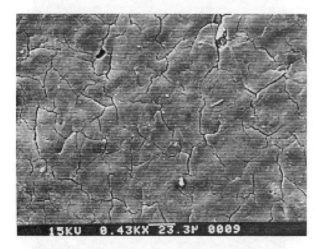

Fig. 5(b). Micro cracking of gel layer on glass surface.

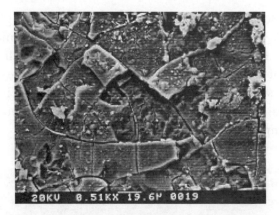

Fig. 5(c). Loss of gel layer parts.

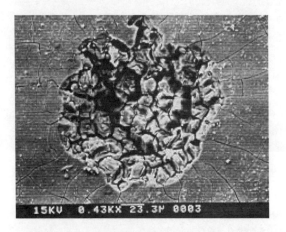

Fig. 5(d). Initial stages of pit formation.

Fig. 5(e). Dense corrosion crust.

2.4. Corrosion Enhancing Factors and Effects

Besides the major influence of composition of the glass, there are other factors which affect the type and rate of glass decomposition:

- *manufacturing process of the glass*
  surface properties
  homogeneity of the glass
- *time of exposure*
- *environmental conditions*
  water (humidity, condensation, precipitation)
  temperature changes
  air pollution
  UV light-causing oxidation processes
  deposition (dust, soot, grease)
- *previous (conservation) treatments*
  cleaning treatments (mechanical and chemical)
  heat treatments for paint consolidation
  inadequate application of coatings
- *attack of microorganisms*
  lichens, fungi, algae, bacteria
- *vibrations*
  road, rail, air traffic, minor earth movements

As already mentioned, water is a primary agent which enhances corrosion, but nevertheless, air pollutants such as sulphur dioxide, nitrogen oxides and ozone in combination with water increase the corrosion rate drastically. On the one hand, they enhance the ion exchange reactions and on the other, they react with the released alkali and earth alkali ions. It is assumed that the first step is the reaction of carbon dioxide with the primary products of corrosion by converting the hydroxides, produced by the attack of water on glass, to carbonates. This step is responsible for the calcite frequently found in the weathering crust of glass. It contains mainly gypsum and syngenite formed through the reaction with atmospheric sulphur compounds. The crust may be up to 1 mm thick, white brownish or even blackish, very soft and powdery, or extremely hard and flinty.

The effect of water and air pollutants are very much linked to time and temperature. The higher the temperature and the time of exposure, the higher the corrosion and its rate. Cycles of temperature change and condensation and drying especially contribute to faster corrosion. Condensation may occur on the inside of church windows especially when heating is installed. These condensation/drying cycles can result in the formation of minute, discrete droplets which then dry out. Any alkali extracted from the glass by the droplets remains as patches on the surface to form nuclei so that the droplets in the next phase of condensation occur at exactly the same spot. Solutions of high pH could then build up at each droplet site causing attack at those points — this may be an explanation of the origin of pitting (see Fig. 5(d) and Sec. 2.2).

Another mechanism for the formation of pits and corrosion phenomena may be the attack by microorganisms (Krumbein et al., 1991; Weissmann and Drewello, 1996; Perez y Jorba et al., 1984; Perez y Jorba and Bettembourg, 1989; Müller, 1992). Much of the action of microbiota could be assigned to water supply via the fungal mycelium. It is suggested that fungi serve as transport agents of water and are thus hydrating forces comparable to the action of a thread of wool. In addition to the supply of water bacteria and fungi act as physical agents and as chemical agents through the formation of various acids, among which citric and oxalic acids are the most prominent ones. Lichen, algae, fungi and bacteria can metabolise, leach, accumulate and redeposit calcium, potassium, magnesium, manganese, etc. Obviously organic growth on both sides of the window glass occurs in warm humid environments where a minimum supply of biological essential elements and nutrients is available. The attack of microorganisms can lead to special etching patterns, biopitting, dealkalisation, staining (darkening) and depositions like calcium oxalate.

The darkening phenomena of windows is not yet understood. It is suggested that the UV light causes oxidation/reduction reactions of manganese and iron oxides which are then deposited along the micro fissures of the gel layer and lead to complete intransparency (Perez y Jorba and Bettembourg, 1989; Müller, 1992).

Inadequate previous conservation treatments, as for example, mechanical or chemical cleaning can damage the glass surface by creating scratches where corrosion processes or enhanced leaching of ions can be initiated (Fuchs *et al.*, 1991).

3. Main Analytical Tools to Examine Corrosion Effects on Glass

The degradation mechanisms and rates depend quite critically on the initial surface and on the evolution of leached layers, gel layers, and reaction product layers as the dissolution process proceeds. Characterisation of these surfaces has brought into play nearly every surface characterisation tool known to material science. For those glasses that develop substantial surface layers, characterisation by scanning electron microscopy, atomic force microscopy, photoelectron spectroscopy, and diffuse reflectance infra-red spectroscopy has proved useful (Salem *et al.*, 1994a and b). Depth profiling techniques such as secondary ion mass spectrometry and sputter induced optical emission spectroscopy allow measurement of diffusion profiles through the glass surface layers. Success in measuring hydrogen concentrations in the hydrated layers has also been obtained by resonant nuclear reactions methods and by Rutherford backscattering.

3.1. SEM/EDX

The Scanning Electron Microscope (SEM) is often the first analytical instrument used when a "quick look" at a material is required and the light microscope no longer provides adequate solution. In the SEM, an electron beam is focused into a fine probe and subsequently raster scanned over a small rectangular area [see Figs. 5(a) to (e)].

3.2. Infra-red and Raman Spectroscopy

Infra-red and Raman spectroscopical methods are gaining importance in investigation of surface layers of glasses. Infra-red reflection techniques allow especially effective examination of the

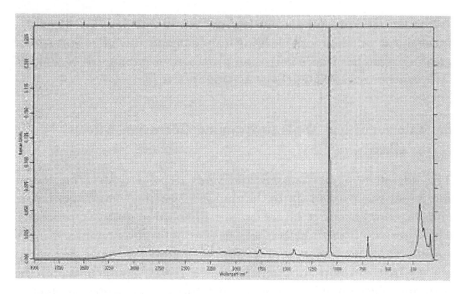

Fig. 6. Raman spectrum of sodium carbonate on a corroded glass surface.

surface. The characteristic vibrations of silicate glasses lie in three regions between 33,000 and 3800 cm^{-1}, 800 and 1200 cm^{-1} and around 500 cm^{-1} and can be recorded by infra-red and Raman spectroscopy. The vibration bands in the infra-red spectra are very strong so that normal glasses are not transparent to infra-red radiation (Fuchs *et al.*, 1988 and 1991a) . Therefore infra-red spectra cannot be recorded in the transmission mode when the glasses are thicker than a few millimetres. A watery gel layer will be formed due to the corrosion process which can be seen by the growth of the OH absorption band at about 3300 cm^{-1}. The corrosion crust also consists of gypsum and syngenite or carbonate crystals (see Fig. 6) which can ideally be identified by the sharp absoprtion bands in the Raman spectrum.

4. Evaluation of Environmental Impact

In conservation research, it is important to evaluate the aggressivity and corrosivity of environments. Basic investigations of the interaction between materials and the environmental stress at the very spot or building structure of interest can render detailed understanding of the deterioration phenomena and corrosion mechanisms. The corrosion relation between specific stress situations and resulting decay processes may be essential for the design of adequate protective measures. The assessment of the combined corrosive effects induced by climatic influences, pollution related parameters and synergetic effects is also of high interest for judgements concerning the protective features of existing conservation concepts.

In the field of stained glass conservation, the state of the art is characterised by transferring standard analytical methods from the fields of climatology and environmental engineering into conservation research. Measurements of single parameters such as temperature, humidity, condensation effects, pollutant concentrations, precipitation and run-off water, examination of microbiota — the spectra of relevant environmental parameters as well as the efforts of analyzing them can be extended at will (Oidtmann *et al.*, 2000).

Up to now, knowledge about the precise meaning of these influences and the resulting impact on the material is fragmentary. To cope with this restrictions, an alternative and integrating concept for total stress level assessments was desirable so the type of sensitive material found in stained glass windows seemed a useful surrogate. Studies using the original materials of the stained glass windows are restricted by the heterogeneity of the assets, the complexity of altered surface layers and, in most cases, the exact history of historic pieces which may not be well documented.

Therefore a solution to these problems is offered by using simulation materials, known as *glass sensors* (Fuchs *et al.* 1988, 1991a and b; VDI-Richtlinie, 1992; Leissner and Fuchs, 1992a), which show qualitatively the same corrosion phenomena, but within a much shorter time. A reliable comparability between different objects and environmental situations can be obtained. Glass sensors are based on low durable potassium-calcium silicate glasses similar to mediaeval stained glasses with a high sensitivity to corrosive stresses. Their easy-to-handle technique enables the detection of combined impact potential. Due to the sensitivity of the dosimeter glass, judgements about corrosive stresses and damage risks can be made within 12 months.

The principle of the *Glass Sensor Method* is the registration of the combined impacts of the environment. The surface of the glass sensor interacts with its immediate atmosphere causing alterations in the surface layer:

- leaching of potassium and calcium
- formation of a gel layer
- formation of a corrosion crust

The degree and kinetics of this alterations as well as the crystallisation of the corrosion crust correspond with the total stress level during the exposition, and intergrates all effective environmental influences as well as their synergetic interactions. The changes can be determined by different analytical routine methods, the data being collected before and after exposure of the glass sensors. The two major corrosion effects, the leached gel layer and the increasing crystalline corrosion crust (K-Ca sulfate hydrates) cause an increase in the intensity of the OH absorption band, in the infra-red spectrum of the exposed glass sensor. The higher the increase in the OH absorption band, the higher the corrosive stress. Additional microscopical investigations allow additional qualitative and semi-quantitative estimations about the degree of corrosion.

The *Glass Sensor Method* is a very valuable tool in the field of stained glass conservation, e.g. to estimate the effectiveness of protective glazings (a widely accepted protection measure). As a case study, three differently ventilated types of protective glazings (internal ventilation, no ventilation and an external ventilation) were investigated with glass sensors (Leissner and Fuchs, 1992b; Romich *et al.*, 1999). The results (Fig. 7) showed clearly that the best outcome is obtained with an internally ventilated protective glazing, whereas the external ventilation shows no protective effect (position 2 has nearly the same corrosive stress as the outdoor position 4). The worst case is with the non-ventilated protective glazing — here the situation for the position 2 is even worse than without protective glazing and will lead to an enhanced corrosion progress. The height of the bars in Fig. 7 represents the corrosive stresses to which the different sides of a stained glass window are exposed (position 1 = front side of original stained

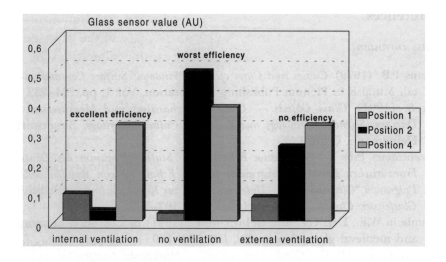

Fig. 7. Efficiency of different ventilation types of protective glazings.

glass window; position 2 = reverse side of original stained glass window, gives information about achieved protective effect; position 4 = reverse side of protective glazing facing outdoor conditions). Up to now, a large number of protective glazings have been investigated by glass sensors which resulted in the general trend, that internally ventilated protective glazings are more effective than externally or non-ventilated types. Nevertheless, every stained glass window and its protective glazing stand in a unique situation concerning overall climate situation, size of slots, indoor climate, etc. and therefore needs its own evaluation of the achieved protective effect.

5. Conclusions

The corrosion process of glass is a very complex mechanism. Many factors which promote corrosion are well known but not all the influences are understood. Since environmental pollution is a major factor for the enhanced corrosion process, it is necessary to protect the historic windows with highly efficient protective glazings and to evaluate the environmental stresses to which they are exposed in order to save them for the future generations.

References

Glass corrosion:

Adams P.B. (1979) *Causes and Cures of Dirty Windows: Surface Contamination,* ed. Mittal K.L. Plenum Publishing Corporation, Vol. 1, pp. 224–232.

Fitz S. (1991) *Glass Objects — Causes, Mechanisms and Measurements of Damage; Science, Technology and European Cultural Heritage.* Butterworth-Heinemann, Oxford.

Gemeinsames Erbe — Gemeinsam Erhalten, 1. Statuskolloquium des Deutsch-Französischen Forschungsprogramms für die Erhaltung von Baudenkmälern, Programm "Glasschäden" — Untersuchungen zur Konservierung mittelalterlicher Glasfenster, Champs-sur-Marne (1993), pp. 197–299.

Krumbein W.E., Urzi C.E. and Gehrmann C. (1991) Biocorrosion of antique and medieval glass, *Geomicrobiol. J.* **9**, 139–160.

Marschner H. (1985) *Glaskonservierung — Historische Glasfenster und ihre Erhaltung, Arbeitsheft 32, Bayer.* Landesamt für Denkmalpflege.

Müller W. (1992) Corrosion phenomena of medieval stained glasses. *Bol. Soc. Esp: Ceram. Vid.* **31-C**, 219–239.

Newton G. and Davison S. (1989) *Conservation of Glass.* Butterworths, London.

Perez y Jorba M., Dallas J.P., Bauer C., Bahezre C. and Martin J.C. (1984) Deterioration of stained glass by atmospheric corrosion and micro-organisms. *J. Mater. Sci.* **15**, 1640–1647.

Perez y Jorba M. and Bettembourg J. M. (1989) *Opacification des verres médiévaux: rôle du manganèse: les arts du verre, actes du Colloque de namur,* pp. 119–125.2.

Salem A.A., Grasserbauer M. and Schreiner M. (1994a) Study of corrosion processes in glass by a multitechnique approach. Part 1. Atomic absorption spectroscopy, atomic emission spectroscopy and scanning electron microscopy. *Glass Technol.* **35**(2) 89–96.

Salem A.A., Kellner R. and Grasserbauer M. (1994b) Study of corrosion processes in glass by a multitechnique approach. Part 2. Infrared spectroscopy, *Glass Technol.* **35**(3), 135–140.

Sanders and Hench L.L (1973) Mechanisms of glass corrosion. *J. Am. Ceram. Soc.* **56**, 373.

Scholze H. (1988) *Glas — Natur, Struktur, Eigenschaften.* Springer-Verlag, Berlin.

Vogel W. (1979) *Glaschemie; VEB Deutscher Verlag für Grundstoffindustrie, Leipzig.*

Weissmann R. and Drewello R. (1996) Attack on glass. In *Microbially Influenced Corrosion of Materials,* eds. Heitz *et al.* Springer-Verlag, Berlin-Heidelberg.

Evaluation of environmental impact:

Fuchs D.R., Patzelt H., Tünker G. and Schmidt H. (1988) Model glass test sensors — A new concept to investigate and characterize external protective glazings. *News Lett. CVMA* **41/41**, 27–29.

Fuchs D.R., Römich H. and Schmidt H. (1991a) Glass sensors — assessment of complex corrosive stresses in conservation research. In *Materials Issues in Art and Archaeology II*, eds. Vandiver P.B., Druzik J., Wheeler G. S., *Mat. Res. Soc. Symp.* **185**, 239–251.

Fuchs D.R., Römich H., Tur P. and Leissner J. (1991b) Konservierung historischer Glasfenster — Internationale Untersuchungen neuer Methoden. *Forschungsbericht UFO-PLAN-Nr.* **108 07 005/03**.

Leissner J. and Fuchs D.R. (1992a) Glass sensors — a European study to estimate the effectiveness of protective glazings at different cathedrals. I congreso internacional rehabilitacion del patrimonio arquitectonico y edificacion. *Islas Canarias* **1**, 285–290.

Leissner J. and Fuchs D.R. (1992b) Investigations by glass sensors on the corrosive environmental conditions at stained glass windows with protective glazings in Europe. In *Materials Issues in Art and Archaeology III*, eds. Vandiver P.B., Druzik J., Wheeler G.S. and Freestone I., *Mat. Res. Soc. Symp.* **267**, 1031–1038.

Oidtmann S., Leissner J. and Römich H. (2000) Schutzverglasungen. In *Restaurierung und Konservierung historischer Glasmalereien*, ed. Wolff A. Verlag Philipp von Zabern, Mainz.

Römich H., Fuchs D.R. and Leissner J. (1999) Glass sensor studies for the evaluation of protective glazings on stained glass windows — a survey of results after 10 years experience. *Proceedings of the CVMA-Colloquium, Cracow, 1999*, pp. 223–231.

VDI-Richtlinie 3955, Blatt 2 (1992) *Bestimmung der korrosiven Wirkung komplexer Umgebungsbedingungen auf Werkstoffe: Exposition von Glassensoren*, Beuth Verlag GmbH, Berlin.

CHAPTER 9

THE EFFECTS OF OZONE ON MATERIALS — EXPERIMENTAL EVALUATION OF THE SUSCEPTIBILITY OF POLYMERIC MATERIALS TO OZONE

D.S. Lee, P.M. Lewis, J.N. Cape, I.D. Leith
and S.E. Espenhahn

1. Introduction

Much of the debate on the deleterious effects of ozone (O_3) in the boundary layer has focused upon vegetation and human health. The effects of O_3 on materials have been largely ignored until recently. Materials can be defined in a number of ways: for convenience, we define materials as the inorganic and organic components of things that humans use in their day-to-day life, for example, clothing, buildings, vehicles, etc. Here, we focus upon polymeric materials and rubbers in particular, and the effects of O_3 on them.

Research into the effects of O_3 on materials is hardly new: some of the older literature goes hand-in-hand with the development of understanding of the generation and abundance of photochemical oxidants (e.g. Bradley and Haagen-Smit, 1951). However, despite the fact that early attempts at limiting O_3 concentrations were targeted at materials effects, further research between the 1960s and 1990s was extremely limited. The research into the effects of O_3 on various materials has been reviewed and this research gap highlighted (Lee et al., 1996; PORG, 1997). The lack of research partly reflects a perception that this is a solved problem. For example, the effects of O_3 on fabrics were understood some time ago and measures put into place in the manufacturing process to mitigate against damage. In the case of polymeric materials and in particular, rubber, antiozonants were incorporated into natural and synthetic rubbers to prevent the well-known effects.

This chapter summarises part of a research programme put into place by the U.K. Department of Environment Transport and the Regions. The research comprised two essential elements: basic research, and analysis of results in terms of economic assessment derived from the basic research and emissions and modelling scenarios of future atmosphere. The latter part, the assessment, is not considered here but is described by Holland et al. (1998). The experimental programme and results described here synthesise a number of particular experiments that will be described in more detail in forthcoming publications.

In Sec. 2, the basic experimental philosophy and design is described. In Sec. 3, the results of accelerated exposure of rubber and elastomeric materials are described and discussed. Finally, in Sec. 4, conclusions are drawn.

2. Experimental Design

2.1. Experimental Philosophy and Strategy

The experimental design of the programme centred around the use of so-called "open-top chambers" or OTCs, which have been used widely in biological research into the effects of air pollutants on plants and trees (e.g. Fowler et al., 1989). Basically, the chambers resemble large greenhouses with open tops within which controlled levels of pollutants are injected. These OTCs provided a means of accelerating exposure of materials to O_3 and are described in more detail in

Sec. 2.3. However, before exposing materials over the course of months in the OTCs, a screening exercise was undertaken to identify those materials that were more susceptible to O_3 by exposure in the laboratory to higher concentrations of O_3. This system is described in Sec. 2.2. Finally, materials were also exposed in the field at national monitoring network sites for O_3 to investigate the uniformity of effect (particularly on natural rubber) for the same dose but for slightly different "pollution climates". For example both urban and rural sites, characterized by different balances and concentrations of gaseous pollutants, were chosen.

2.2. The High Ozone Exposure Chambers

Accelerated tests were carried out at the Tun Abdul Razak Research Centre (TARRC). A nominal O_3 concentration of 500 ppb was maintained in a darkened aluminium lined test room operating at a temperature of 20–25°C and a relative humidity of 50–80%. O_3 levels were measured up to two times a week using electrochemical and ultraviolet (UV) methods. Tests at much higher concentrations of 5000 ppb O_3 were conducted in a commercial Hampden 603 O_3 cabinet operating at 25°C.

2.3. The Open-Top Chambers

The controlled exposure of materials to a range of O_3 concentrations was carried out in OTCs at the Centre for Ecology and Hydrology (CEH), Edinburgh (shown in Fig. 1). The octagonal glass-sided OTCs create a controlled environment in which conditions are close to those of the field. Impregnated charcoal filters remove the major pollutant gases, nitrogen dioxide (NO_2), sulphur dioxide (SO_2) and O_3, from the air supplied to the chambers via ducted fan units. This large flow rate of air (30 m^3 min^{-1}) minimises the warming effect of the chamber to approximately +1°C relative to ambient. The horticultural 3 mm glass normally used was replaced in one of the chambers with UV transmitting glass (Sanalux) on the southeast to northwest facing panels for the "4× ambient +UV chamber". The chambers were fitted

Fig. 1. Open-top chamber facility at ITE, Edinburgh.

with detachable roofs to prevent any precipitation reaching the experimental material. All the paint panels and the rubber frames were mounted on a free-standing framework (Dexion) within the chambers. The rubber experiments were shaded from direct sunlight using horticultural green netting (50% shading).

Rather than injecting known amounts of O_3 in order to determine dose-response relationships, a system was devised to mimic the concentration fluctuations in the real environment by "tracking" the ambient concentrations and enhancing these concentrations in the chambers by 2× ambient, 4× ambient and 4× ambient +UV with one chamber assigned to each O_3 concentration. In addition, one chamber was run at ambient concentrations of O_3 as a control. The computer controlled system monitored the O_3 concentration in the ambient chamber, then adjusted the O_3 concentration to achieve the multiple of the measured ambient concentration, i.e. 2× and 4×.

After development of a computer controlled feedback system, and testing to ensure pre-defined target O_3 concentrations could be achieved, the system was commissioned.

O_3 was generated by electrical discharge in oxygen and fed via electronic mass flow controllers under computer control to the individual chambers. O_3 was injected into the filtered air flow, then into the OTC via a polyethylene manifold. The large air flow-rate produced effective mixing of the O_3 within the chamber.

Air was continuously pumped from each of the OTCs via a series of solenoid valves. Each chamber was individually sampled in rotation; during each 90-second sampling cycle a 45-second purging of the air was followed by a 45-second monitoring period during which the O_3 concentrations were recorded every 0.1 seconds and an average calculated.

The average O_3 concentration in each OTC was then compared to the appropriate multiple of the measured ambient concentration enabling the computer programme to calculate the required flow rate for the mass flow controller to track the pre-defined multiple concentration. O_3 concentrations were monitored using a Thermo Electron O_3 analyser (TECO 49C). The system included appropriate fail-safe controls that automatically stopped the programme if the O_3 analyser failed, in the event of power failure, and if the O_3 concentrations were outside the pre-set maximum and minimum concentrations.

3. Exposure of Rubber and Polymeric Materials to Ozone

3.1. Introduction

Rubbers having an unsaturated carbon-to-carbon main polymer chain are highly susceptible to O_3 degradation, so that scission across the double bond may easily take place. On stretched surfaces, this attack results in the formation of cracks, the density of which increases with the applied extension. The reaction is rapid with many general purpose diene rubbers and, without protection, surface cracks can become evident within even a few days exposure to atmospheric levels of O_3. The damage detracts from appearance and more seriously can reduce product service life through premature tearing, fatigue, loss of sealing stress and diminished impermeability.

O_3 attack is generally accepted as a major weakness of commodity elastomers such as natural rubber (NR) and its closest synthetic competitor, styrene-butadiene rubber (commonly known as SBR), but satisfactory protection can be provided by the addition of antiozonant materials or by blending with other rubbers having a higher inherent resistance to O_3 cracking. However, there are many applications, for example in healthcare, where antiozonants cannot be used and in these cases protection, often during storage, depends on suitable packaging of the product. Furthermore, an increasing number of applications, including many non-tyre automotive components, are now based largely on O_3-resistant rubbers in order to avoid the use of antiozonants and guarantee crack-free resistance, irrespective of the severity of exposure.

Work conducted at the TARRC and elsewhere has demonstrated that O_3 resistance can be defined in terms of two distinctive physical characteristics (Lake, 1970): the first is a threshold strain, typically around 5% in most unprotected rubbers, below which no cracking will occur irrespective of the O_3 concentration used; the second is that the rate of growth of an isolated crack above the threshold is constant with time and linearly dependent on the O_3 concentration. Antiozonant materials raise the threshold strain or reduce the rate of crack growth, or in some instances, both. These findings were obtained at O_3 concentrations considerably higher than those experienced in product service, not only to obtain results more quickly but also to ensure constant exposure conditions and to avoid interaction from other forms of environmental deterioration. Qualitative agreement has been found with ambient O_3 conditions but the relationship between crack growth and O_3 concentration has not been previously examined quantitatively at low and more typical severities, mainly because of the absence of test chambers able to function reliably at O_3 concentrations around ambient.

The principal objective of this part of the work was to assess the effect of controlled changes in O_3 concentration upon crack growth behaviour, the resistance of a range of antiozonant-protected rubbers and the behaviour of a small group of rubber products.

3.2. Test Materials

A series of rubber mixes of known composition were prepared at TARRC. All ingredients were based upon 100 parts by weight of raw rubber (phr). These were as follows:

Control compounds for crack growth measurements

The first compound examined (later known as Compound 2) was an unfilled (gum) natural rubber vulcanised to contain polymerised 2,2,4-trimethyl-1,2-dihydroquinoline as an antioxidant; this is not an antiozonant for NR but is necessary to guard against thermal oxidative degradation. The rubber also contained 5 phr of N330 type carbon black (an effective UV absorber) to reduce the risk of photo-oxidation. Compounds without antioxidant and with an alternative non-staining phenolic antioxidant (Compound 3) were also prepared.

Protected rubbers to assess efficacy of added antiozonant materials

Protection was conferred by a commonly used antiozonant, NN'-dioctyl-p-phenylenediamine (DOPD) added at 0, 1, 1.5, 2, 2.5, 3, 3.5 and 4 phr; the small increments were used to determine the level of protection needed for different O_3 concentrations. Three other compounds were prepared with a commonly used antiozonant system of N-1,3-dimethylbutyl-N'-p-phenylenediamine (6PPD) and a paraffin wax, in this case Sunproof Improved (Uniroyal). The level of wax was held at 2 phr and the level of 6PPD varied at 1, 2 and 3 phr. 6PPD is an example of a crack growth rate-reducing antiozonant that needs the additional assistance of a surface wax bloom to ensure static O_3 resistance, whereas DOPD is the best known antiozonant capable of enhancing threshold strain. The proprietary materials used were Santoflex 6PPD (Flexsys) and, for DOPD, UOP 88 (Universal Oil Products).

The base NR formulation, which in this instance contained 45 phr N330 black for reinforcement is representative of compound used in a wide range of applications, including tyre components, bearings and mountings, and belting covers. NR has been used as the base polymer because of its widespread use in these products but it is also representative of other ozone-susceptible elas-tomers in need of antiozonant protection, including SBR and polybutadiene.

Synthetic rubber

A chloroprene rubber (CR) formulation was examined because CR (an unsaturated rubber) has an inherent ozone resistance that is derived from a very low rate of crack growth (compared with NR or SBR). It can undergo cracking but is sufficiently resistant to be used in many outdoor applications calling for a measure of weathering resistance.

In one compound, 2 phr octylated diphenylamine was used as antioxidant. In the other, this was joined by 3 phr of a diaryl-p-phenylenediamine (Wingstay 100, Goodyear).

The base black-filled formulation, with suitable protection, is typical of CR compounds used in applications ranging from bearings to weather stripping.

NR/EPDM blends

Ethylene-propylene-diene copolymer (EPDM) is the leading example of a completely O_3-resistant rubber because of the absence of main chain unsaturation. Black-filled NR/EPDM blends were prepared and compared with unprotected NR and EPDM controls. EPDM and its blends are used in applications calling for O_3 resistance that does not depend on surface protection. Applications range from hose covers and tyre sidewalls to gaskets and seals.

All test mixes under (a) to (d) were moulded into standard 2 mm thick vulcanised sheets from which test pieces could be cut. The vulcanising conditions for all NR compounds were 40 minutes at 140°C in a steam-heated press. The conditions for CR were 30 minutes at 153°C.

3.3. Test Piece Evaluation

Test pieces were inspected at various intervals during exposure.

Crack lengths, within the surface and along the cut edges, were measured with a graticule having a resolution of 0.1 mm. An average was taken of five to ten crack lengths on both faces of test strips unless fewer cracks were involved. The mean of the two averages was only used where the two faces had similar crack appearances. Difficulties in measurement increase with crack size and crack density through interference and the merging of cracks, but by this time there should be sufficient crack growth to establish the rate of cracking.

Attempts were made to measure the depth of cracking because this is often more critical to the life of the product than crack length within the surface. It was found to be very difficult and necessarily involved destruction of the test pieces. Therefore, as an alternative, stress retention was measured on exposed dumb-bells. The cracked test pieces were stretched to a test extension of 100% and the stress measured after 1 minute relaxation. The stress, or relaxed modulus as it is commonly called, is a measure of the residual effective cross-section of the test piece and therefore falls as crack depth increases. Previous work had shown that the method is not suitable for test pieces having completely isolated cracks and therefore is best used for test pieces exposed at extensions above the threshold strain of the rubber, for example 10% in the case of the unprotected test compound. Tensile strength measurements are more sensitive to isolated cracks than stress retention but were not used here because several duplicate test pieces are necessary to allow for the higher inherent scatter of strength measurements.

For protected rubbers, time to first crack was determined by ×7 magnification with, where necessary, a distinction is made between surface and edge cracks.

A typical test frame for assessment of crack growth is shown in Fig. 2. Two test strips are held at 7.5% extension, one further dumb-bell is held at 10% extension (for modulus measurements) and another is held at 120% extension. The latter is used to measure the effect of any significant differences in air flow-rate over the rubber surface. At high extensions, the crack density is high and previous experience with pieces indoors has shown that the rapid consumption of O_3 may deplete the concentration of O_3 at the surface layer, such that crack growth rate falls unless the O_3 is replenished. Relaxed

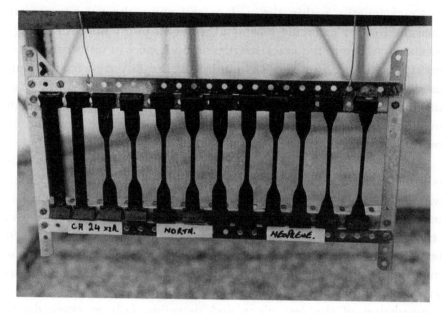

Fig. 2. Frame of rubber test strips (in this case neoprene from Expt. 4) at elongations of 20, 40 and 100%.

modulus measurements showed no discernible effect of flow rate on crack growth in the OTC exposures, as might be expected, where turbulent diffusion is large and the surface resistance (i.e. r_c) is not constrained in an overall deposition velocity term.

3.4. Results

Crack growth measurements

Crack size is not solely dependent upon O_3 concentration and time of exposure. Crack growth is most rapid just above the characteristic threshold strain; in this region only a few surface flaws are large enough to reach the tearing energy needed for crack initiation and therefore the resulting crack density is low, with much reduced chance of interference between neighbouring crack sites. Crack growth is also sustained because the rate of O_3 depletion in the air layer above the surface is much lower than it is when crack density is high (Lake, 1970). Exposure time is significant too because cracks will eventually merge irrespective of the crack density and some may be sufficiently long to rupture the test piece. O_3 cracking may also be affected by stress relaxation if the rubber is in the vicinity of the threshold strain. In addition, surface cracking can result from other forms of deterioration, including surface hardening or embrittlement, photo-oxidation and, in some elastomers, static fatigue. Furthermore, nonlinear crack growth behaviour has been found in some rubbers.

It is therefore important that the test rubbers and straining conditions used in the study are fully characterized, initially under accelerated laboratory conditions.

Figure 3 confirms that a linear relationship initially exists between crack length and exposure time. The test extension of 7.5% is just above the threshold strain of the main test rubber (compound 2), which is estimated to be between 3 to 5% extension. For the 10 mm wide strip used for most of the work, the linearity continues up to a crack length of 5 mm (i.e. half the width): thereafter the rate of growth decreases markedly because of crack interference. Crack density increases markedly above the threshold strain and interference reduces the maximum crack length for linearity.

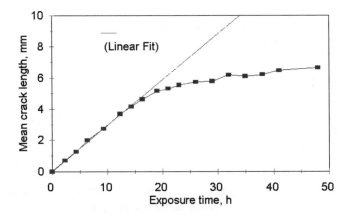

Fig. 3. Dependence of crack growth on exposure time (compound 2, 10 mm strip at 7.5% strain, 500 ppb O_3).

Exposure in the open-top chambers

Two series of tests were conducted in the four CEH OTCs. The first was primarily to assess consistency in crack growth at three positions within each chamber. This exercise demonstrated that there were no significant differences provided steps were taken to eliminate direct sunlight. The rates of crack growth in the chambers operating at ×2 and ×4 ambient O_3 concentration were also closely consistent with predictions from accelerated test measurements, clearly demonstrating that the linearity between crack growth and O_3 concentrations extended to near-ambient severities (Fig. 4).

The crack growth rate in the control chamber (No. 20) was lower than predicted and despite the presence of carbon black there was evidence of photo-oxidation, although other factors might also be involved in the extra time period needed for reasonable crack growth in this chamber. If surface modification or ageing had occurred in this instance, exposure of the returned test piece to an accelerated O_3 test might be expected to show differences with the behaviour of fresh test pieces. This is clearly the case in Fig. 4, with the crack growth rate of the previously exposed test piece (shown as "X" in the figure) only being a quarter of that of a new test piece.

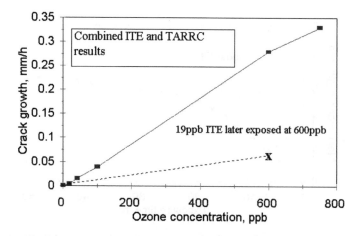

Fig. 4. Correlation between rate of crack growth and O_3 concentration up to 750 ppb (compound 2 at 7.5% strain, first series).

Better agreement among the four test chambers was found in the second series of tests, both for mean crack length (Fig. 5) and for relaxed modulus (Fig. 6); the latter figure showing combined results from the first series of exposures in the OTCs and from the TARRC O_3 room. The dependence of relaxed modulus on the product of O_3 concentration and time is also shown for compound 3 in Fig. 7. Light

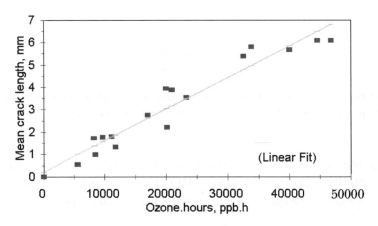

Fig. 5. Correlation between cracking and exposure in CEH (ITE) chambers. First series, compound 2 at 7.5% strain.

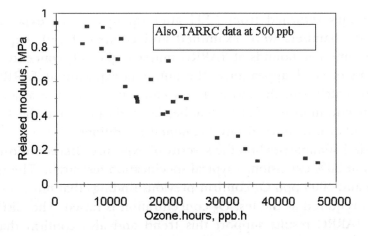

Fig. 6. Correlation between relaxed modulus and O_3 time product for CEH chambers (compound 2 exposed at 10% strain, first series).

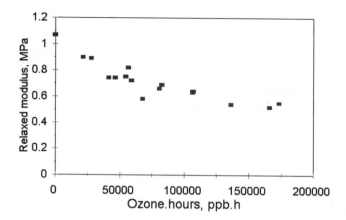

Fig. 7. Correlation between relaxed modulus and O_3 time product for CEH chambers (compound 3 exposed at 10% strain, first series of exposures).

ageing was not evident for compound 3 and the difference suggests that trimethyldihydroquinoline (TMQ) may be capable of promoting photo-oxidation despite the presence of a UV absorber.

Tests with protected and ozone-resistant rubbers

Antiozonant-protected vulcanizates

Test frames returned from CEH after up to a year's exposure in the test chambers were examined and compared with exposures of the same compounds at TARRC under higher O_3 concentrations. In terms of crack appearance, the outdoors exposure at TARRC was a good match with the more closely controlled exposure at ambient O_3 concentration in CEH's Chamber 20, indicating that the OTC is representative of normal environmental conditions.

Table 1 summarises how the severity of exposure affects antiozonant activity at 20% extension, a typical specification test strain. The results at 500 and 5000 ppb O_3 confirm previous findings that more antiozonant is required as the test O_3 concentration is raised. The CEH and other TARRC results support this trend and also confirm that the relationship is nonlinear. Even indoors at O_3 concentrations that are much less than those outdoors, 3 phr DOPD is required for protection

Table 1. Effect of O_3 concentration on antiozonant level for crack-free protection at 20% extension for one year.

Antiozonant (phr)	Indoors (office)	Outdoors at TARRC	CEH ×1 ambient	CEH ×2 ambient	CEH ×4 ambient	500 ppb at TARRC	5000 ppb at TARRC*
Control	X	X	X	X	X	X	X
DOPD 1	X	X	X	X	X	X	X
1.5	X	X	X	X	X	X	X
2	X	X	X	X	X	X	X
2.5	X	X	X	X	X	X	X
3	X	–	–	X	X	X	X
3.5	–	–	–	–	–	X	X
4	–	–	–	–	–	–	X
6PPD, 1; wax, 2	–	–	–	–	–	–	X
6PPD, 2; wax, 2		–	–	–	–	–	X
6PPD, 3; wax, 2	–	–	–	–	–	–	X

*: Test terminated after 24 hours.

X: Cracked.

–: No cracking.

whereas only 4 phr becomes necessary for 500 ppb O_3 and less than 5 phr for 5000 ppb O_3 (Fig. 8). The difference between Chamber 20 and Chamber 24 (with its four-fold O_3 concentration) is equivalent to no more that 0.5 phr antiozonant. This is also apparent for the enhancement of threshold strain (Table 2).

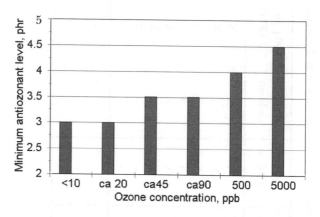

Fig. 8. Estimated level of DOPD antiozonant for crack-free protection at 20% strain for at least one year (combined CEH open-top chamber and TARRC results).

Table 2. Effect of O_3 concentration on estimated threshold strain of antiozonant protected natural rubber.

Antiozonant (phr)	Outdoors at TARRC	CEH ×1 ambient	CEH ×2 ambient	CEH ×4 ambient	500 ppb at TARRC
Control	5–10	5–10	5–10	5–10	5–10
DOPD 1	<10	<10	<10	<10	<10
1.5	<10	<10	<10	<10	<10
2	10–20	10–20	10–20	10	10
2.5	10–20	10–20	20	10–20	10
3	20–40	20–40	20	20	10–20
3.5	>40	>40	>40	40	20
4	>40	>40	>40	>40	20–30
6PPD, 1; wax, 2	>40	>40	>40	>40	10–20
6PPD, 2; wax, 2	>40	>40	>40	>40	20–30
6PPD, 3; wax, 2	>40	>40	>40	>40	30–40

6839496475y port,

Further tests were made on a reduced time scale in the CEH open-top chambers (Table 3). This second batch was included to assess the influence of the time of year in the efficiency of protection since it is known that antiozonant protection can be "conditioned" by prior exposure to a relatively low O_3 concentration (as might occur in Winter). However, the O_3 levels at the start of the two series of tests were about the same. Even so, the level of resistance was slightly lower during the second series and one likely contributing factor is the known loss in antiozonant activity during sample storage.

The 6PPD/wax protective system was effective at all three 6PPD levels in the first series of exposures. Wax blooms are less sensitive to comparatively small levels of O_3 than they are to changes in temperature, which in these tests were small with temperatures always below the solubility limit of Sunproof Improved.

Table 3. Estimates of antiozonant level for one year's crack-free protection of natural rubber.

Severity	DOPD (phr)
For 10% extension	
Indoors	2.5
Outdoors	3
60 ppb O_3	3
100 ppb O_3	3
500 ppb O_3	3
5000 ppb O_3	3.5
For 20% extension	
Indoors	3
Outdoors	3.5
60 ppb O_3	3.5
100 ppb O_3	3.5
500 ppb O_3	4
5000 ppb O_3	5
For 40% extension	
Outdoors	3.5
60 ppb O_3	3.5
100 ppb O_3	4

Chloroprene rubber

An initial exercise was to determine whether the crack growth of the unprotected CR compound was linearly related to the time of exposure. This was confirmed for a test extension of 10% at approximately 500 ppb O_3 (Fig. 9). The rate of crack growth determined at this concentration and a slightly higher one showed a linear dependence on O_3 concentration (Fig. 10). This plot indicates that at 100 ppb O_3 (the concentration region for Chamber 24 over the

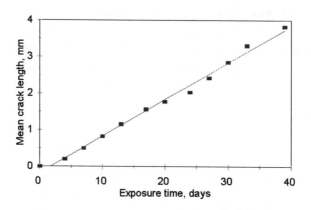

Fig. 9. Linear dependence of crack growth in chloroprene rubber (unprotected CR compound, 10% strain, 500 ppb O_3).

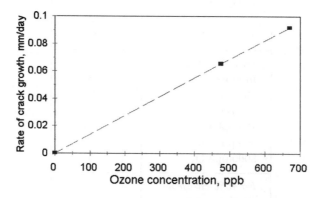

Fig. 10. Dependence of crack growth of chloroprene rubber on O_3 concentration (unprotected CR compound exposed at 10% strain).

year's exposure) the mean crack size should be at least 3 mm. Yet no cracks were detected on the test pieces returned to TARRC, with crack-free protection also being evident at 20% strain. Only at 40% extension did the CR compound suffer any serious cracking. This greater-than-predicted resistance suggests that the octylated diphenylamine antioxidant may be able to exert a small antiozonant effect at low severities and thus give rise to a nonlinear relationship. Assuming the linearity of crack growth with time is maintained, the results at 40% strain also suggest nonlinearity because the crack length in Chamber 24 is at least 20 times longer than in Chamber 20 and five times longer that in Chamber 23. An indication of a small antiozonant activity is a marked conditioning effect at low O_3 concentration. Test pieces exposed for a year outdoors at TARRC continued to show some resistance when inserted in the O_3 room at 500 ppb, a reduction being seen in both the rate of cracking and more significantly in the density of cracks.

The fully protected CR compound was completely resistant at all O_3 concentrations examined (including 5000 ppb).

NR/EPDM blends

None of the NR/EPDM compounds underwent any cracking in the CEH and TARRC tests. Resistance was even found at 100% extension after several months at 500 ppb O_3.

4. Conclusions on the Exposure of Rubber and Polymeric Materials to Ozone

(1) The rate of crack growth of unprotected natural rubber, measured in terms of crack length, is directly proportional to O_3 concentration at the severities used in the CEH test chambers, i.e. between ambient and four times ambient. The increase in crack length is also linearly related to the time of exposure until crack interference occurs. These findings confirm earlier work conducted at very much higher O_3 concentrations far removed from those experienced around rubber products. The results for crack length and for relaxed modulus or stress (as a measure of crack depth)

can be plotted as a function of the product of O_3 concentration and time; this parameter is therefore a convenient means of expressing severity and may be a suitable basis for a dose-response function. These tests are the first to be undertaken at conditions much closer to ambient O_3 concentrations and show that the linear relationships found at very much higher concentrations still apply at close to ambient conditions of O_3.

(2) A more complex relationship exists between resistance and O_3 concentration for fully or partially protected rubbers, a consequence of the separate activity of antiozonants and the ability of some of these to enhance the threshold strain for crack initiation. A significant level of antiozonant is needed even for the very low O_3 concentrations found indoors, but only a relatively small increase is required for protection at outdoor concentration. In terms of crack-free protection for at least one year, the difference between the three O_3 conditions in the OTCs is equivalent to about 0.5 phr of DOPD antiozonant.

(3) There is evidence that the proportionality between crack growth and O_3 concentration extends to unprotected rubber products but resistance is also strongly influenced by packaging, variations in strains and the risk of photo-oxidation in light coloured surfaces. If light ageing is evident, any increase in O_3 concentration will be proportionately more severe than predicted since crack growth can occur before photo-oxidation ensues. Similar conclusions might apply to a rubber that may become more resistant to O_3 cracking simply through stress relaxation.

Acknowledgments

This work was funded by the U.K. Department of the Environment, Transport and the Regions, Air and Environmental Quality Division, and is published in this form with their kind permission. The programme would not have taken place without the enthusiasm and encouragement of Dr. Trudie McMullen, the Project Officer, and we are deeply grateful to her for this. Other colleagues contributed to the overall programme on the assessment of surface coatings (Mr. Norman Falla and Dr. Keith Mower, Paint Research Association) economic assessment (Dr. Mike Holland, AEA Technology) and analysis of future atmospheres (Dr. Dick Derwent, Meteorological Office) and we are appreciative of their comments during the course of the experimental programme.

References

Bradley C.E. and Haagen-Smit A.J. (1951) The application of rubber in the quantitative determination of ozone. *Rubber Chem. Technol.* **24**, 750–755.

Fowler D., Cape J.N., Deans J.D., Leith I.D., Murray M.B., Smith R.I., Sheppard L.J. and Unsworth M.H. (1989) Effects of acid mist on the frost hardiness of red spruce seedlings. *New Phytol.* **113**, 321–335.

Holland M.R., Haydock H., Lee D.S., Espenhahn S.E., Cape J.N., Leith I.D., Derwent R.G., Lewis P.M., Falla N. and Mower K.G. (1998) *The Effects of Ozone on Materials.* Final report, Report no. AEAT-4401, AEA Technology, Culham.

Lake G.J. (1970) Ozone cracking and protection. *Rubber Chem. Technol.* **43**, 1230–1254.

Lee D.S., Holland M.R. and Falla N. (1996) The potential impact of ozone on materials in the U.K. *Atmos. Env.* **30**, 1053–1065.

PORG (1997) *Ozone in the United Kingdom.* Fourth Report of the United Kingdom Photochemical Oxidants Review Group, Institute of Terrestrial Ecology, Edinburgh.

CHAPTER 10

THE SOILING OF BUILDINGS
BY AIR POLLUTION

J. Watt and R. Hamilton

1. Introduction

When Blake wrote of "dark, satanic mills", he evoked a widespread
perception of an industrial cause and effect — buildings blackened by
atmospheric pollution due in large part to industrialisation. The "green
and pleasant land", in contrast, is mostly associated with an equally
strong impression of a clean rural environment. Britain's early
industrialisation and development of urban manufacturing centres
meant that atmospheric pollution had reached a level that was noticed
by contemporary scientists by as early as the second half of the
19th century (Camuffo, 1992). These effects persisted or intensified
for the next hundred years, until the 1950s, when a series of intense
London smogs galvanised Parliament into producing a series of
Clean Air Acts, though by this time the frequency of London fogs
had decreased significantly from the numbers reported in Victorian
times (Brimblecombe, 1987). The acts ended the burning of coal in
homes and factories in most large towns and cities, with a switch to
electric power produced in large rural power stations, together with
greater use of oil and gas burners, which were much cleaner.

In the years that followed, many buildings and cities were cleaned
and a wide variety of colours and textures have re-emerged from under
the layers of blackness. Yet it has become apparent that in spite of the

huge reduction in smoke resulting from the prohibitions placed on burning coal in cities, for both domestic and industrial purposes, buildings are continuing to become dirty. The massive reductions in the levels of suspended particulate material and sulphate clearly do not contain the whole story. Other mechanisms, especially the role of traffic pollution, must now be investigated. This chapter reviews the work undertaken to date. It looks at experimental and theoretical approaches to estimating the rate of soiling for a number of materials, at the trends in air quality and at their implications for the future, including economic and cultural factors.

Soiling is a visual nuisance resulting from the darkening of exposed surfaces by deposition of atmospheric particles. The size of this nuisance depends strongly on the perception of the individual and on the general conditions of the local environment. Generally, people are much more sensitive to the soiling effect in relatively clean environments than in areas with large dust loads, such as in industrial regions.

To give this nuisance a more objective character, the Workshop on Research Strategies to Study the Soiling of Buildings and Materials (Haynie and Spence, 1984) defined soiling as:

> *A degradation that can be undone by cleaning, washing and painting and that can be measured as the reflectance contrast of a surface with deposited particles by comparison to the reflectance of the base substrate.*

Soiling results mainly from the deposition of atmospheric particles and may be caused by both natural and anthropogenic mechanisms. Increased frequency of cleaning, washing, or repainting of soiled surfaces becomes a considerable economic cost and can reduce the useful life of the soiled material. By requiring surface degradation, the above definition would exclude deposited ambient dust, which although a major cause of complaint due to its high visibility, does not produce a permanent effect. It is caused in general by coarse particles, which deposit by gravitational settling, most efficiently on horizontal surfaces. These particles do not travel far and are relatively easily removed by wind and rain. Various methods exist to quantify dustfall, such as BS Dust gauges and glass-slide soiling (Brooks and Schwar, 1987; Moorcroft and Laxen, 1990; Vallack and Shillito, 1998), and there is perhaps some overlap with studies of soiled horizontal surfaces. Terrat and Joumard (1990) showed that a simple plate method (a measurement of the number of particles deposited on a flat plate of inert material), as well as the measurement of reflectance and transmission of light, were capable of quantifying soiling deposit in a town. These simple plates would seem to be more suitable for areas with high particle pollution and the optical methods may be more suitable for less polluted areas. Martin and Souprounovich, (1986) reported an exposure study to examine the soiling of building materials in Melbourne, Australia. They used reflectance measurements on a range of materials exposed both horizontally and vertically and attempted to address the difference between what has been defined as dustiness and soiling by making two measurements, before and after removal of loose grime with a brush and water. There was no attempt to derive a soiling rate, however and particle composition was not determined (apart from a determination of % solubles).

There have been a number of interesting recent papers that have examined the sources and effect of coarse particles within museums (Brimblecombe *et al.*, 1999; Yoon and Brimblecombe, 2000). These have shown that many particles emanate from visitors to the museums, often brought in by clothing and shoes and re-suspended within the museum. The role of this material in damaging the collections housed within the galleries are not well understood but would seem to be largely analagous to the nuisance dusts described in this section and they have therefore been excluded from the soiling discussions.

2. Soiling and Material Damage

The separation of soiling from other atmospheric damage to buildings and materials is, to some extent, artificial. There is a considerable amount of available information that supports the fact that exposure to acid-forming aerosols has been found to limit the life expectancy of paints by causing discolouration, loss of gloss, and loss of thickness of the paint film layer. Various building stones and cement products are known to be damaged from exposure to acid-forming aerosols.

However, it should be noted that the extent of the damage to building stones and cement products produced by pollution, beyond that expected as part of the natural weathering process, is uncertain. Some investigators have suggested that the damage attributed to acid-forming pollutants is over-estimated and that stone damage is predominantly associated with relative humidity, temperature, and, only to a lesser degree, air pollution. One area in which stone damage and soiling do overlap considerably is the formation of black crusts, especially on limestone and sandstone buildings (DelMonte et al., 1981). These encrustations often develop in urban stone structures in areas protected from rainfall. They have been shown to be combinations of carbon and gypsum, the latter believed to be formed by interaction with the parent building stone. The consequent roughening of the surface has been postulated to increase soot deposition (Camuffo, et al., 1983). These crusts are therefore a considerable part of the blackening of many buildings, and yet they are also very destructive since they eventually spall off and destroy the surface of the stone.

There is in practice no real boundary between decay and soiling, although the study of the various effects of air pollution has often had to be separated for analytical convenience. Black crusts may therefore be considered to form an area of study on their own and are excluded from the study of soiling mechanisms discussed in this chapter, which relates to processes that take place on flat, non-reactive surfaces. Here the reflectance of the surface is related to the reflectance of the deposited particles and the fraction of the surface covered by particles. One model also considers the depth of cover.

Attempts have been made to calculate damage functions for the erosion of buildings (Lipfert, 1989) but these do not include the effects of spalling black crusts as they are based on calculation of steady-state loss mechanisms from various materials and erosion regimes.

One other area of potential overlap between soiling and building damage concerns conservation practice (Torraca, 1988.) Cleaning is the first step in conservation of stone surfaces and it is vital to the success or failure of the whole process. Stone fabric may often be in poor condition due to lack of proper maintenance in previous years and rapid mechanical cleaning, for example by sand blasting, may exacerbate the damage by increasing the number of cracks and introducing salts into fissured and porous surfaces. Specialised cleaning techniques, though more expensive, should be used to reduce this damage.

3. The Nature of Atmospheric Particles

The major contribution to all processes of soiling of buildings is the deposition of particulate matter. In contrast to individual gaseous pollutants, which are considered in other chapters, particles are not single, well-defined substances but are a complex mixture arising from a large number of sources. This is not the place to attempt to review the enormous literature that already exists in the field but it is necessary to discuss the major characteristics of particulate pollution as it affects the soiling of buildings. The major influences on soiling processes are determined by the size and chemistry of the particulate matter, which are largely determined by its source. Techniques to measure particle concentration and to estimate the source contributions are briefly discussed, with special reference to carbonaceous particles because of their importance to soiling.

There is no single comprehensive method of characterisation of airborne particles, which can be assessed by many methods which determine one or more characteristic. It is unlikely that there will ever be a single comprehensive study of airborne particulate pollution, which completely measures their number, size distribution, morphology and composition, thus making it difficult to accurately quantify the impacts of particulate air pollution. There is still great uncertainty in the assessment of background concentrations of particles, and spatial and temporal patterns of concentration are not well understood.

In spite of these difficulties, there are some broad classifications that can be made in terms of size, which are also helpful when considering the origin and fate of airborne particulate material. Three groups of particles (sometimes called "modes") have been defined.

The *nucleation mode*, the group with the smallest particle size, consists of ions and nuclei (often of the dimensions of molecular clusters), and the particles into which they grow as a consequence of the condensation of vapours upon them. Particles arising from gas to particle conversion (e.g. sulphuric acid droplets from the oxidation of sulphur dioxide) are initially formed by condensation onto a nucleus. The size range of particles in this mode extends from that of molecular clusters — say, 0.001 μm in diameter — to about 0.1 μm. Condensation nuclei are usually present in very large number concentrations in urban atmospheres, but because of their small size they make a relatively small contribution to the total mass concentration.

The *accumulation mode* consists of particles which have grown from the nucleation mode by further condensation of vapours upon them or by coagulation. Their size range is usually taken to be from about 0.1 μm to about 1 μm. They are relatively persistent, since the processes which remove particles from the atmosphere (e.g. diffusion, washout and sedimentation) are least efficient for particles in this size range.

The *coarse* mode consists of particles greater than about 1 μm in diameter. Many of them originate from mechanical processes such as erosion, re-suspension and sea spray. Soil dust and most industrial dusts come within this category, as do pollens, mould spores and some bacterial cells.

The *fine particle mode* is a term used to indicate the combined nucleation and accumulation modes.

The composition of particulate matter is dependent on source, chemical transformations in the atmosphere, long-range transport effects and removal processes. Measurements tend to be of average concentrations, which may conceal important extreme or episodic events. Diurnal and seasonal trends may be apparent but a great deal of research remains to be undertaken before a complete understanding is achieved.

3.1. Primary Particulate Matter

Primary particulate material, including anthropogenic emissions (especially significant in the current context), is emitted directly to the atmosphere from a source. The major sources of primary particulate material are:

- Petrol and diesel vehicles, the latter being the source of most black smoke.
- Controlled emissions from chimney stacks.
- Fugitive emissions. These are diverse and uncontrolled, and include the re-suspension of soil by wind and mechanical disturbance; the re-suspension of surface dust from roads and urban surfaces by wind, vehicle movements, and other local air disturbance; and emissions from activities such as quarrying, road and building, construction, and the loading and unloading of dusty materials.

Emission inventories have been developed and are a useful technique in the assessment of the relative contributions of specific sources to the overall particulate load. Most useful in estimating anthropogenic emissions of particulate material, emission inventories provide an indication of trends in emissions and the proportional importance of activities that result in particulate release. Black carbon is very important in the context of soiling of buildings and is almost entirely emitted by vehicles (Hamilton and Mansfield, 1991).

3.2. Secondary Particulate Matter

Secondary particles result from the reaction of two gases or vapours, forming a substance that condenses onto a nucleus. The major sources are the atmospheric oxidation of sulphur dioxide droplets to sulphuric acid, and the oxidation of nitrogen dioxide vapour to nitric acid. Hydrochloric acid vapour (refuse incineration and coal combustion are major sources) is also present in the atmosphere, and both this and nitric acid vapour react reversibly with ammonia (largely associated with agricultural sources) to form ammonium salts (Allen *et al.*, 1989). Sulphuric acid reacts irreversibly in two stages to form either ammonium hydrogen sulphate or ammonium sulphate. Such ammonium salts are formed continuously by the oxidation of sulphur and nitrogen dioxide, whenever there is sufficient ammonia for neutralisation. They are therefore part of a large-scale pollution phenomenon affecting both urban and rural areas. Most of the accumulation mode particles in the United Kingdom atmosphere consist of ammonium salts.

3.3. Particle Mass Concentrations

A number of techniques have been used to measure particle mass concentration and the soiling experiments described below have related to several of them:

- Black smoke is associated with the smoke stain technique (BSI 1747 Part 11, 1993) developed in the 1960s.
- Total suspended particulate (TSP) or suspended particulate matter (SPM) describes all airborne particles. TSP is determined gravimetrically using a high-volume sampler and is expressed as $\mu g/m^3$.
- PM_{10} defines the fraction of particulate matter which passes through a selective inlet of specific size with a 50% cut-off at 10 μm aerodynamic diameter [10 μm is therefore the mass median aerodynamic diameter (MMAD)]. $PM_{2.5}$ has a MMAD of 2.5 μm.

The black smoke method has been the longest established particle measurement technique in the U.K. The reflectance of the stain produced by air drawn through a white filter paper is measured using a reflectometer. A standard calibration curve is used to convert the reflectometer measurement into a nominal mass concentration of the airborne particles (known as the "black smoke concentration"). The method is cheap and simple, but it has a number of disadvantages, the worst of which is that the standard calibration curve was established for particulate material of the type that existed in U.K. urban areas up to the early 1960s. The validity of the results depends upon the fraction of carbonaceous (i.e. black) material in the sample, which is now very different (Muir and Laxen, 1995; Muir and Laxen, 1996; Horvath, 1996). The amount of particulate elemental carbon (PEC), which is emitted at proportionally different concentrations from different sources (Hamilton and Mansfield 1991), has changed dramatically. Indeed, the proportion of each component has changed with time, with diesel vehicles becoming the main contributors of PEC in cities that have controlled domestic and industrial emissions from coal. Figure 1 shows that the contribution of coal to urban smoke in the U.K. has decreased substantially since the introduction of smoke-free zones in the Clean Air Act (1956), and diesel emissions have greatly increased in significance. Despite this, the Black

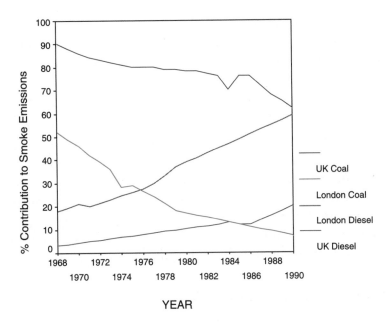

Fig. 1. Contributions of coal and diesel fuel to smoke in the U.K. (Kendall *et al.*, 1994).

Smoke method is currently being used at 222 sites in the U.K. (Broughton *et al.*, 1998) to provide a long-term dataset. There have been attempts to re-calibrate the Black Smoke results to the modern aerosol (Muir and Laxen, 1995; Muir and Laxen, 1996; Horvath, 1996). The greater blackness of the current aerosol is, of course, one of the factors that is contributing to the continued soiling of buildings.

Measurements of TSP or SPM have succeeded and/or complemented the measurement of black smoke. The most widely used method collects particles using a high-volume sampler onto a pre-weighed glass fibre filter for gravimetric analysis in the laboratory. Variable characteristics of the sampler inlet and variable collection efficiencies of larger particles in different wind speeds and conditions have led to the development of a number of further methods of mass determination.

• Dichotomous samplers, which separate the fine and coarse fractions, are used in the United States.

- In 1992, the U.K. Automated Urban Network (AUN) started to use tapered element oscillating microbalance (TEOM) instruments to provide a continuous reading of PM_{10} (using a pre-heating to 50°C, which influences some of the particles captured).

The relationship between these different measures of particle mass is not straightforward, since the situation is complicated by the changing nature of the particulate pollution. PM_{10} mass concentrations are higher than those of black smoke at co-located AUN sites, and remain seasonally variable and highly site-specific. Measured particle mass concentrations are also dependent on the measuring technique employed. Results from different samplers operating simultaneously or under different conditions do not always agree even where they are nominally capturing the same particles.

3.4. Chemical Composition of Particles

Airborne particles comprise three main categories of compounds:

- Insoluble minerals or crustal material,
- Hygroscopic inorganic salts, and
- Carbonaceous material.

Background particulate concentrations are strongly dependent on the local geology and soil type, distance from the sea, amount and type of vegetation cover, season and local weather conditions.

It is possible to make some general statements about the classification of the type of aerosol being considered, e.g. urban, rural or marine, which contain common chemical components at relatively similar concentrations. The UK Quality of Urban Air Review Group produced an estimate of the typical urban airborne particulate matter composition which synthesised the analytical work available at that time for the U.K., shown in Fig. 2 (Quality of Urban Air Review Group, 1993).

3.5. Carbonaceous Compounds

Carbonaceous matter represents approximately 40% of total particulate matter and represents at least half of the fine fraction. There are

Total

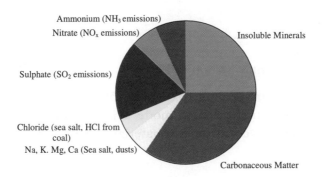

Fine Fraction

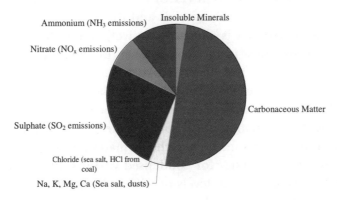

Coarse Fraction

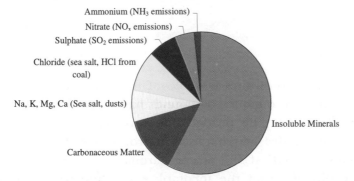

Fig. 2. Typical composition of fine and coarse fractions of particulate matter in the U.K. (Quality of Urban Air Review Group, 1993).

three forms in which carbon compounds are found in particulate matter:

- Carbonate
- Organic carbon
- Particulate elemental carbon (PEC).

Carbonate constitutes approximately 5% of the total mass of particulate, mainly in the coarse mode, and is mainly comprised of soil-derived minerals, construction material and re-suspended dust.

Elemental carbon

Particulate elemental carbon (PEC) is the black component of smoke responsible for soiling of materials and the absorption of light. Light absorption by particles is almost exclusively caused by PEC (Horvath, 1993) and is the most influential factor in the measurement of smoke concentration using the black smoke reflectance method. PEC also has strong adsorptive properties due to its large surface area and the availability of one extra valence electron per exposed carbon atom at particle surfaces. Elemental carbon provides an effective catalytic site for atmospheric reactions such as the formation of sulphuric and nitric acids and acts as a carrier for condensed vapours. PEC measurements taken in the U.K. indicate levels of ~6% in Leeds (Clarke *et al.*, 1984) and ~3% in Birmingham (Harrison, and Jones, 1995). It mainly derives from vehicle emissions, especially diesel combustion (Hamilton, and Mansfield, 1991).

Organic carbon

Organic carbon constitutes 60–80% of the total particulate carbon. A large number of organic compounds have been detected in urban particles. These may be divided into two major source groups — primary condensates and oxidised hydrocarbons. Primary condensates — alkanes (C17–C36), alkenes, aromatics and polyaromatics — originate directly from the incomplete combustion of fossil fuels and are adsorbed onto the surface of particulate matter. Oxidised hydrocarbons may either be attached to the particles as primary condensates or may be produced as the result of atmospheric oxidation reactions. Such compound groups include acids, aldehydes, ketones, quinones, esters, phenols, dioxins or dibenzofurans. These compounds may have an important role in source apportionment of inputs of carbon to soiled layers and black crusts on the built environment and they may also be important in increasing adhesiveness of particles once deposited. As far as direct measurement of soiling is concerned, however, it is the light absorption of elemental carbon that is most important (Hamilton, and Mansfield, 1991).

Source contributions to atmospheric fine carbon particle concentrations

Carbon particles in the atmosphere may be derived from more than 70 different air pollution source types (Gray and Cass, 1998; Hamilton and Mansfield, 1991). The main sources arise from the burning of fossil fuels, with additional contributions from contemporary carbon particles from woodsmoke, cooking and even cigarettes. There are also carbon components in some fugitive emissions such as tyre and brake dusts. Gray and Cass (1998) developed a Lagrangian particle-in-cell air quality model to predict primary carbon particle concentrations over long periods of time. An extensive inventory of fine particle carbon emissions was assembled for the Los Angeles area, which is summarised in Table 1. The air quality model was used to determine source class contributions to both fine primary total carbon and fine elemental carbon concentrations at seven receptor (air monitoring site) locations. It was found that diesel engine applications, including both highway and non-highway applications, dominated black elemental carbon particle concentrations at most monitoring sites. Proportions varied with location.

Miguel *et al.* (1998) confirmed the importance of heavy-duty diesel vehicles to the production of black carbon aerosol. They measured polyaromatic hydrocarbon (PAH) and black carbon emissions in the Caldecott tunnel in the San Francisco Bay Area, making comparisons between different bores of the tunnel, which permitted mixed traffic or light vehicles only. The black carbon measurements in the truck-influenced bore were about five times greater than those found in the tunnel that had only light vehicles, despite the latter

Table 1. Fine particle (< 2.1 micron) carbon emissions in central portion of California coast air basin surrounding Los Angeles, 1982.

Source type	% Total carbon	% Elemental carbon
Mobile sources		
Petrol powered vehicles	13	11
Diesels : Highway	19	49
: Ships, rail, etc.	7	18
: Aircraft and others	1	3
Fugitive sources		
Tyre and brake wear, paved road dust	21	7
Fugitive combustion	24	5
Stationary sources		
Fuel combustion	4	3
Industrial sources		
Point sources	11	4

Source: From Grey and Cass (1998).

having higher traffic volumes. Calculations of emission rates gave values of 30 ± 2 mg/kg of fuel burned for light vehicles and 1440 ± 160 mg for heavy-duty diesel vehicles.

3.6. Particle Deposition

There are many processes by which air pollution is transported to the surface, which can be gathered together into wet and dry deposition, with the former being processes of scavenging by cloud and rainwater. Dry deposition, which occurs by adherence or physical absorption of the pollutant to the surface, is most important for soiling. The term is somewhat misleading since humidity and surface moisture are important controlling mechanisms, including deposition from fog and mist. Water also has an important role in the removal of particles from surfaces.

Dry deposition is a term that includes a number of processes by which particles and gases transfer to surfaces by impaction, sedimentation and Brownian motion. Air is rapidly mixed by turbulent motions generated by the friction of the wind at the surface ensuring roughly uniform particle concentrations throughout this layer, except for a millimetre or so layer immediately adjacent to the surface within which air movements are retarded (the boundary layer). Particles deposit from this layer by a combination of gravitational settling, Brownian diffusion, interception and impaction. Brownian motion is the erratic movement observed in small particle suspensions resulting from the variations in the number and directions of impacts of air molecules on the surface of the particles. The smallest particles show the greatest response to molecular impacts, and the effects decrease with increasing particle size. This phenomenon causes the slow dispersion of particles in still air and increases collisions of particles with each other and with surfaces.

3.7. Deposition Velocities

The flux density of particles deposited on a surface depends on the concentration of the particles in the air and on the deposition velocity. Deposition velocity can be defined as the net velocity with which the particles move to the surface, i.e. the ratio of the flux density of particles depositing per unit time and area, and the number of particles per unit volume in the airborne state.

The size of the soot particles in an urban environment is determined by their sources, which are mainly combustion and automobile exhaust. In both cases, the main fraction (> 80%) of the particles is in the diameter range between 0.05 and 0.5 µm. In this size range, both the velocity under the influence of gravity and the inertia of the particles is sufficiently small that the particles reach the surface by diffusion due to Brownian motion, since sedimentation and impaction are unimportant as deposition mechanisms. The deposition depends on the gradient of the particle concentration in the vicinity of the surface, which varies with the thickness of the surface layer (which depends on the flow characteristics around the surface).

Deposition velocities have been determined by different authors (Milford and Davidson, 1985; Vawda *et al.*, 1989; Nicholson, 1988; Crump and Seinfeld, 1981); values differ by up to an order of magnitude, which may depend on the assumed flow conditions and

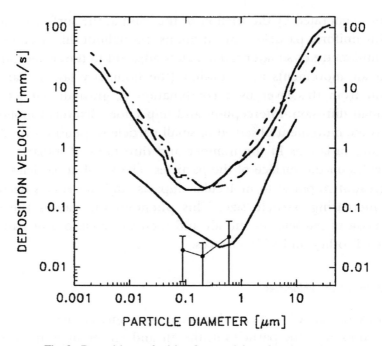

Fig. 3. Deposition velocities for particles of various sizes.

the laminar surface layer thus formed. As examples, two sets of curves obtained by theoretical considerations by different authors are presented in Fig. 3. The upper set of curves is obtained for a model of a very structured surface (grass), the lower for a fairly flat surface. The difference between the two is an order of magnitude for the size range between 0.05 and 0.5 µm particle diameter.

The upper set of curves is taken from Milford and Davidson (1985), the lower set of curves is taken from Hollander and Pohlmann (1991). The data points with the error bars were determined experimentally by Horvath *et al.* (1996) (see below) for deposition to building surfaces.

4. Soiling Models

Soiling is produced by deposition of particulate material onto a surface over a period of time. The rate at which soiling occurs depends on the deposition velocity together with the colour and chemistry of the particles. The nature (especially roughness) and orientation of the receptor surface remains important. All of these factors combine to produce a very site-specific set of deposition and removal processes (re-suspension, washout). A number of field studies have provided data describing the rate at which the soiling process occurs in a given situation. Some of these studies have attempted to clarify the nature of the particles retained on the deposition surface by subjecting the surface to physical and/or chemical analysis, particularly by electron microscopy. By their very nature, these field studies are capable only of describing the soiling process at the location and time at which the studies occurred; a deeper theoretical understanding is required if the results are to be generalised to other sites and other times. This is particularly important if the effect of predicted trends and scenarios for airborne particulate emissions and concentrations is to be evaluated in terms of the consequence for soiling. Two types of model have been produced — empirical models in which the soiling rate is deduced from examination of the relationships produced by a series of measurements (of dose/response over time) and theoretical/physical models derived from consideration of the underlying processes.

4.1. Field Studies

The earliest field studies were largely descriptive. Parker (1955) reported the occurrence of black specks on the newly painted surface of a building in an industrial area. The black specks were not only aesthetically unappealing, but also physically damaged the painted surface. Depending on the particle concentration, the building required repainting every two to three years. Spence and Haynie (1972) published data on the effects of particles on the painted exterior surfaces of homes in Steubenville and Uniontown, OH, Suitland and Rockville, MD, and Fairfax, VA. There was a direct correlation between the ambient concentration of particulate matter in the city and the number of years between repainting. The average repainting time for homes in Steubenville, where particulate matter concentrations averaged 235 $\mu g/m^3$, was approximately one year. In the less polluted city, Fairfax where the particulate matter concentrations only reached 60 $\mu g/m^3$ (arithmetic means), the time between repainting was four years.

The first attempt to produce a general dose-response relationship for particles was produced by Beloin and Haynie (1975) of the U.S. Environmental Protection Agency. They introduced the idea that soiling could be measured using the contrast in reflectance of a soiled surface to the reflectance of a bare substrate. Six different types of material (painted cedar siding, concrete block, brick, limestone, asphalt shingles, and window glass) were monitored over a two-year period at five sites with TSP concentrations ranging between 60 and 250 µg/m^3. The five locations included rural, urban, industrial and mixed-use areas. The effects of direction of exposure, use of preservative to prevent mildew on painted surfaces and sealant on concrete, brick and limestone were also tested. Rainfall, temperature and dewfall were not different between the sites at the 0.05 probability level. One site had higher relative humidity, which led to some problems with mildew formation. Analysis of variance of each of the controlled sources of variability (location, exposure time, exposure direction, paint preservative, surface film and colour) and the interactions between them, revealed that location and months of exposure were significant for all materials. The interaction between these sources of variance was significant for all materials except glass. Some effects of the other sources were seen and discussed. The correlation index (R^2) values for the totalled effects of location, time and the interaction between them were highest for unsheltered acrylic emulsion paint (0.972) and white asphalt shingles (0.903).

The results were expressed as regression functions of reflectance loss (soiling) directly proportional to the square root of the dose. These relationships took the form:

$$\Delta R = A + B\sqrt{C_{TSP}t}$$
$$= A + K \sqrt{t} \tag{1}$$

where ΔR = change in reflectance

  C_{TSP} = concentration of total suspended particulate

  A, B and K were constants obtained by best-fitting the data to the equation.

Haynie and Lemmons (1990) conducted a soiling study at an air monitoring site in a relatively rural environment in Research

Triangle Park, NC. The study was designed to determine how various environmental factors contribute to the rate of soiling of white painted surfaces. White painted surfaces are highly sensitive to soiling by dark particles and represent a large fraction of all man-made surfaces exposed to the environment. Hourly rainfall and wind speed data, weekly data from dichotomous sampler measurements and TSP concentrations were collected. Gloss and flat white paints were applied to hardboard house siding surfaces and exposed vertically and horizontally for 16 weeks, either shielded from or exposed to rainfall. Particle mass concentration, percentage of surfaces covered by fine and coarse mode fractions, average wind speed and rainfall amounts, and paint reflectance changes were measured at two, four, eight and 16 weeks. Scanning electron microscopy stubs, that had been flush-mounted on the hardboard house siding prior to painting, were also removed and replaced with unpainted stubs at these intervals. The unsheltered panels were initially more soiled by ambient pollutants than the sheltered panels; however, washing by rain reduced the effect. The vertically exposed panels soiled at a slower rate than the horizontally exposed panels. This was attributed to additional contribution to particle flux from gravity. The reflectivity was found to decrease faster on glossy paint than on the flat paint.

Based on the results of this study, the authors concluded that:

(1) coarse mode particles initially contribute more to soiling of both horizontal and vertical surfaces than fine mode particles;

(2) coarse mode particles, however, are more easily removed by rain than are fine mode particles;

(3) for sheltered surfaces, reflectance change is proportional to surface coverage by particles, and particle accumulation is consistent with the deposition theory;

(4) rain interacts with particles to contribute to soiling by dissolving or desegregating particles and leaving stains; and

(5) very long-term remedial actions are probably taken because of the accumulation of fine rather than coarse particles (Haynie and Lemmons, 1990).

Similar results were also reported by Creighton *et al.* (1990). They found that horizontal surfaces, under the test conditions, soiled faster than did vertical surfaces, and that large particles were primarily responsible for the soiling of horizontal surfaces not exposed to rainfall. Soiling was related to the accumulated mass of particles from both the fine and coarse fractions. Exposed horizontal panels stain because of dissolved chemical constituents in the deposited particles. The size distribution of deposited particles was bimodal, and the area of coverage by deposited particles was also bimodal with a minimum at approximately 5 μm.

In one of the first studies in the U.K., Mansfield and Hamilton exposed white wood tablets for 250 days in a road tunnel. These authors also placed the same type of white wood plates on the top of a roof in the London urban area over a period of 110 days. The results of both exposure experiments related the rate of soiling to the atmospheric levels of black smoke. During the various experiments performed, meteorological variables such as wind intensity, wind direction and rainfall were monitored in an attempt to relate the rate of soiling to meteorological conditions. Physical models, which took into account deposition processes, were developed to explain soiling processes; empirical models allowed for meteorological conditions by their best fit approach to real data (Hamilton and Mansfield 1992; Mansfield and Hamilton, 1989).

This early work was extended into two major research contracts sponsored by the European Commission. The first of these, under-taken with partners in Portugal, Austria and the U.K., examined the effect of airborne particulate matter on building surfaces (Contract Number: STEP-CT90-0097).

The programme was concerned with describing both the rate at which airborne particulate pollution is soiling buildings and the influence which acidic aerosols may have on stone deterioration. The nature of the particulate pollution was investigated in the course of the programme by collection of particulate matter using several different techniques. Theoretically derived emission inventories for elemental carbon and trace elements provided data to estimate soiling rates and apportion contributing sources to the collected particulate pollutants.

Filter samples using high volume samplers were collected from London, Vienna, Oporto and Coimbra. For comparative purposes, additional samples were collected at a suburban site near London and a rural site near Vienna. A portion of each filter was used to measure black carbon, organic carbon and carbonates, a second portion to analyse water soluble cations and anions and a third portion was extracted with acids and analysed for trace element content. Black smoke measurements were carried out.

Soiling and corrosion experiments were performed using a previously tested exposure protocol and materials provided by the Building Research Establishment in the U.K. Two different types of tablets were used for soiling measurements: white painted wood and calcareous Portland stone. One sample of each material was placed vertically in the exposure system facing the four cardinal directions, north, south, east and west, either in an uncovered position on the top of a mast where it was affected by wind and rainfall or under a roof, which protected the samples from rain but left them open to the wind. The main conclusions of this research programme were:

- Sheltered stone tablets showed a weight gain.
- Unsheltered stone tablets showed a weight loss and this loss showed a strong correlation with atmospheric SO_2 level.
- The input of sulphur and nitrogen via the aerosol form represented a small fraction of the total input.
- Soiling appeared to proceed in two stages; an initial stage which can be well represented by an exponential relationship and a later stage which can be well represented by a square root relationship.

The study was extended in Portugal by the Portuguese research team (Pio *et al.*, 1998), who concluded that sheltered surfaces had a continuous decrease in reflectance, which followed a square root equation on exposure time. Their data predicted a 30% decrease in reflectance would take between 5.5 and 8.8 years. They attributed 70% of the black carbon particles responsible for this soiling to vehicle emissions.

The second experimental study undertaken by Middlesex University and its co-workers was sponsored by the European Union (Contract: EV5V-CT94-0519) and ran from 1 July 1994 to 31 October 1996.

The project was established to examine the ways in which the rate of soiling of buildings from our cultural heritage can be directly related to the composition of depositing particles and their deposition velocity. The study was again undertaken in the three European cities — Oporto, Vienna and London, building on the previous collaborative research at the same locations. For comparative purposes, additional samples were again collected at a suburban site near London and a rural site near Vienna.

This project extended the examination of the nature and properties of particulate pollution likely to affect buildings in three ways. Analysis of organic and elemental carbon was undertaken to provide data on carbon levels sampled from the air by low volume filtration and to compare them to analyses of crust samples. TSP was monitored at two sites in central and north London. The distribution of some important PAHs in both air and crust was studied using gas chromatography/ mass spectrometry.

Monitoring of TSP and subsequent analysis of the particulate matter for total organic carbon, particulate elemental carbon, 16 PAH compounds and 23 *n*-alkanes revealed higher concentrations of all of these compounds at the central London location. Clear signatures associated with transport sources were identified in both atmospheric and crust samples. Weather factors were found to be important. Good inter-site relationships existed between the two sites for TSP, total organic carbon (TOC) and PEC. In Oporto too, multivariate statistical analysis lead to the conclusion that the predominant alkanes and PAHs were characteristic of urban dust and urban area emissions (especially diesel and fossil fuel combustion).

Particle characterisation by scanning electron microscopy (SEM) described the individual chemistry of particles deposited at St Paul's Cathedral, London. The SEM also revealed some very interesting examples of some of the structures associated with early development of black crusts. Different stages of crust development or structure were tentatively identified, including apparent growth stems, which protrude from the crust surface. The visual correspondence of these structures to those predicted by fractal modelling (next paragraph) for early simple disk aggregation provides great encouragement for further development in this area. SEM examination of different crusts from London's St Paul's cathedral revealed large numbers of particles of anthropogenic origin, probably emitted from oil and coal combustion sources, which accords well with other published research but may largely reflect earlier deposits.

As discussed above, it is generally accepted that, in modern cities, the unpleasant appearance of many stone surfaces of old buildings is mainly due to the deposition of soot particles. These particles cause blackening of the surface, and can also react with atmospheric gases, especially SO_2, which consequently produce sulphate and $CaSO_4$, deteriorating the stone. The manner in which the structure and porosity of particle agglomerates develop is known to affect the physical properties of materials and may also control the growth rate of the crusts formed (Massey, 1993). The U.K. Building Research Establishment examined the growth and structure of the agglomerates in more detail using a computer model based on a variation of the Diffusion Limited Aggregation algorithm. Computer generated pixels or pixels agglomerated into disks were deposited, one by one, onto a line to simulate the deposition of particles on to a planar surface in two dimensions. Fractal analysis, which is ideally suited to describe irregular surfaces such as those encountered in particulate deposition and the agglomerates formed, was used to characterise the structure (Watt *et al.*, 2000).

Conclusions from the fractal studies were that aggregation models provide a controlled way to examine the dynamics of particle growth. It was clear that disk models provide a more intuitive representation of agglomeration than pixel models. These simulations were used to make an estimate of the rate at which soiling might occur based on the model. Preliminary calculations were made of how soiling would build up over time based upon realistic values of atmospheric concentration. The calculations from the model gave a figure of about seven years for soiling to occur to a level that equated to a fairly dense covering of sub-micron particles approximately 0.5 μm thick over the surface of the material. Although this accords well with estimates from soiling rate models described above, it must be stressed that the fractal model under discussion is very preliminary and this result has not been fully substantiated. It has been shown that useful information concerning the internal structure of the agglomerate can be gained from measurements taken from a surface profile. This information can potentially be used to assess the strength of the agglomerate.

4.2. A Theoretical Framework

Following the initial field studies, which yielded empirical relationships between soiling rate and air quality, such as equation (1) above, the EPA 1983 "Workshop on Research Strategies to Study the Soiling of Buildings and Materials" highlighted the need for a theoretical model of soiling of surfaces by airborne particles. Haynie (1986) reported such a model, which provided an explanation of how ambient concentrations of TSP are related to the accumulation of particles on surfaces and ultimately the effect of soiling by changing reflectance. Soiling is assumed to be the contrast in reflectance of the particles on the substrate to the reflectance of the bare substrate given by:

$$R = R_0 (1 - X) + R_p X$$

where R = reflectance of soiled surface
          R_0 = initial reflectance
          R_p = reflectance of deposited particles
          X = fraction of surface area covered by particles.

Assuming a deposition surface area A_0, of which an area A is covered ($X = A/A_0$), the mass deposition rate to the uncovered surface will be given by:

$$\frac{dm}{dt} = (A_0 - A) C V_d \qquad (2)$$

where m = mass of particulate matter deposited, C = concentration of particulate matter in the atmosphere, and V_d = deposition velocity.

A simple relation between mass deposited and area covered results from considering the properties of n particles, each of radius r and density ρ:

$$m = 4/3 \ \pi \ r^3 \ \rho \ n \quad : \quad A = \pi \ r^2 \ n$$

It follows that:

$$\frac{dA}{dt} = \frac{3}{4r\rho} \frac{dm}{dt}$$

Substituting into equation (2) gives

$$\frac{dA}{dt} = (A_0 - A) \quad \text{where} \quad k = (3\,C\,V_d)/(4\,r\,\rho).$$

Under constant conditions, the rate of change in fraction of surface covered is directly proportional to the fraction of surface yet to be covered. Therefore, after integration:

$$A/A_0 = X = 1 - \exp\,(-kt)$$

where *k* is a constant, known as the soiling constant, and is a function of particle size distribution and dynamics and t is time.

It follows that reflectance will change according to:

$$R = Rp + (Ro - Rp)\,\exp\,(-kt).$$

In most situations, $R_p \ll R_0$, so the above equation reduces to:

$$R = R_0\,\exp\,(-kt) \tag{3}$$

An alternative approach (Lanting, 1986) assumes the specific absorptivity of PEC is responsible for soiling, eliminating the assumptions about particle size. This model assumes that the exposed surface is covered at a uniform rate and that reflectance decreases with increased light absorption by deposited particles, which is determined by the thickness of the soiling layer. The theory is similar to that for X-ray attenuation and results in an exponential dependence as described by equation (3), but in this case

$$k = 2\alpha\,C\,V_d$$

where α is the specific absorptivity of PEC and the factor 2 is included to account for the fact that light passes through deposited particles twice on a smooth surface. Specific absorptivity is assumed to be $1 \times 10^4\,\text{m}^2\,\text{kg}^{-1}$. Attempts to model the soiling process, using the results from field studies, have normally employed equations (1) or (3).

Haynie (1986) examined the soiling of a gloss painted vertical surface and described his results as:

$$R = R.\,\exp\,(-0.0003\,[0.0363C_f + 0.29C_c]\,t)$$

where R and RO are reflectance and original reflectance, respectively; C_f and C_c, are fine and coarse mode particle concentrations in $\mu g/m^3$, respectively, and t is time in weeks of exposure.

The road tunnel experiments carried out in England by the Middlesex University team (Mansfield and Hamilton, 1989) were well described by the equation:

$$\Delta R = 100 \ (1 - e^{-1.92t}), \tag{a}$$

where t is time in years.

The level of black smoke in the tunnel was estimated as 246 $\mu g/m^3$.

These authors also related the rate of soiling with the atmospheric levels of black smoke using their results for white wood plate exposures on the top of a roof in the London urban area either using the empirical equation of Beloin and Haynie (1975), equations (c) and (d), or an exponential fit, equation (b):

$$\Delta R = 100 \ (1 - e^{-0.044 \ C_{BS} \ t}) \tag{b}$$

$$\Delta R = 2.8 \ (C_{TSP} \ t)^{1/2}, \quad \text{for exposed surfaces} \tag{c}$$

$$\Delta R = 4.2 \ (C_{TSP} \ t)^{1/2}, \quad \text{for sheltered surfaces} \tag{d}$$

where C_{BS} is the concentration of black smoke in the atmosphere during the exposure period, in $\mu g/m^3$.

Comparison of equations (a) and (b) shows that, in these two sets of experiments, the soiling constant is not proportional to black smoke concentrations.

Recent experiments (Pio *et al.*, 1998) have used multiple regression techniques in an attempt to identify which air quality parameter (TSP or black smoke or PM_{10} concentration, humidity, rainfall etc.) has most effect on the soiling constant but no clear picture emerged. At this time, soiling rate can be described by either model but the soiling constant is site-specific.

5. Deposition Velocity

Camuffo and Bernardi (1993) demonstrate the importance of microclimatic factors on the deposition of particulate matter and resulting damage due to air pollution on the Trajan Column in Rome. They are able to show that Rome has two daily maxima of air pollution concentrations, and that these occur at times when the deposition regime at the column is very different. The authors examine the effects of temperature, humidity and wind conditions on deposition and show that the column frequently has different processes operating at different locations vertically and around the circumference. Such studies demonstrate the dangers of generalising deposition rates from average values and show that the latter only provide a first order approximation of what actually occurs.

Horvath *et al.* (1996) considered the need to devise a new experimental method for examining the deposition velocity for particles in the size range of atmospheric soot particles, which is in the order of fractions of millimetres per second. Direct determination of the deposition velocity of naturally occurring soot particles by weighing samples is impossible, since any mass increase by deposition is masked by other processes occurring simultaneously (such as gypsum growth). Deposition could be measured in an environment with an extremely high aerosol concentration or by using a method where the aerosol particles deposited on the surface can be determined independently from any other processes causing a mass increase or decrease of the sample. Since the mass of one single aerosol particle of a diameter of 0.1 µm is 5.23×10^{-16} g, the deposit formed in one hour in the above example represents 17 million particles on a surface of 25 cm^2 or 60 particles on a 100 µm by 100 µm field of view in a microscope. This is easily countable if the particles of this size can be made visible. Spherical latex particles, which contain a fluorescent substance, can be obtained in sizes between 0.025 and 3.0 µm as liquid suspensions and from 7 to 165 µm as powders. When using these particles for the deposition investigations, the fluorescent light emitted by the particles is used as a yes-no decision, i.e. seeing a luminous dot simply means that a particle is present at the surface which is inspected in the fluorescent light microscope. Thus even particles having diameters far below the detection limit of the microscope can easily be seen. Any other particles, which do not show fluorescence, are not seen. Preliminary laboratory results were presented in Fig. 3 above.

For experimental measurements outdoors at monuments, special surrogate surfaces were prepared consisting of 3×3 cm sheet metal plates coated with a mixture of cement and glue, in order to imitate the wall-structure of stone as closely as possible. They were magnetically affixed to the six sides of a 3 cm cube. This set-up allowed the investigation of the effect of different orientation of the surrogate surface with respect to the wind on the deposition velocity.

Four atomisers were used to generate the aerosol, consisting of fluorescent latex spheres of diameter 0.6 µm, which was introduced to a tube of approximately half a metre length. A fine wire grid attached at the end of the tube produced a uniform veil of aerosol. The velocity of the aerosol in the tube was about 10 cm per second. When leaving the tube, the particles almost immediately gain the velocity of the surrounding air.

The cube carrying the surrogate surfaces was exposed to the aerosol for a time of 2.5–4 hours. Afterwards the deposited particles on the surrogate surfaces were counted by means of a light microscope provided with an optical filter.

In contrast to other methods such as weighing or washing off the deposit, this procedure permits a direct and unequivocal identification of the examined 0.6 µm particles. The accuracy of the measurements is solely statistically determined when comparing particle deposition on different locations of the surface relative to each other. This fact allowed a reasonably precise determination of the deposition pattern on the cube sides. The deposition velocity is calculated from

$$v_d = v_f \frac{n_s}{n_f}, \quad v_t = \frac{V}{r_f^2 \pi \cdot T_m}$$

where n_s and n_f are the counted number of particles on the filter and on the surrogate surface, respectively, and v_f the face velocity to the filter with r_f as the filter radius, V is the volume of aerosol drawn through the filter, and T_m is the exposure time of the cube to the aerosol.

The deposition experiments in the open were performed in winter and spring at the town hall of Oporto, St. Paul's cathedral in London, and the University of Vienna. All measurements were carried out on the roof of these buildings. The range of measured deposition velocities was within one order of magnitude at each side of the cube. The ratio between the deposition velocities on the six sides stayed remarkably constant. A careful analysis of the deposit showed that the particles were distributed non-homogeneously on the surface. The deposition velocity showed a maximum at the upwind and downwind edges and a minimum somewhere in the middle of all laterally orientated surrogate surfaces (on the top, bottom, left and right sides of the cube). In most cases, the maximum was higher at the upwind edges. In order to study this effect, the cube with the six surrogate surfaces was positioned in the symmetrical axis of a wind tunnel. The deposition velocities were lower than the results from the field measurements and the differences between the deposition velocities of the six sides of the cube were lower because of the altered pressure gradients due to the presence of the walls of the wind tunnel. Nevertheless, the distinct deposition pattern (highest deposition at the edges and higher deposition on the front side compared to the back one) was reproduced in the deposition measurements performed in the wind tunnel. A complicated flow field around, and especially within a few millimetres of the sidewalls of the cube, is evident from these data. Therefore, care must be taken not to over-generalise the conclusions about deposition characteristics from this single example to those found in other experiments. Though a higher deposition was not always found on the upwind edge compared with that on the downwind edge, there was always a minimum in the middle of the surrogate surface. The reason for a higher deposition on the downwind edge is the existence of small variations in the alignment of the cube with respect to the airstream.

The rougher surfaces showed a higher deposition velocity, as could be expected. The lower deposition velocity on the back compared to the smoother surface is within the statistical error. On the other hand, since the back is in the wake of the cube, it may be the case that the surface structure plays only a minor role in the deposition process.

Particles with diameter between about 0.1 and 1 μm have the lowest deposition velocities, because all deposition mechanisms are least effective. For the 0.6 μm particles, the only possible deposition mechanism (assuming no electrophoretic or thermophoretic forces) is turbulent diffusion, even for the lower turbulence. The deposition rate experiments undertaken by Pesava *et al.* (1999) have demonstrated that, although there are important differences in conditions, existing literature values for deposition to rough surfaces accord very well with measured rates to surrogate building surfaces.

Using a cube for deposition experiments, one can determine deposition velocities in all main orientations with respect to the wind. Usually, however, an object or an ornament is attached to a plane surface, which can be the floor or the wall of a building. The presence of a plane surface like a wall alters the flow field by introducing a boundary layer adjacent to the wall.

To study this special case, a box of 3.0 cm width, 2.3 cm height, and 8.6 cm length was placed on the floor of the wind tunnel, with the cross-section of 3.0, 2.3 cm facing the wind. The height of the box was approximately equal to the boundary layer thickness at the measuring station. The front was tilted to the left at an angle of 1.5° to the channel axis. This set-up allows the investigation of an alignment effect on the deposition pattern, which was suspected to produce the higher deposition that sometimes occurred on the downwind edge of the cube.

The deposition velocities on the left side of the box are lower than on the right side at both the upwind and downwind edges. Due to the tilt, the left side experiences an adverse pressure gradient, which extends the separation region, compared to the right side. On the right side the edge effect is more pronounced, hence the deposition is higher, particularly at the downwind edge and the right edge of the front surface of the box.

The deposition velocity on all sides, except the top surface, decreases from top to bottom. This is the result of the boundary layer: the rate at which particles are transported to the surface diminishes with decreasing velocity relative to the location where the particle concentration of the free airstream is sampled.

Turbulence intensities in the boundary layer of the atmosphere are dependent on height, velocity component, and eddy scale (Hall and Emmott 1994). The lateral and vertical components of the large-scale turbulence intensity have the same effect as fluctuations in wind direction. Only the small scale eddies act in the same way as the velocity fluctuations in the wind tunnel. This is the reason that the deposition velocities in the experiments performed in the open were higher compared to the wind tunnel. Both the alignment effect and the higher turbulence intensity enhance the deposition. In addition, there are considerable microscale variations due to complexities in the flow field of bluff objects. Thus, one has to proceed with caution when deposition velocities from laboratory data are used to calculate particle deposition rates for real buildings in the open.

6. Indoor Soiling

Ligocki *et al.* (1993) studied potential soiling of works of art. Concentrations and chemical composition of suspended particles were measured in both the fine and coarse size modes inside and outside five museums in Southern California in both summer and winter. The seasonally averaged indoor/outdoor ratios for particle mass concentrations ranged from 0.16 to 0.96 for fine particles and from 0.06 to 0.53 for coarse particles, with lower values observed for buildings with sophisticated ventilation systems that include filters for particle removal. Museums with deliberate particle filtration systems showed indoor fine particle concentrations generally averaging less than 10 µg/m. One museum, a historic house in Los Angeles with no environmental control system showed indoor fine particle concentrations that averaged 60 µg/m^3. Fine particle concentrations at this site were almost identical to outdoor concentrations and therefore significant fractions of the dark-coloured fine elemental carbon and soil dust particles present in the outdoor environment may constitute a soiling hazard to displayed works of art. Analysis of indoor versus outdoor concentrations of major chemical species indicated that indoor sources of organics may exist at all sites, but that none of the other measured species appear to have major indoor sources at the museums studied. This finding is interesting since smoking and use coal or wood burners are forbidden, and these are the usual source of raised indoor particulate matter. All sites do appear to have detectable black carbon values indoors. This finding that fine particle chemistry inside a museum was similar to the results outside was also reported by DeBock *et al.* (1996) for the Correr Museum in Venice. They demonstrated differences in the coarse particles but reported great similarities in the fine particulate.

A simplified steady state material balance model (Nazaroff and Cass, 1991) was used to explore the relationship between indoor and outdoor fine particle components to attempt to reveal indoor source contributions. Given that similarly sized particles ought to have the same contribution from outside, since their dynamics will be similar, it was interesting to see that modelled outdoor contributions for elemental carbon were sometimes higher than for organic matter or sulphate, implying that filtration efficiency was lower for these particles. Average ratios for elemental carbon in the indoor fine fraction compared to outdoor values ranged from 0.2 to 0.5 for buildings with fine particle filter systems and it is therefore clear that there is a risk of indoor soiling wherever elemental carbon values are high outdoors, even if filtration is undertaken.

The Nazaroff and Cass (1991) model applied above was also used to examine possible control strategies at the historic house museum, Sepulveda House. According to model results, the soiling rate can be reduced by at least two orders of magnitude through practical application of methods that included reducing the building ventilation rate, increasing the effectiveness of particle filtration, reducing the particle deposition velocity onto surfaces of concern, placing objects within display cases or glass frames, managing a site to achieve lower outdoor aerosol concentrations, and eliminating indoor particle sources. Combining improved filtration with either a reduced ventilation rate for the entire building or low-air-exchange display cases would be likely to extend the time taken to achieve noticeable soiling from years to centuries at a site like Sepulveda House, which might be considered to be analogous to many European cultural buildings, being in relatively polluted environments and with little or no existing ventilation control.

7. Economics of Soiling

Several types of economic costs can be associated with material damage and soiling, which may include the reduction in service life of a material and additional required maintenance, including cleaning. There may be a decrease in utility, a need to substitute a more expensive material and additional costs of protecting susceptible artefacts.

The relative importance of these various losses will vary between materials and between locations. The cost of materials is only one component in assessment of the value of the building, and clearly the same materials will have different apparent values when a modern dwelling is being considered compared to when the damage is occurring on an ancient monument. There are a number of methods that have been used to estimate the magnitude of the effects of air pollution.

The first of these is known as the damage function method, which in the context of the effects of air pollution on materials, is most advanced in the consideration of corrosion, especially from SO_2, NO_x and O_3. This field has progressed to the point where sensible first order calculations of associated costs can be made, at least for modern materials. A recent evaluation of the current situation was undertaken by the workshop on Economic Evaluation of Air Pollution Abatement and Damage to Buildings including Cultural Heritage, which took place in Stockholm on 23–25 January 1996. The workshop was organised jointly by the International Co-operative Programme on Effects of Air Pollution on Materials, including Historic and Cultural Monuments (1CP Materials) and the Task Force on Economic Aspects of Abatement Strategies.

One of the conclusions adopted by the workshop was that a methodology for corrosion cost assessment of the built environment had been established. It is based on dose-response functions, damage functions, modelling of the most important environmental degradation factors, statistical building inventories to establish the stock-at-risk and an estimation of the costs. It has been developed and used in the Czech Republic, Germany, Norway, Sweden, the United Kingdom and the United States. Attempts have been made to extrapolate these results to the European level (ApSimon and Cowell, 1996).

The damage function approach requires a number of steps to be elucidated:

- A *dose-response function* relates the dose of pollution, measured in ambient concentration and/or deposition, to the rate of material corrosion. As far as corrosion is concerned, ICP Materials has developed dose-response functions for a number of pollutants and materials. Even taking corrosion effects in isolation, there are as yet no satisfactory dose-response functions for particulate matter and, as discussed above, the soiling models are not yet sufficiently advanced to produce reliable functions to relate reflectance change to atmospheric concentration.

- A *physical damage function* links the rate of material damage (i.e. only corrosion damage at this point; soiling damage could perhaps be broadly estimated) to the time of replacement or maintenance of the material. There are additional difficulties when assessing the soiling damage function since the requirement to clean a structure is not a direct function of damage alone but relates to a perception on the part of the local population or the building manager. Different levels of soiling will be tolerated in different locations and at different times.
- A *cost function* links changes in the time of replacement, cleaning or repainting to monetary cost.
- An *economic damage* function links cost to the dose of pollution by an amalgamation of the previous steps.

Once the economic damage functions have been established, cost calculations can be performed after identification of the stock-at-risk. There are a number of methods for drawing up building inventories, assessing the relative proportions of each different material and for evaluating the degree of damage based on lifetime assessments. The following categories can be distinguished — residential, industrial and other buildings; infrastructure; and cultural monuments. Stock-at-risk data can be collected at different levels of detail: (i) building types; (ii) construction materials; and (iii) repair/replacement materials. The identification of groups of buildings or parts of a building with statistically the same mixture of material (known as "identikits") is an important part of the methods developed. These identikits have to be adapted to local conditions when applied in different locations.

There are very large gaps in the stock-at-risk data as far as cultural heritage is concerned. The biggest difficulty in estimating damage to cultural monuments is related to the economic valuation. However, there are also problems in identifying the stock-at-risk and determining which buildings should be considered as cultural monuments. In

Italy, a national risk mapping exercise for cultural heritage is under way which singles out the risk from air pollution damage among a list of other factors.

The US EPA Air Quality Criteria Document (United States Environmental Protection Agency, 1996) reviews some of the literature that covers economic losses associated with soiling. The following section is based on this review, since much of the material covered refers to EPA reports that have not been seen in the original. The references are retained in the text. Haynie (1990) examined the potential effects of PM_{10} non-attainment on the costs of repainting exterior residential walls due to soiling in 123 counties in the United States. The analysis was based on a damage function methodology developed for an earlier risk assessment of soiling of painted exterior residential walls (Haynie, 1989). The database was updated with 1988 and 1989 data. The costs of soiling to exterior residential walls were estimated to be US$1 billion.

An experimentally determined soiling function for unsheltered, vertically exposed house paint was used to determine painting frequency (Haynie and Lemmons, 1990). An equation was set up to express paint life in integer years because the painting of exterior surfaces is usually controlled by season (weather). Different values for normal paint life without soiling and levels of unacceptable soiling could be used in the equation. If four was taken as the most likely average paint life for other than soiling reasons, then painting because of soiling would be likely to be done at one-, two-, or three-year intervals.

Soiling costs by county were calculated and ranked by decreasing amounts and the logarithm of costs plotted by rank. The plot consisted of three distinct straight lines with intersections at ranks 4 and 45. The calculated cost values provide a reasonable ranking of the soiling problem by county, but do not necessarily reflect actual painting cost associated with extreme concentrations of particles. Households exposed to extremes are not expected to respond with average behaviour. The authors concluded that repainting costs could be lowered if:

(1) Individuals can learn to live with higher particle pollution, accepting greater reductions in reflectance before painting;

(2) Painted surfaces were washed rather than repainted; and
(3) If materials or paint colours that do not tend to show dirt were used.

Extrapolating the middle distribution of costs to the top four ranked counties reduces their estimated costs considerably. For example, Maricopa County, AR, was calculated to rank first at US$70.2 million if all households painted each year as predicted, but was calculated to be only US$29.7 million based on the distribution extrapolation.

Based on these calculations and error analysis, the national soiling costs associated with repainting the exterior walls of houses probably were within the range of US$400 to US$800 million a year in 1990. This sector represents about 70% of the exterior paint market, so that extrapolating to all exterior paint surfaces gives a range of from US$570 to US$1140 million (Haynie and Lemmons, 1990).

Math Tech Inc. (1990) assessed the effects of acidic deposition on painted wood surfaces using individual maintenance behaviour data. The effects were a function of the repainting frequency of the houses as well as pollution levels.

The loss of amenity or direct financial losses can also be estimated econometrically in studies that do not use the physical damage approach to derive monetary economic damages. These approaches have been used to relate changes in air pollution directly with the economic value of avoidance or mitigation of damages. Such studies are prone to errors in accounting for all factors that affect cost other than air quality. In general, all approaches to estimating costs of air pollution effects on materials are limited by the difficulty in quantifying the human response to damage based upon the ability and the incentive to pay additional costs (Yocom and Grappone, 1976).

McClelland *et al.* (1991), conducted a field study valuing eastern visibility using the contingent valuation method. Given the problem of embedding between closely associated attributes, the survey instrument provided for separation of the visibility, soiling, and health components of the willingness-to-pay estimates. Households were found to be willing to pay US$2.70 per $\mu g/m^3$ change in particle pollution to avoid soiling effects.

8. Costs to Cultural Heritage Caused by Soiling

As described above, many past studies that have tried to relate air pollution damage to construction or cultural materials have often used the avoided maintenance cost approach to estimate the benefits of reduced emissions, i.e. they have calculated the savings implied from a reduction in maintenance costs due to reduced damage rates. However, maintenance costs are probably not the correct measure of the benefits derived by society from reduction of damage to cultural resources. An estimate of economic value is given by the public's willingness to pay (WTP) for reduced damages or, equivalently, their willingness to accept (WTA) these damages. WTP to prevent damage may be larger, smaller or equal to maintenance or mitigation costs. Original valuation studies are expensive but, in the case of cultural resources, calculation of maintenance costs may seriously underestimate the damages, since there are likely to be substantial non-use values. Valuation studies must also examine the effects of maintenance practices themselves, since they may not prevent injuries from occurring, with part of the value being irreversibly lost when the original material is altered or replaced.

Mourato, (1997) presented an overview of the existing literature on the economic valuation of cultural goods with a special focus on damages arising from air pollution. She found that cultural heritage valuation studies were scarce and limited in scope and content. With a few exceptions, they were mostly confined to finding a price for the goods in question, without properly describing the extent of the quality change that occurred, without any systematic attempt to test for validity or reliability of the estimates produced or to fully explore the nature of people's preferences towards cultural goods. Only three papers dealt specifically with the estimation of air pollution damage.

Navrud (1992) reported results from a contingent valuation study to assess the value of losses to visitors from corrosion of Nidaros Cathedral in Trondheim, Norway, from air pollution. A sample of 163 Norwegian and foreign visitors were asked to value two alternative options: a "restoration option" where all the cathedral's original parts were replaced, versus a "preservation option" where the remaining original parts were preserved through reduced air pollution. There was no statistically significant difference in the mean annual WTP for each option: 318 NOK (US$49) per person for the preservation option versus 278 NOK (1992 US$43) for the full restoration scenario. This is a fascinating result, in that it implies that, on average, those surveyed did not seem to derive any extra benefits from having the original, as opposed to a restored, building. However, 65% of respondents said that the original meant more to them than the restored church and this group had a significantly higher WTP for preservation than for restoration. Mourato concludes that the results from this study indicated that there was a significant WTP for preservation of the Nidaros cathedral from air pollution damage — although no extra benefit is attached to preserving original parts rather than replacing them. However, the authors did not report: (i) what payment vehicle was used (i.e. tax or donations); (ii) what the damages from air pollution were; (iii) what exactly did both preservation and restoration options imply in terms of future rates of decay; or (iv) any validity tests on the WTP estimates. Hence, the accuracy of their results cannot be established.

Grosclaude and Soguel (1994) estimated the social cost of damages from air pollution from traffic to historical buildings in Neuchatel (Switzerland). Two hundred residents were asked about their willingness to make a voluntary contribution to a fund to be established to maintain 16 historical limestone buildings. Respondents were shown pictures of the 16 buildings and asked to identify those in need of immediate conservation measures. An average of six buildings was selected. The elicitation procedure included an open-ended WTP question followed by a bidding game to determine the maximum WTP amount. Forty-three per cent of respondents declared a zero WTP — of those, 25% did not consider the damage to historical buildings to be a problem (the "indifferents" or "true zeros"), 19% could not afford the payment (the "non-solvents") while the remaining 56% were considered "false zeros" and hence classified as "free riders". The average observed annual WTP to repair damages to historical buildings was SFr. 172 (US$123) for the total sample and SFr. 192 (US$137 with the exclusion of "true zeros"; the average predicted WTP is SFr. 121 (US$86) for this latter group. On aggregate, the annual social cost of damage caused to historical buildings by road traffic air pollution is estimated to be SFr. 1.7 million (US$1.2 million) with a cost SFr. 283,000 per building (US$202,000).

Mourato pointed out that no indication was given of what the maintenance works would imply in terms of aesthetic improvements or future rates of corrosion, the link between damages to historic buildings and road traffic air pollution was extremely tenuous, an essentially arbitrary procedure was adopted to screen valid responses by differentiating between respondent's motivations and no mention was made to other cultural resources to which Neuchatel citizens might also attribute some value.

Morey *et al.* (1997) conducted a comprehensive stated preference valuation study of damage from acid deposition to marble monuments and historical buildings in Washington, DC. Considerable effort was put into specification the pollution damage scenarios: respondents were presented with descriptions, photos and a map of outdoor marble monuments in Washington DC; written descriptions of erosion and chemical alteration with supporting illustrations were given; digitally enhanced photos in a time line were used to suggest damage levels from air pollution now, in 75 and in 150 years time; the same procedure was used to depict the reduction in the rates of deterioration that would arise from specified preservation programmes. Four possible evolution patterns were considered: the current injury levels "baseline" and three alternative damage reduction programmes. The three latter options delayed the appearance of weathering effects in the time line: 25% later in Option A, 50% in option B and 100% in Option C. The proposed scenario involved hypothetical preservation programmes based on (imaginary) chemical coating procedures.

A random sample of 272 paid respondents was interviewed in two locations (Boston and Philadelphia), with the chosen procedure being group presentations followed by individual answering of a questionnaire. The chosen elicitation format was pair-wise choices with different combinations of two attributes: price (a one-time household payment) and preservation option, with nine and four different levels, respectively (including a zero priced baseline option). A follow-up payment card was included as a comparator. The payment mechanism was not specified.

Predicted mean household WTP (from "male" respondents — women were found to be willing to pay twice as much as men; instead of averaging out responses, the authors adopted the "conservative" approach of treating every respondent as male) was US$26 for a 25% shift in the injury time line (Option A), US$38 for a 50% shift (Option B) and US$56 for a 100% shift (Option C). Assuming that the WTP estimates were sufficiently representative of the populations from which the samples were taken (Boston and Philadelphia), mean aggregate estimates were calculated to be US$50, US$72 and US$105 million, respectively for preservation Options A, B and C.

Mourato considered this study to be the best available for estimating economic damages from air pollution on cultural resources. It was the first attempt to link the changes in the goods, as perceived by the consumer, to the source of the damage, as described in the scientific literature. The survey comprehensively described the goods to be valued and the proposed changes with an extensive use of visual aids. It recognised the fact that the relevant valuation issue is to achieve a decline in the rate of deterioration and not an unattainable goal of complete preservation.

The currently available literature is insufficient to provide overall estimates of the value of cultural resources in general, and of the damages caused by air pollution in particular, with any degree of reliability or precision. However, people do seem to place a value on cultural heritage and these early estimates indicate that this may amount to a considerable sum. A lot of work clearly remains to be done, with many more surveys needed. Any attempt to generalise from particular case studies needs to examine the possibilities for benefit transfer. Here there will be many difficulties, for example differences in materials and use make it difficult to generalise the way in which changes in air pollution damage will affect values as perceived by consumers. Further research will also need to examine how measured WTP can be transferred between societies. In the case of cultural heritage buildings, many of which are unique, this has additional problems (it is important not to seem to imply that Stonehenge is somehow more important than, say, the Acropolis). A direct and uncritical comparison of the Grosclaude and Soguel (1994) and Navrud (1992) studies seems to imply that an average of six buildings in Neuchatel has, for some reason, a considerably larger value to the local population than the Nidaros cathedral has to Norwegians. In relative terms, the former estimate corresponds to 0.32% of Switzerland's per capita income while the latter amounts to only 0.17% of Norway's per capita GNP. Navrud (1992) has demonstrated the difficulty of transferring health impact of air pollution values measured in one country for use in another.

Nevertheless, valuation studies are valuable because they remain the only way to account for benefits that may be unrelated to any specific use but simply arise from the knowledge that cultural assets exist (existence value) and will be available for the enjoyment of other people (altruistic motive) and future generations (bequest value) or for one's own possible future use (option value). These benefits are usually known as non-use values and are thought to be a significant proportion of total value in the case of cultural heritage.

In the case of damage due to soiling, where the dose response functions are not well elucidated as discussed above, valuation studies may be particularly effective. Soiling damage would seem to be particularly suitable for translation into descriptions of visible effects on cultural resources that are detectable to the untrained eye.

9. Conclusions

Soiling is an integrating process, which occurs over a number of years and there need to be ongoing studies to relate soiling to air quality and meteorology. Modelling is clearly important and we see scope for further theoretical approaches to develop a fundamental understanding of the empirical relationships developed in these studies.

The aim of achieving improved models describing the deposition and accumulation of pollutants under real conditions has been addressed by a number of innovative experimental studies. The fractal model work showed that realistic time-based soiling curves could be constructed using the model. These ideas can be readily developed, and as the fractal model becomes more complex with additional data from further interpretation of the experimental results on deposition rate, size distribution and adhesiveness of particles, soiling curves will be produced which will form the basis for prediction of effects of different air pollution and remediation strategies.

Analysis of the formation and structure of particle agglomerates may also assist in the treatment of the crusts regarding cleaning and protection of surfaces. Given that the strength of the agglomerate is determined by particle contacts, then selection of a cleaning regime that utilised the least effort to break these contacts can be specified. The fractal model is still a relatively simple one and its initial finding, that early cleaning may be the most efficient strategy for mono-sized particles, will need to be further investigated to include such factors as adhesion, stickiness and size distributions. These factors may have a significant effect on the number of grain contacts and the compactness of the agglomerate, which will have a corresponding influence on any recommended cleaning strategy.

Despite a large amount of qualitative evidence that air pollution can cause substantial harm to culturally important buildings and monuments, there remains little quantitative data to link pollutant exposure with material damage, and trends are inconclusive. Clearly, more quantitative evidence is required if control strategies are to be effective in reducing the impacts of particle soiling in urban areas.

It seems reasonable to suggest that economic costs associated with atmospheric pollution can be described and that households are willing to pay positive amounts to reduce material damage and soiling. However, perception of nuisance varies and, although blackening of buildings has been shown by surveys of public attitudes to traffic nuisance to be a factor that people resent (McCrae and Williams, 1994; Williams, and McCrae, 1995,) there have been no studies that examine these effects in relation to concentration, particle size and chemical composition. Indeed it would be difficult to imagine how this might be achieved and yet without such information it is very difficult and highly uncertain to quantify the relationship between ambient particle concentrations, soiling and associated economic cost.

Acknowledgements

We would like to acknowledge the help of all of our colleagues in the Urban Pollution Research Centre and all of our co-workers on our various soiling projects. The funding for the work was largely from the EU Framework programmes, to which we are very grateful.

References

Allen A.G., Harrison R.M. and Erisman J.W. (1989) Field measurements of the dissociation of ammonium nitrate and ammonium chloride aerosols. *Atmos. Env.* **23**, 1591–1599.

ApSimon H.M. and Cowell D. (1996) The benefits of reduced damage to buildings from abatement of sulphur dioxide emissions. *Energy Policy* **24**, 651–654.

Beloin N.J. and Haynie F.H. (1975) Soiling of building materials. *J. Air Pollut. Control Assoc.* **25**, 399–403.

Brimblecombe P. (1987) The big smoke: a history of air pollution in London since medieval times. Methuen and Co.

Brimblecombe P., Blades N., Camuffo D., Sturaro G., Valentino A., Geysels K., van Grieken K., Busse H.-J., Kim O., Ulrych U. and Wieser M. (1999) The indoor environment of a modern museum building, the Sainsbury Centre for Visual Arts, Norwich, UK. *Indoor Air* **9**,146–164

Brooks K. and Schwar M.J.R. (1987) Dust deposition and the soiling of glossy surfaces. *Env. Pollut.* **43**, 129–141.

Broughton, G.F.J., Bower J.S., Clark H. and Willis P.G. (1996) *Air Pollution in the UK*, AEAT-2238.

Camuffo D., DelMonte M. and Sabbioni C. (1983) Origin and growth mechanisms of the sulphated crusts on urban limestone. *Water Air Soil Pollut.* **19**, 351–359.

Camuffo D. (1992) Acid rain and deterioration of monuments — how old is the phenomenon? *Atmos. Env. Part B Urban Atmos.* **26**, 241–247.

Camuffo D. and Bernardi A. (1993) Microclimatic factors affecting the Trajan Column. *Sci. Total Env.* **128**, 227–255.

Clarke A.G., Willison M.J. and Zeki E.M. (1984) A comparison of urban and rural aerosol composition using dichotomous samplers. *Atmos. Env.* **18**, 1767–1775.

Creighton N.P., Lioy P.J., Haynie F.H., Lemmons T.J., Miller J.L. and Gerhart J. (1990) Soiling by atmospheric aerosols in an urban industrial area. *J. Air Waste Manag. Assoc.* **40**, 1285–1289

Crump J.G. and Seinfeld J.H. (1981) Turbulent deposition and gravitational sedimentation of an aerosol in a vessel of arbitrary shape. *J. Aerosol Sci.* **12**, 405–415.

DeBock L.A., van Grieken R.E., Camuffo D. and Grime G.W. (1996) Microanalysis of museum aerosols to elucidate the soiling of paintings: case of the Correr Museum, Venice, Italy. *Env. Sci. Technol.* **30**, 3341–3350.

DelMonte M., Sabbioni C. and Vittorio O. (1981) Airborne carbon particles and marble deterioration. *Atmos. Env.* **15**, 645–652.

Gray H.A. and Cass G.R. (1998) Source contributions to atmospheric fine carbon particle concentrations. *Atmos. Env.* **32**, 3805–3825.

Grosclaude P. and Soguel N. (1994) Valuing damage to historic buildings using a contingent market: a case study of road traffic externalities. *J. Env. Planning Manag.* **37**, 279–287.

Hall D.J. and Emmott M.A. (1994) Some comments on the effects of atmospheric turbulence on ambient particle samplers. *J. Aerosol Sci.* **25**, 355–366.

Hamilton R.S. and Mansfield T.A. (1991) Airborne particulate elemental carbon — its sources, transport and contribution to dark smoke and soiling. *Atmos. Env. Part A Gen. Topics* **25**, 715–723.

Hamilton R.S. and Mansfield T.A. (1992) The soiling of materials in the ambient atmosphere. *Atmos. Env. Part A — Gen. Topics* **26**, 3291–3296.

Harrison R.M. and Jones M. (1995) The chemical composition of airborne particles in the UK atmosphere. *Sci. Total Env.* **168**, 195–214.

Haynie F.H. and Spence J.W. (1984) Air pollution damage to exterior household paints. *J. Air Pollut. Control Assoc.* **34**, 941–944.

Haynie F.H. (1986) Theoretical model of soiling of surfaces by airborne particles in aerosols: research, risk assessment and control strategies. In *Proceedings of the Second US-Dutch International Symposium*, ed. Lee S.D. Lewis Publishers, Williamsburg, VA.

Haynie F.H. (1989) *Risk Assessment of Particulate Matter Soiling of Exterior House Paints.* EPA/600/X-89/304, US Environmental Protection Agency.

Haynie F.H. (1990) *Effects of PM$_{10}$ Non-Attainment on Soiling of Painted Surfaces.* US Environmental Protection Agency Research Triangle Park, NC, USA.

Haynie F.H. and Lemmons T.J. (1990) Particulate matter soiling of exterior paints at a rural site. *Aerosol Sci. Technol.* **13**, 353–367.

Hollander W. and Pohlmann G. (1991) Measurement of the influence of directed particle motion on the turbulent particle deposition velocity by means of laser doppler anemometry. *Particle and Particle Systems Characterization* **8**, 12–15.

Horvath H. (1993) Atmospheric light absorption — a review. *Atmos. Env. Part A Gen. Topics* **27**, 293–317.

Horvath H., (1996) Black smoke as a surrogate for PM(10) in health studies. *Atmos. Env.* **30**, 2649–2650.

Horvath H., Pesava P., Toprak S. and Aksu R. (1996) Technique for measuring the deposition velocity of particulate matter to building surfaces. *Sci. Total Env.* **190**, 255–258.

Kendall M., Hamilton R., Williams I.D. and Revitt D.M. (1994) Smoke emissions from petrol and diesel engined vehicles in the UK. In *Proceedings of the Dedicated Conference on the Motor Vehicle and the Environment — Demands of the Nineties and Beyond, Aachen, Germany, 1994.*

Lanting R.W. (1986) Black smoke and soiling in aerosols: research, risk assessment and control strategies. In *Proceedings of the Second U.S.–Dutch International Symposium*, ed. Lee S.D. Williamsburg, VA.

Ligocki M.P., Salmon L.G., Fall T., Jones M.C., Nazaroff W.W. and Cass G.R. (1993) Characteristics of airborne particles inside Southern California museums. *Atmos. Env. Part A Gen. Topics* **27**, 697–711.

Lipfert F.W. (1989) Atmospheric damage to calcareous stones: comparison and reconciliation of recent experimental findings. *Atmos. Env.* **23**, 415–429.

Mansfield T.A. and Hamilton R.S. (1989) The soiling of materials: models and measurement in a road tunnel in Man and his ecosystem. In *Proceedings of the 8th World Clean Air Congress*, eds. Brasser L.J. and Mulder W.C. Elsevier Science Publishers B.V., Amsterdam, The Netherlands, pp. 353–357.

Martin K.G. and Souprounovich A.N. (1986) Soiling of building materials about Melbourne — an exposure study. *Clean Air* **20**, 95–100.

Massey S.W. (1993) Fractal analysis of stone fabric. In *3rd International Colloquium "Materials and Restoration"*. Expert Verlag, pp. 1712–1719.

Math Tech Inc. *Economic Assessment of Materials Damage in the South Coast Air Basin: A Case Study of Acid Deposition Effects on Painted Wood Surfaces Using Individual Maintenance Behaviour Data.* California Air Resources Board Sacramento, CA, USA.

McClelland G., Shulze W., Waldman D., Irwin J., Schenk D., Stewart T., Deck L. and Thayer M. (1991) *Valuing Eastern Visibility: A Field Test of the Contingent Valuation Method.* Draft report to the USEPA, Co-operative agreement no CR-815183-01-3, Washington, DC, USA.

McCrae I.S. and Williams I.D. (1994) Road traffic pollution and public nuisance. *Sci. Total Env.* **147**, 81–91.

Miguel A.H., Kirchstetter T.W., Harley R.A. and Hering S.V. (1998) On-road emissions of particulate polycyclic aromatic hydrocarbons and black carbon from gasoline acid diesel vehicles. *Env. Sci. Technol.* **32**, 450–455.

Milford J.B. and Davidson C.I. (1985) The sizes of particulate trace elements in the atmosphere — a review. *J. Air Pollut. Control Assoc.* **35**, 1249–1260.

Moorcroft J.S. and Laxen D.P.H. (1990) Assessment of dust nuisance. *Env. Health*, 215–217.

Morey E., Rossmann K., Chestnut L. and Ragland S. (1997) *Valuing Acid Deposition Injuries to Cultural Resources.* Report for the National Acid Precipitation Assessment Program.

Mourato S. (1997) *Effects of Air Pollution on Cultural Heritage: A Survey of Economic Valuation Studies.* Prepared for the United Nations Economic Commission for Europe, Task Force on Economic Aspects of Abatement Strategies, Madrid.

Muir D. and Laxen D.P.H. (1995) Black smoke as a surrogate for PM_{10} in health studies. *Atmos. Env.* **29**, 959–962.

Muir D. and Laxen D.P.H. (1996) Black smoke as a surrogate for PM_{10} in health studies — reply. *Atmos. Env.* **30**, 2648–2648.

Navrud S. (1992) Norway. In *Pricing the European Environment*, ed. Navrud S. Scandinavian University Press, Oslo.

Nazaroff W.W. and Cass G.R. (1991) Protecting museum collections from soiling due to the deposition of airborne particles. *Atmos. Env. Part A Gen. Topics* **25**, 841–852.

Nicholson K.W. (1988) The dry deposition of small particles — a review of experimental measurements. *Atmos. Env.* **22**, 2653–2666.

Parker A. (1955) The destructive effects of air pollution on materials. National Smoke Abatement Society, London pp. 3–15.

Pesava P., Aksu R., Toprak S., Horvath H. and Seidl S. (1999) Dry deposition of particles to building surfaces and soiling. *Sci. Total Env.* **235**, 25–35.

Pio C.A., Ramos M.M. and Duarte A.C. (1998) Atmospheric aerosol and soiling of external surfaces in an urban environment. *Atmos. Env.* **32**, 1979–1989.

Quality of Urban Air Review Group (1993) *Urban Air Quality in the United Kingdom*. First report of the Quality of Urban Air Review Group, HMSO.

Spence J.W. and Haynie F.H. (1972) *Paint Technology and Air Pollution: A Survey and Economic Assessment*. AP-103 U.S. Environmental Protection Agency, Office of Air Pollution Research Triangle Park, NC, USA.

Terrat M.N. and Joumard R. (1990) The measurement of soiling. *Sci. Total Env.* **93**, 131–138.

Torraca G. (1988) Air pollution and the conservation of building materials. *Durab. Build. Mater.* **5**, 383–392.

United States Environmental Protection Agency (1996) *Air Quality Criteria for Particulate Matter*. EPA/600/P-95/001bF, United States Environmental Protection Agency Washington, DC, USA.

Vallack H.W. and Shillito D.E. (1998) Suggested guidelines for deposited ambient dust. *Atmos. Env.* **32**, 2737–2744.

Vawda Y., Colbeck I., Harrison R.M. and Nicholson K.W. (1989) The effects of particle size on deposition rates. *J. Aerosol Sci.* **20**, 1155–1158.

Watt J., Massey S.W. and Kendall M. (2000) Fractal modelling of particulate deposition in the development of black crusts on stone. In *9th International Congress on Deterioration and Conservation of Stone*, ed. Fassina V. Elsevier Science B.V., pp. 637–646.

Williams I.D. and McCrae I.S. (1995) Road traffic nuisance in residential and commercial areas. *Sci. Total Env.* **169**, 75–82.

Yocom J.E. and Grappone N. (1976) *Effects of Power Plant Emissions on Materials*. EPRI/EC-139, Electric Power Research Institute, Palo Alto, CA, USA.

Yoon Y.H. and Brimblecombe P. (2000) Contribution of dust at floor level to particle deposit within the Sainsbury Centre for Visual Arts. *Studies in Conservation* **45**, 127–137.

CHAPTER 11

CHANGES IN SOILING PATTERNS OVER TIME ON THE CATHEDRAL OF LEARNING

W. Tang, C.I. Davidson, S. Finger, V. Etyemezian,
M.F. Striegel and S.I. Sherwood

1. Introduction

Air pollution has been responsible for increasing the deterioration rate of structures made of limestone and marble. These calcareous stones are vulnerable to attack by several natural processes, including dissolution by rain, physical stresses such as freeze-thaw cycles, and microbial activity on the stone surface. Anthropogenic pollutant emissions may accelerate the natural erosion, resulting in pitting, cracking and discolouration (Sherwood *et al.*, 1990).

One major cause of anthropogenic degradation is the formation of gypsum. This is the product of the reaction between calcium carbonate and acidic forms of sulphur, such as sulphuric acid. Gypsum occupies a greater volume than calcium carbonate, causing the stone to crack when gypsum forms. Furthermore, gypsum is more soluble in rain water than calcium carbonate, and thus rain may wash off the gypsum deposits, leaving pits in the stone. Gypsum is also more porous than the original stone, and can serve as an effective surface for the deposition of particles such as soot carbon. This can lead to discolouration of the stone, which is well-documented for limestone buildings (Sherwood *et al.*, 1990).

In previous work (Etyemezian *et al.*, 2000; Davidson *et al.*, 2000), we hypothesised that soiling on a tall limestone building in Pittsburgh, Pennsylvania has been the result of two competing processes. The first is the deposition of pollutants on the stone, especially on sections of stone where gypsum has formed. The second process is washoff of soiled material by rain. Soiling patterns change when the relative rates of pollutant deposition and rain washoff vary over time.

In this paper, the changes in soiling patterns over time on the same limestone building have been studied based on archival photographs, analysis of soiling on architectural features, and computer modeling of horizontal rain flux. The results are used to support the hypothesis that soiling is determined mainly by the two competing processes.

2. Changes of Soiling Patterns

2.1. Background

The structure of interest is the Cathedral of Learning, a National Historic Landmark located in the densely populated Oakland area of Pittsburgh. This is a 42-storey Indiana limestone building on the University of Pittsburgh campus, constructed between 1926 and 1937. Two sides of the Cathedral have extensive soiling, particularly on the lower half of the building. Since the time of construction, soiling has been evident as a result of numerous air pollutant sources within a few kilometres of the building. These include steel manufacturing plants that employ coke ovens and blast furnaces, a coal-burning steam heating plant, motor vehicle traffic, coal-burning railroads and riverboats, and a large number of domestic coal combustion sources such as home furnaces.

The Cathedral of Learning has attributes which lend themselves to this type of study. The location of the Cathedral in an urban setting with detailed records of pollutant sources and concentrations allows the study of changes in soiling over time. Archival photographs of the building are available to permit comparisons between observed soiling and pollutant levels. The Cathedral is the tallest structure in the area, and thus prevailing wind and weather patterns will not be altered much due to surrounding structures, at least on the upper levels. There are certain architectural features repeated at many locations on the walls of the Cathedral, which can be used to quantify the amount of soiling at different elevations. The Cathedral has never been cleaned, except by natural rainfall. Finally, the Cathedral has historic and cultural value in its own right.

Since the time of construction of the Cathedral, Pittsburgh has experienced substantial changes in air pollution concentrations (Davidson, 1979). During the 1930s and 1940s, coal burning was responsible for the city's notorious smoke levels. In the late 1940s and throughout the 1950s, enforcement of smoke control ordinances reduced pollutant emissions. Stricter county ordinances in 1960 and 1970 as well as new federal regulations resulted in continued decreases in air pollution levels. Figure 1 presents annual average dustfall in or near the downtown area over an 85-year period. The continued decrease through recent decades is evident, especially the rapid

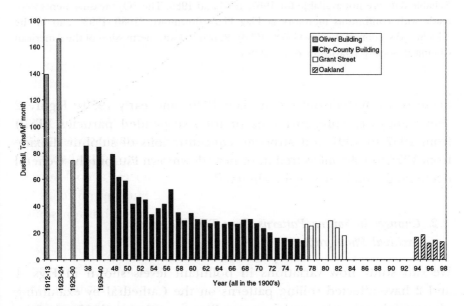

Fig. 1. Annual average dustfall at four different sites in or near downtown Pittsburgh. No data are available for 1980 or for 1984–1993. Data are taken from archival and recent records at the Air Quality Program of the Allegheny County Health Department.

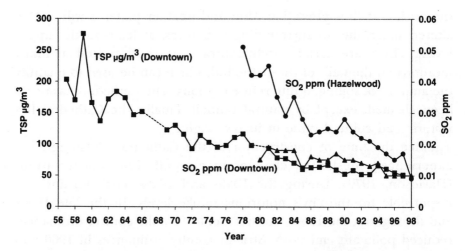

Fig. 2. Annual arithmetic average concentrations of total suspended particles (TSP) and sulphur dioxide (SO₂) in or near downtown Pittsburgh. The TSP measurements were made with high volume samplers at two downtown locations: the County Office Building (1957–1982) and Flag Plaza (1983–1997), as part of the National Air Sampling Network and the Air Quality Program of the Allegheny County Health Department. Reliable data are not available for 1967, 1968 and 1980. The SO₂ measurements were made with continuous monitors at Flag Plaza downtown (1980–1998) and in the Hazelwood section of the city (1978–1998). Reprinted with permission of the American Chemical Society (Davidson *et al.*, 2000).

decrease in dustfall during the late 1940s and early 1950s. Figure 2 shows airborne concentrations of total suspended particles (TSP) from 1957 to 1997 and airborne concentrations of sulphur dioxide from 1978 to 1998 measured in or near downtown Pittsburgh. General decreasing trends are again observed.

2.2. Changes in Soiling Patterns Over Time Based on Archival Photographs

We can study how variations in pollutant levels shown in Figs. 1 and 2 have affected soiling patterns on the Cathedral by examining photographs taken in previous years. For convenience, the faces of the Cathedral have been labelled with names of nearby streets. These are Bigelow Boulevard (southwest side of the building), Fifth Avenue (northwest side), Bellefield Avenue (northeast side), and Forbes Avenue (southeast side).

The first pair of photos in Fig. 3 shows the Bigelow Boulevard side of the building. The photo from 1937 shows heavy soiling from approximately the fourth floor to the roof, except for the very top floor. An interesting feature of the building is that between 1929 and 1931, stonework was installed from the fourth floor up to the top. After that, work was stopped due to financial problems. It was not until the mid-1930s that stones for the lowest four floors were added

1937   1995

Fig. 3. The Bigelow façade of the Cathedral of Learning in 1937 and 1995. (Sources: 1937 — University Archives, University of Pittsburgh; 1995 — Justin Parkhurst.)

and the top floor was reconstructed (Brown, 1987). Because of this fact, white stones at the top of the Cathedral (visible in the 1937 photo on the right side) have been a reference point to distinguish soiled sections from white ones. Using this reference point suggests that a significant amount of soiling occurred during 1931 to 1937. This coincides with heavy smoke in the 1930s throughout the region. In contrast, the photograph from 1995 shows that the entire Bigelow face of the building is almost free of soiling. Since no cleaning or renovation has been done since the completion of construction, it is likely that the reduction in soiling has been influenced by natural processes over time.

The photographs of the Forbes façade from 1930 to 1995 in Fig. 4 are useful for observing changes in soiling patterns over several time intervals. The photo from 1930 shows that the surface was relatively white shortly after laying the exterior stonework began in 1929. By the late 1930s, however, the surface had become highly soiled. The soiling progressively decreases in the three later photographs. Generally, the decrease of soiling on this face is not as dramatic as that on the Bigelow Boulevard side. However, the later photos show that the top one-third of the building on the Forbes Avenue face has been virtually free of soiling since 1989.

In addition to observations of the whole building, smaller scale changes on individual sections provide insight into the rain washing process. The location marked with an arrow on the first two photographs points out where there have been notable changes in soiling. On the photo from the late 1930s, the region below the arrow shows a demarcation between soiled and white areas on the left side of the Forbes Avenue face. The white region appears as a "notch" in the soiling, which has become enlarged in the downward direction over time. The bottom of the notch reaches the fourth window from the top of the section. By 1962, the notch has reached the middle of the fifth window. The photo from 1989 shows that the notch now reaches between the fifth and sixth windows. By 1995, the notch extends to the sixth window.

These archival photographs suggest that the Cathedral of Learning has been washed by natural rainfall over time, which supports the hypothesis that soiling on building surfaces is the result of a competitive

Fig. 4. Archival photographs of the Forbes Avenue (southeast-facing) side of the Cathedral of Learning on the University of Pittsburgh campus. Changes in soiling patterns, such as those in the region below the arrow in the first two photographs, are apparent by comparison with the later photos. [Sources: 1930 — University Archives, University of Pittsburgh; Late 1930s — Carnegie Library of Pittsburgh; 1962 — University Archives, University of Pittsburgh; 1989 — Ferguson Photographic Enterprises; 1995 — Justin Parkhurst. Reprinted with permission of the American Chemical Society (Davidson *et al.*, 2000).]

process between pollutant deposition and rain washoff. The overall trend of annual precipitation in Pittsburgh has been roughly constant over these decades (Etyemezian, *et al.*, 1998). However, airborne concentrations of SO_2 and particles have decreased steadily over the same time period (Davidson, 1979). Thus, those areas of the façade that were soiled in the late 1930s have become white in recent years because the rate of removal of soiled material by rain washing is greater than the rate of soiling by pollutant deposition and chemical reaction. The opposite was true in the 1930s when air pollutant concentrations were considerably greater than at present.

The rates of washoff of soiling have been different on the four faces of the Cathedral. During the early years when pollutant deposition was dominant, soiling was almost uniform on each face, as shown by numerous archival photographs from the 1930s. However, it is likely that the Bigelow face has received a greater rain flux than the Forbes face, as will be discussed below, so that the decrease of soiling on the Bigelow face is much more significant.

2.3. Analysis of Soiling on Architectural Features

To assess quantitatively the patterns of visible damage that have occurred on the Cathedral, the soiling patterns of repeated architectural features have been documented. One such repeated feature is a stone carving 0.56 m × 0.75 m in the shape of a large "X", hereafter referred to as a "cross". There are 226 crosses scattered on all four faces of the Cathedral at different elevations. The soiling patterns on each cross have been sketched and scanned into a computer, and the percentage of discoloured area has been determined (Gould *et al.*, 1993; Lutz *et al.*, 1994; Etyemezian *et al.*, 1995).

Figure 5 shows examples of four sketches with different percentages of soiled area. One sketch shows an ideal, unsoiled cross, while the other three sketches show soiled crosses on different floors on the Bigelow face of the building.

By examining data for all 226 crosses, we can find strong evidence for the hypothesis that pollutant deposition and rain washoff determine soiling patterns on the building. Most sharp edges of the carvings have been cleaned because they are exposed to raindrop impact and

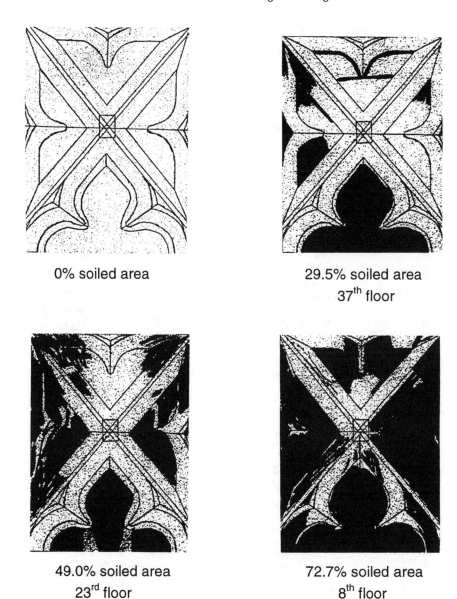

0% soiled area

29.5% soiled area
37th floor

49.0% soiled area
23rd floor

72.7% soiled area
8th floor

Fig. 5. Soiling patterns on crosses carved into the stone, a repeated architectural feature on the Cathedral of Learning. The upper left sketch shows an ideal, "unsoiled" cross, used as a blank in the computations of percent soiled area. The other three sketches are examples of soiled crosses taken from different floors on the Bigelow face of the building.

downward dripping of rainwater. In contrast, the sheltered areas below the edges show more soiling. Even for the crosses with a small percentage of soiled area, the lower centre regions are black. The reason is that very little rainwater can flow over this area because it is sheltered by the edges. All crosses are at least 20% soiled, even in those areas where the flat sections of the wall are nearly entirely white.

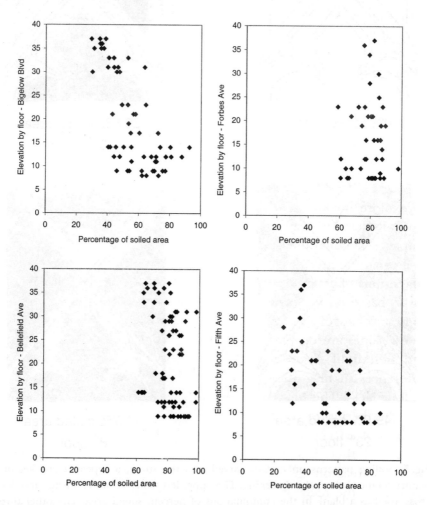

Fig. 6. Elevation versus percentage of soiled area for decorative crosses on the four faces of the Cathedral of Learning. [Reprinted with permission from the American Chemical Society (Davidson *et al.*, 2000).]

Data from the crosses are presented in Fig. 6 as plots of percentage of area soiled versus elevation. From this figure, a negative correlation between percent area soiled and elevation by floor is observed, especially for the Bigelow Boulevard and Fifth Avenue faces where more soiled areas have been washed off. However, in the 1930s, a vertically uniform soiling pattern had been present for each face of the Cathedral as suggested by Fig. 3 and other archival photographs. Furthermore, sampling of pollutants at the Cathedral has suggested that the distributions of airborne concentrations and deposition rates are roughly uniform with height at the building (Etyemezian *et al.*, 1998). This implies that differences in the amounts of soiled area as a function of height observed today are the result of differences in rain flux rather than differences in pollutant levels.

2.4. Comparison of Soiling Patterns with Modeling of Rain Impingement

To explore further the role of rain washoff, the delivery of rain to the walls has been approximated by modelling the Cathedral as a simple rectangular block. Each face has been divided into 15 sections (3 horizontal by 5 vertical sections), with each section having dimensions of 10 m × 32 m. The modelling results are presented in Fig. 7, based on the original data from Etyemezian *et al.* (2000). The highest values of rain flux are on the Fifth Avenue face while the lowest values are on the Forbes Avenue face; the Bigelow Boulevard and Bellefield Avenue faces have intermediate values. Despite differences in magnitude, patterns of rain delivery are similar for all four faces. The top sections of each face receive the greatest rain flux. Furthermore, the amount of rain delivered to the individual sections of a face increase with distance from the vertical centreline. In general, there is a reasonable, although not exact, correspondence between areas on the surface of the Cathedral that are white and sections of the rectangular block in the model that receive the most rain. Thus, the rain modelling results are consistent with both the observations of the crosses and with overall soiling patterns on the building.

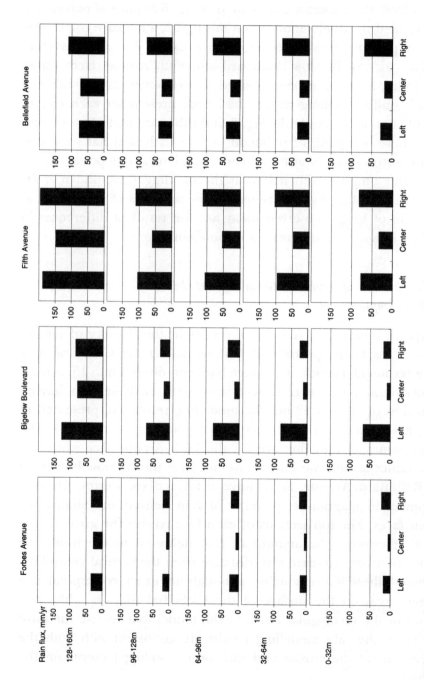

Fig. 7. Modelling results of rain fluxes on each face of the Cathedral of Learning. (Each face is divided into 3 × 5 = 15 sections with dimensions of 10 m × 32 m.)

3. Conclusions

We hypothesise that soiling on calcareous stone buildings is the result of two competing processes: deposition of pollutants and washoff by rain. We have explored this hypothesis for the Cathedral of Learning, a 42-storey limestone building constructed in the 1920s and 1930s in Pittsburgh, Pennsylvania. Several approaches have been used in this effort.

Comparison of archival with recent photographs shows that the Cathedral developed extensive soiling shortly after the completion of construction, and the soiling has decreased over the past several decades. This is consistent with decreasing trends in airborne pollutant concentrations and deposition rates since smoke control began in Pittsburgh in the late 1940s. Rainfall was roughly constant over the 60-year history of the building, and thus it is likely that the process of decrease of soiling began when pollutant levels had fallen sufficiently.

We have studied architectural features on the Cathedral to assess quantitatively the washoff of soiling on the building. By examining crosses carved into the stone at over 200 locations on the building, we have found that those carvings with the highest percentages of soiled area occur at the lowest elevations on the building. This is true despite airborne concentration and deposition data, suggesting a roughly uniform distribution of pollutants with elevation at the building. The findings are consistent with the result of modelling raindrop impingement: the lower floors of the building surface receive a smaller rain flux than the higher floors. Overall, these results suggest that soiling on buildings in polluted areas is determined largely by both pollutant deposition and by delivery of rain to the building surface.

Acknowledgments

We are grateful to the University of Pittsburgh for permission to conduct this research. We acknowledge the help of graduate students T. Gould, M. Lutz and R. Strader as well as the help of the following undergraduate students in the work reported here: N. Barabas,

B. Caster, W.R. Chan, D. Deal, J. Doan, B. Fontaine, D. Iorio, S. Jindal, S. Jutahkiti, J. Lee, S. Luckjiff, F. Molfetta, G. Mumpower, M. Nicholson, J. Parkhurst, K. Pinkston, A. Puttaiah and K. Vance. We also acknowledge the assistance of R. Savukas and R. Westman in providing data from the Air Quality Program of the Allegheny County Health Department. The help of D. Fagnelli of the Facilities Management Department at the University of Pittsburgh was valuable in obtaining information about the Cathedral of Learning. Additional assistance and suggestions on the research were provided by J. Andelman, L. Cartwright, A. Fiore, S. Pandis, A. Robinson, M. Small and J. Tarr. Assistance in obtaining and copying photographs was obtained from the Carnegie Library of Pittsburgh, the University of Pittsburgh Archives, Ferguson Photographic Enterprises, and Photography and Graphic Services at Carnegie Mellon. This work was funded by the United States National Park Service Cooperative Agreements CA042419005 and 1443CA00196035.

References

Brown M.M. (1987) *The Cathedral of Learning: Concept, Design, Construction.* Catalog of the University Art Gallery, Henry Clay Frick Fine Arts Building, University of Pittsburgh, pp. 17–18.

Davidson C.I. (1979) Air pollution in Pittsburgh: a historical perspective. *J. Air Pollut. Control Assoc.* **29**, 1035–1041.

Davidson C.I., Tang W., Finger S., Etyemezian V., Striegel M.F. and Sherwood S.I. (2000) Soiling patterns on a tall limestone building: changes over sixty years. *Env. Sci. Technol.* **34**, 560–565.

Etyemezian V., Davidson C.I., *et al.* (1995) *Influence of Atmospheric Pollutants on Soiling of a Limestone Building Surface.* Progress Report for the National Park Service.

Etyemezian V., Davidson C.I., Finger S., Striegel M.F., Barabas N. and Chow J. (1998) Vertical gradients of pollutant concentrations and deposition fluxes at a tall limestone building. *J. A. I. C.* **37**, 187–210.

Etyemezian V., Davidson C.I., Zufall M., Dai W., Finger S. and Striegel M.F. (2000) Impingement of rain drops on a tall building. *Atmos. Env.* **34**, 2399–2412.

Gould T.R., Davidson C.I., *et al.* (1993) *Influence of Atmospheric Pollutant Concentrations and Deposition Rates on Soiling of a Limestone Building Surface.* Progress Report for the National Park Service.

Lutz M.R., Davidson C.I., *et al.* (1994) *Influence of Atmospheric Pollutant Concentrations and Deposition Rates on Soiling of a Limestone Building Surface 1991–1994.* Progress Report for National Park Service.

Sherwood S.I., Gatz D.F., Hosker R.P. *et al.* (1990) *Progress of Deposition to Structures.* National Acid Precipitation Assessment Program, Acidic Deposition: State of Science and Technology, Vol. 3, Report 20.

CHAPTER 12

EXPOSURE OF BUILDINGS TO POLLUTANTS IN URBAN AREAS: A REVIEW OF THE CONTRIBUTIONS FROM DIFFERENT SOURCES

D.J. Hall,[*] A.M. Spanton,[*] V. Kukadia[†] and S. Walker[†]

1. Introduction

The high density of human activities in urban areas leads to a related high density of emitted air pollutants. As a result, urban areas tend to be amongst both the major sources of and sufferers from pollutants. Both pollutant sources and their effects are multifarious. Pollutants carried by the wind disperse to cover steadily increasing areas and those from different sources overlap and combine to generate the overall level of exposure that is experienced at particular sites.

From the point of view of the recipient, whether human or inanimate, the individual sources that form this total exposure may not be readily distinguishable. For example, a given level of exposure may come from a relatively small polluting source close at hand or a large source at a much greater distance. Similarly, the contribution of individual sources to the total cannot be readily distinguished at the point of reception unless they have markedly different characteristics.

[*]Envirobods Ltd., 13, Badminton Close, Stevenage, Herts SG2 8SR, UK.
[†]BRE Ltd., Garston, Watford, Herts WD25 9XX, UK.

This distinction is of more than academic interest. There is a natural and practical desire to avoid the effects of air pollution where possible. From a local point of view there may be no effective possibility of avoiding large polluting sources at long distances, which may pervade the whole urban area. However, there are greater possibilities for avoiding and controlling local sources so that their effects can be diminished by planning and regulation. This has been long recognised and the collection of urban emission inventories and their use in modelling pollution levels in urban areas became common practice as soon as it was a practicable activity (see for example, Benarie, 1980). Similarly, it is normal for static polluting discharges from combustion plant and industrial processes to be regulated by way of both emission controls and minimum heights of discharge stacks. In more recent years, there has been an additional growing control over mobile emission sources (Schwela and Zali, 1999).

Apart from the desire for pollution control in a public sense, there is also a more individual element of interest, for example in the choice of preferred sites for buildings or of the characteristics of buildings designed to suit particular sites. A particular example is the choice of ventilation systems and the placement of ventilation inlets and exhausts in order to minimise internal contamination problems. Dealing with such matters raises queries about the nature of the exposure to pollutants on specific sites and how it may vary from one site to another, over the area of the site or with the passage of time. Similarly, it may be useful to know whether pollutants are mainly due to local sources or otherwise. A further question may be the degree to which pollution levels on a particular site can be predicted from data collected elsewhere within the urban area.

It is difficult to deal with problems of this sort without some understanding of the character of discharged pollutants from different sources, especially those at varying distances, and the way in which their dispersion contributes to the overall pollution levels at a particular site. This is, of course, the usual subject of numerical dispersion studies of urban pollution. However, such studies do not cover all urban areas and it is unusual to find them using small resolution or short time scales. The smallest unit of area likely to be found in such a study is a kilometre square and the shortest time scale about an hour. At these scales, vertical and lateral gradients in the pollutant concentration are likely to be small and many models do not, in any case, consider such matters. However, individual interests are often on significantly smaller scales than this. The site of a single building, for example, may occupy scales of only 100 m or less. Within these smaller scales, there can be quite large variations in pollution levels in space and time which are significant from this smaller scale point of view.

A major factor in the dispersion of pollutants in urban areas is the severe topography due to the large number of surface obstacles, mainly buildings but including a variety of other structures. The effect of the urban topography on wind effects is well known and much discussed (see for example, Cook, 1985 and 1990), but its effects on the dispersion of pollutants at short ranges within the "urban canopy" is less well understood. The requirements for simple short range dispersion models in urban areas have recently been reviewed by Hall *et al.* (1996b) who discuss these problems in more detail.

The relationship between pollution levels at these different scales is the subject of discussion in the present paper. It is dealt with in a generic way, rather than with reference to specific pollutants or pollution problems and illustrated with examples from dispersion studies, both at full and model scales, and with pollution monitoring data from urban areas. One of the most important features that distinguishes the character of different pollution sources is their distance from the point of interest, so that in the discussion that follows, the characteristics of pollution sources are mainly treated in terms of increasing distance.

2. Dispersion Over Different Scales in Urban Areas

2.1. The Definition of Scales and Spatial Variability

The term "scale" of air pollutants can apply to both space and time. Both are important for various reasons and they are connected by way of the mean windspeed, which sweeps pollutants over a specific area in a given time while dispersing them. In his thorough and very interesting monograph on "The Design of Air Quality Monitoring Networks", Munn (1981) covers a number of aspects of urban pollution that are relevant to the present discussion. In considering time and space variability, he defines some characteristic scales of spatial air pollution patterns, which in order of increasing size are:

Microscale (0–100 m)
Neighbourhood scale (100–2000 m)
Urban scale (5–50 km)
Regional scale (100–1000 km)
Continental, hemispheric and global scales.

The ranges are as defined by Munn and do not overlap. However, as the values are approximate this is not important. These are roughly the different scale orders that have been used here, as they correspond fairly well to the different types of dispersion patterns that occur. Only the first three scale orders are of direct interest as variables within the scale of urban areas. Pollution levels at the larger two scales would show no significant variation over the scale of an urban area and would class as contributors to the "background level" of pollutants in the area. Quite what should be classed as "background" levels of pollution is considered later in the paper.

In a similar way, Munn defined a number of characteristic time scales associated with pollutants:

Minute to minute variations
The daily (diurnal) cycle
Large scale weather fluctuations (3–5 days)
Weekly emission cycles
Annual emission and weather cycles.

It can be seen that these are associated either with natural meteorological cycles or with patterns of human activity. For the shorter ranges mainly of interest here, it is useful to subdivide the shortest time scale into two further divisions related to the stochastic (i.e. the unsteady) nature of dispersing pollutant plumes:

Times below which the fluctuating characteristics of dispersing plumes are apparent (typically seconds)

Times beyond which the time-averaged concentrations in dispersing plumes are stable (typically minutes).

The time scales associated with this subdivision are a little arbitrary as the time scales for stable time-averaged concentrations do not have finite limits and are also affected by large scale wind disturbances, of the sort readily generated by windflows around buildings in urban areas.

It was noted earlier that the spatial and time scales are related to some extent by the windspeed. Thus, taking typical UK windspeeds around the mean, say 3–5 m^{-1} per second, the microscale and shorter neighbourhood scales are associated with time scales of seconds, the longer neighbourhood scales and lower urban scales with time scales of minutes and the upper urban scales and regional scales with time scales of hours. The continental and larger scales correspond to time scales approaching days and beyond.

It is the combination of the multiplicity of discharged polluting sources from this range of distances in the generally upwind direction that produces (simply by the summing of instantaneous pollutant concentrations) the overall pollution level that is experienced at the point of interest. In principal, the contributions of these sources to the overall level are not distinguishable and one of the major reasons for dispersion modelling is as a means of making this distinction. However, there are differences in the character of the contribution from sources at different distances, both in their spatial and tem-poral characteristics, that help to identify them and their contribution to the total. The ensuing discussion attempts to characterise these differences and the ways in which they affect the combined pollution levels experienced at the point of interest.

2.2. Dispersion at Short (Microscale) Ranges

The definition of "short ranges" is a little arbitrary, but implies here sources which are mostly within direct line of site of the point of interest or where dispersing plume widths are relatively small compared with the scale of the surface obstacles. The practical range may be anywhere from 10 m to 1 km, and in exceptional cases further. Thus, this may embrace both the microscale and neighbourhood scales defined by Munn. For low level sources in urban areas, the limiting distances are usually of a few hundred metres and are thus closer to Munn's microscales.

The critical characteristics of dispersing plumes within this scale range are their high pollutant concentrations, small footprint, rapidly fluctuating intensities and (especially within urban areas) meandering qualities. The majority of polluting discharges are from "point" sources, that is their cross-section at discharge is small compared with the plume cross-section even at short distances. This applies, for example, to most combustion and process plant discharges, vehicle exhausts and many ventilation discharges. In these cases, undisturbed discharge plumes are highly concentrated; most of the pollutant material is contained within a subtended angle from the source of about 10°. Thus within the small area of the dispersing plume, there are high concentrations of pollutant, with little pollutant material elsewhere. Also, because of the stochastic (unsteady) nature of dispersion, there will be large variations of concentration within the plume itself. Figure 1 (taken from Hall and Kukadia, 1994) is a visualised dispersing plume which shows clearly the intermittent nature of the concentrations locally within the plume. Because of the stochastic properties of dispersing plumes, at distances within a few hundred metres, it is possible for small regions of undiluted source material to exist and this has been observed in field experiments by Jones (1983).

Fig. 1. Visualised dispersing plume showing variable pollutant content.

Besides the internal variability of pollutant concentration in the plume, the wind environment near the ground in urban areas usually shows a high degree of variability in speed and direction due to the aerodynamic disturbances from buildings and other large structures. This introduces an additional variability in the plume path, usually described as plume "meandering". Thus the overall characteristic of exposure to a dispersing pollutant at short ranges is usually of relatively infrequent, highly intermittent exposure over short periods (of the order of seconds) to relatively high pollutant concentrations. The importance of rapid fluctuations in pollutant levels in a number of applications has recently been discussed by Jones (1996).

In urban areas one strongly modifying factor to this description is the ability of buildings and other large structures to generate rapid dispersion of discharged pollutants over large areas in the aerodynamic wake regions behind their downstream faces. This generally results in a much larger area of exposure to the pollutants, though at lower concentrations than with the slender plume, and in a more persistent and time-continuous form.

Figure 2 shows an illustration (taken from Hall and Kukadia, 1994) of these two types of exposure to a dispersing plume at short ranges. It shows measurements made in the field by Helen Higson of the Environmental Technology Centre (UMIST), of a pollutant plume approaching and dispersing around a rectangular building in otherwise smooth terrain. The plots are of pollutant concentration against time and show two cases, in the undisturbed plume upwind of the building (Case a) and in the region of rapid dispersion in the wake region behind the building (Case b). On both plots, the mean concentration is shown as a broken line. The traces are typical of the two types of dispersion. That in the undisturbed plume shows large variations in concentration, with fluctuations well in excess of the mean, over periods of seconds, and a high degree of intermittency (i.e. there are significant periods when no pollutant is present in the plume). The concentration/time trace downwind of the building shows the continuous presence of pollutant with relatively low levels of fluctuation, so that levels of concentration remain close to the mean. If the plume were dispersing in an urban area, the concentration trace upwind of the building would show additional intermittency

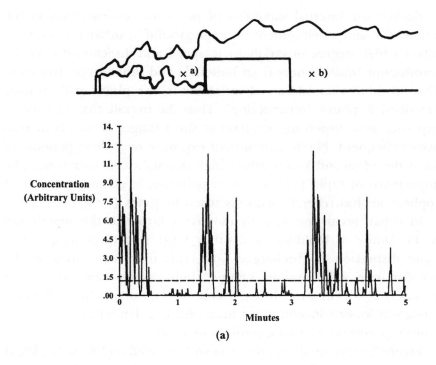

(a)

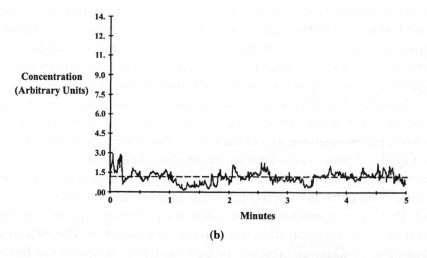

(b)

Fig. 2. Concentration/time measurements of the pollutant in a plume dispersing at short ranges near a rectangular building. (a) In the undisturbed plume just upwind of the building. (b) In the wake region in the lee of the building.

due to meandering of the plume and thus longer periods without the presence of any pollutant.

Exposure to pollutants at short ranges in urban areas is a combination of these two types of dispersion pattern, depending upon the siting of the discharge and the presence of buildings and other surface structures. The form of the exposure need not be one type or the other, it is also possible for both regimes to occur together, so that the exposure is the sum of the two traces (Fig. 2). Figure 3 shows a sketch of a situation in which this will occur, with a polluting discharge on the downwind side of a building. Here, over the longer term the discharge is dispersed in the building wake region, as in Fig. 2(b). However, since the point of discharge is also in the building wake region, in the shorter term there is an additional exposure to the highly concentrated meandering plume in the disturbed flow in the building wake, as in Fig. 2(a).

A practical example of this is shown in Fig. 4 (taken from Hall *et al.*, 1996a), which shows small scale wind tunnel measurements of concentration/time traces at the ground behind a building of similar shape to that of Fig. 2, for a pollutant source just behind the building at different heights ranging from ground level to 1.5 building heights. With the source on the ground, the mean wind carried the dispersing plume away from the point of measurement, so that only the longer

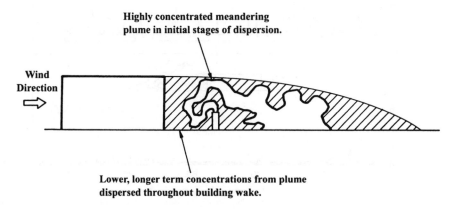

Fig. 3. Sketch of a situation in which the dispersion patterns of Fig. 2(a) and (b) can occur together.

term concentrations in the building wake, as in Fig. 2(b), were observed. However, for the next two pollutant source heights the discharge was within the wake region and the sampling station recorded both the longer term concentrations in the re-circulation regions

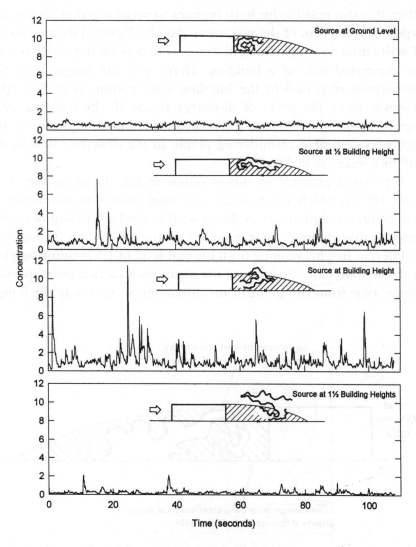

Fig. 4. Concentration/time traces at the ground in the lee of a building due to contaminant sources at different heights also in the lee of the building.

together with shorter term intermittent high concentrations from occasional exposure to the meandering plume. The final source height, at 1.5 building heights, discharged pollutant above the wake region, but part of the plume dispersed downwards and was entrained into the building wake further downwind. The sampling point was again then exposed to the two types of dispersion pattern, but concentrations were lower than before as only part of the dispersing plume was entrained into the wake. In urban areas, where the occupational density of buildings and other structures is high, dispersion patterns like those in Fig. 4 are likely to be the most frequently occurring at short ranges.

Polluting sources at short ranges generate high levels of spatial as well as temporal variability in urban areas. The disturbed windflows that are a feature of urban areas generate a spatial variability that is not only high but which is also sensitive to source position and to the meteorological parameters, especially wind direction. There can, for example, be very large variations in pollutant concentrations across the corners of a street intersection or between the windward and lee

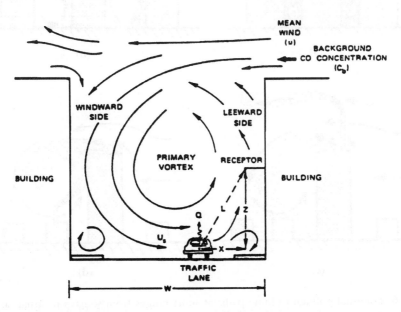

Fig. 5. Simplified flow pattern in a street canyon (from Dabbert *et al.*, 1973).

faces of a building. It is not proposed to discuss this complex subject in detail here, but a few examples of small scale plume behaviour at short distances are given which show the effects clearly enough. Figure 5, taken from Dabbert *et al.* (1973), shows a simplified representation of the flow pattern in the space between buildings in an urban area, frequently described as a "street canyon". Figure 6, from Oke (1987), shows some mildly misleading representative dispersion patterns in urban areas covering a variety of source positions and shows the complex plume paths that can occur. Diagrams like those in Figs. 5 and 6 usually imply a wind square on to the layout of buildings and streets and fail to take into account the additional complexities introduced by the more commonly occurring wind directions, which are skewed diagonally across the street array. Figure 7, from Hoydysh and Dabbert (1994), shows a more realistic example of the large variations in mean concentration that can occur at a street intersection for

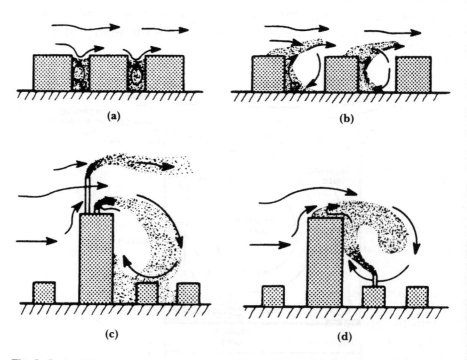

Fig. 6. Potential pollutant plume paths at short ranges from sources in urban areas. (from Oke, 1987).

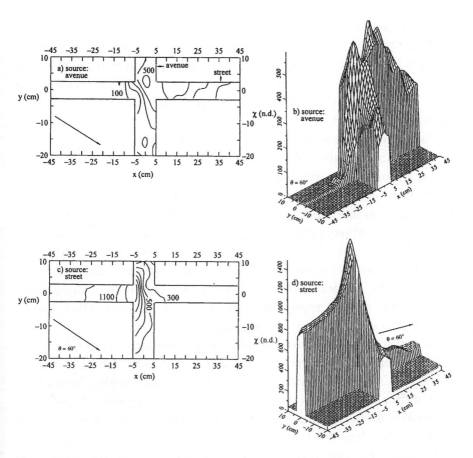

Fig. 7. Mean pollutant concentration patterns at a street junction with a skewed wind direction (from Hoydysh and Dabbert, 1994).

a fixed wind direction skewed across the street array (the measurements were made using a small scale wind tunnel model).

2.3. Dispersion at Neighbourhood Scales (100–2000 m)

Within the greater distances of neighbourhood scales, the high levels of spatial and temporal variability that mark out microscale dispersion patterns reduce. Though spatial and temporal variability in concentration remain, the associated time and distance scales increase. Apart from distance itself, the most important factor influencing this change remains the surface topography, mainly the buildings and other surface obstacles. At these greater ranges, however, it is the size, layout and packing density of the structures in urban areas that are the most important features, rather than the shapes of individual structures and their immediate surroundings, as is more the case with microscale dispersion patterns.

As a rule of thumb, once a source of pollutant is out of line-of-sight in an urban area, the effects of building wakes on dispersion become more important, so dispersion patterns become more stable and the short term variations in concentration should pass from a state like that of Fig. 2(a) to one more resembling that of Fig. 2(b). With increasing distance the diffusing effects of larger numbers of buildings come into play and the variability further reduces. There remain additional complex types of behaviour related to the surface topography. For example, the tendency of pollutants to be channelled along clear paths in the surface topography due to street patterns, which causes higher concentrations in this region.

The spread of pollutants at neighbourhood scales is not presently a well researched subject, though this is a rapidly growing interest (Hall et al.'s (1996b) review discusses most of the existing literature). It is difficult, therefore, to provide a clear description of pollutant dispersion at these distances. As noted above, it is the size, layout and packing density of the buildings that most affect the spread of pollutants. The most recent research suggests that, of the characteristics of the urban form affecting the spread of pollutants, it is the mean height and across-wind widths of the surface structures which has the greatest effect. The typical characteristic of the spread of pollutants at neighbourhood scale is a rapid vertical mixing over (and a little above) the heights of the surface structures in distances covering three to four rows of buildings in the direction of the wind. Beyond this distance, there is a slower rate of vertical spread to greater heights. Lateral spreading is fairly rapid over the individual building widths and further lateral spreading at greater distances depends upon the relationship between building widths and spacing.

It is within the neighbourhood scales and microscales that significant vertical gradients of pollutants can occur. This is a matter of practical interest, for example for human exposure and for the placement of ventilation intakes. As a broad generalisation, it might be considered that pollutant sources at or close to the ground would produce falling levels of pollutant concentration with increasing height, and that pollutant sources at or above the building heights would produce rising pollutant concentrations with increasing height. In urban areas, the dominant near-ground pollutant source is vehicular traffic. A recent emission inventory for the UK West Midlands area has suggested that vehicle emissions are now the major urban polluters (Anon, 1996); similar estimates have been made for Copenhagen and Milan (Vignatti *et al.*, 1996). The dominant pollutant sources at or above building height are mainly discharges from combustion plant and industrial process, for most of which activities there are regulatory requirements in the UK that discharges should be above their immediate surroundings. However, there are in addition a variety of other pollutant discharges at intermediate heights, for example ventilation exhausts (which are often associated with odour problems), standby generators and discharges from some types of gas-fired heating plant.

This broad generalisation for vertical gradients of contamination is of limited reliability. The plume paths sketched in Fig. 6 indicate that at the shorter scales there may be substantial short term local variations in the vertical pollutant gradient for sources at any height from changes in the dispersion patterns due to local aerodynamic effects. The longer term mean of the vertical gradient of pollutants can also vary. Figure 8 shows measurements of the vertical variation of pollutant from sources at the ground in a small scale simulation of an urban area in a wind tunnel, using arrays of cubes set in rows (from Hall *et al.* 1998). Measurements are shown for the surface obstacles occupying 16 and 69% of the ground surface area (the "Area Density"), typical values respectively for a suburban housing estate and the more densely packed central region of an urban area. The pollutant profiles are at distances covering one, three and about ten rows of the surface obstacles, which is a more important criterion than the absolute distance. For a typical urban area with buildings 10–20 m high, the equivalent range of distances was from 100 m to 2 km. The height of the obstacles is marked as a "building height" on the plots, below which the plot is shaded. For the greatest distance, both plots show quite

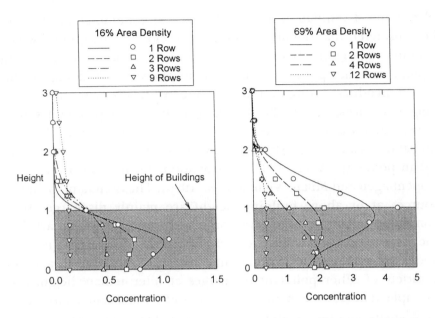

Fig. 8. Vertical profiles of pollutant concentration at different distances through simulated urban arrays with two different densities of building occupation (16% and 69%).

uniform gradients of pollutant to heights well above the obstacles. However, at the shorter distances of a few rows of the obstacles there are sharper gradients, both positive and negative. The measurements show positive gradients near the ground at the shortest distances (behind a single row of obstacles) for both area densities. However, with increasing distances covering additional rows of the obstacles, the gradients become more uniform at the lower area density but pass from positive to negative at the higher area density. In all the cases of distances up to a few rows, the gradient near the top of the obstacles was negative, so that there was a distinct reduction in the level of pollutant at the tops of the obstacles. The implication is that pollutant levels from nearby ground-based sources may well be significantly lower at roof level than lower down, but that below roof level, pollutant levels can vary in complex ways with no certainty as to the vertical gradient of pollutant. It can also be seen in Fig. 8 that, at a given distance from the source, pollutant concentrations are significantly greater for the higher density of obstacle packing than for the lower density.

There seem to be few field measurements of the vertical gradient of pollutants in urban areas. Figure 9 shows some early measurements by Georgii *et al.* (1967) of carbon monoxide (CO) on either side of a street. Since vehicular traffic is the major source of CO, the measurements are largely for a distributed, ground-based source. In this case, the gradient is of reducing concentration with increasing height, but it is interesting that measurements either side of the street are markedly different, the windward side having the lower values. Figure 10 (taken from QUARG, 1993) shows the results of a scan by a remote sensing device (a LIDAR) of the contours of nitrogen dioxide (NO_2) in a London street with dense traffic. The concentrations fall with increasing height above the ground, but show another maximum above building level. It was remarked in this context that the higher maximum might be due to further oxidation of nitrogen monoxide (NO) to NO_2 above the buildings, where more ozone was present. This argument seems a little tenuous for such short time and distance scales, but the higher maximum could equally well be due to

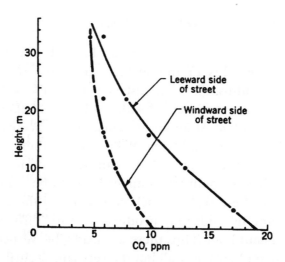

Fig. 9. Vertical profiles of CO concentration on the windward and leeward sides of a street (from Georgii *et al.*, 1967).

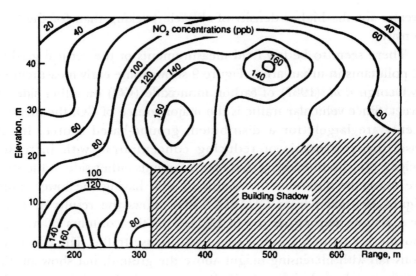

Fig. 10. Remotely sensed contours of NO_2 in a London street with dense traffic (from QUARG, 1993).

a high level discharge from a local combustion plant or some feature of the local dispersion pattern.

2.4. Dispersion at Urban Scales (5–50 km)

Pollutant sources at these distances and beyond, disperse to heights well above the heights of the surface structures and spread over relatively large widths. At 5 km distance, the bulk of the pollutant from a single source is contained within a height of about 250 m and a width of about 1 km. At 50 km distance, the respective heights and widths are about 600 m and 8 km. Pollutants are then starting to mix uniformly within the depth of the surface boundary layer. Also, the large numbers of pollutant sources likely to be contained in the upwind fetch at these scales, which contribute to the total pollution level at a point by addition, produce a more diffuse and slowly changing pollutant level. Thus there are negligible vertical and lateral concentration gradients over all but the very largest surface structures. This is also the regime for which pollutant residence times in the atmosphere are sufficient for chemical processes to occur, for example the oxidation of NO to NO_2, and the generation of photochemical smog.

Most urban modelling studies are carried out at these and greater scales, using mapped inventories of emission data from various pollutant sources and suitable meteorological data. Traditionally, these have been concerned with predicting pollutant levels over relatively long averaging periods, from a day to a year. However, with the recent recommendations of shorter term exposure limits for human exposure to sulphur and nitrogen oxides (EPAQS, 1995 and EEC, 1985) there is a growing interest in making shorter term predictions. Two examples of the results of modelling studies are shown here. The first, due to Ott (1977), in Fig. 11, is an example of the pollutant concentration pattern to be expected from vehicular traffic occupying a grid of streets. It shows the higher levels of pollutants at the junctions and the rapid fall-off of pollutant levels away from the road that appear to occur in practice. This can be compared with the measurements of pollutant concentration at a road junction shown in Fig. 7 from a small scale model experiment, which also show the high concentrations at the junction. The second is due to Timmis and Walker(1989), in

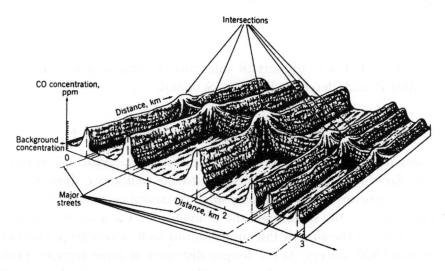

Fig. 11. Map of urban pollutant levels from vehicular traffic on a grid of streets (from Ott, 1977).

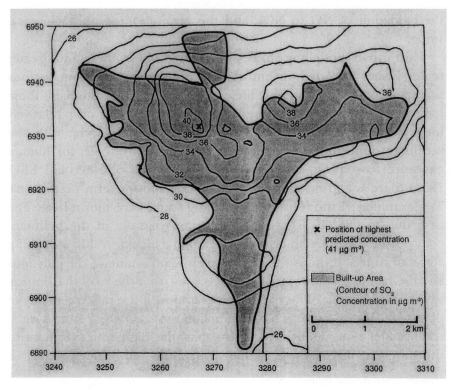

Fig. 12. Contour map of calculated winter mean SO₂ levels in Kirkaldy (from Timmis and Walker, 1989).

Fig. 12, and shows contours of calculated estimates of winter mean sulphur dioxide levels in Kirkaldy, Scotland.

2.5. Dispersion at Regional and Continental Scales (100 km +)

Pollutant sources at these distances uniformly pervade the surface boundary layer, and thus to heights usually well above those of the surface structures, and show only small variations over large areas and long times. At the same time, the pollutant level at a point is usually composed of the sum of the contributions from a very large number of individual sources. At the longer distances of these regimes, even diurnal variations in pollutant discharges are smoothed out and pollutant levels mainly vary with changes in the weather pattern or long term patterns of use. The "upwind" pollutants may have followed complex wind trajectories generated by the weather pattern. These are also the scales at which longer term atmospheric chemical processes occur, such as the further oxidation of nitrogen and sulphur oxides to nitrate and sulphate.

Pollutants at these distances thus constitute the true "background" concentration levels of urban areas in that they pervade the whole area at a uniform level which changes only slowly. They cannot be controlled or avoided within the scales of an urban area. Otts' diagram in Fig. 11 includes a base level of background pollutant concentration to which the local sources additionally contribute.

2.6. The Overall Pollutant Concentration Level Due to the Contribution of Sources at Varying Scales and "Background" Concentrations

It was remarked earlier in this paper that it is the sum of the pollutant concentrations from the multiplicity of pollutant sources at the whole range of upwind distances that produces the overall pollutant concentration level at a point of interest. From the descriptions of the temporal and spatial character of the pollutant levels from sources in the different distance regimes, it will be appreciated how the overall pollutant level is built up from components with a variety of characteristics. Figure 13 shows a hypothetical example of this, with the components from the different distance scales summing to produce the total pollutant level at some point in an urban area. The level of temporal fluctuation and its frequency increases as the scales of the pollutant source distances increase. Thus, the high frequency component of the overall pollutant level is due to the microscale and neighbourhood scale components and the stable long term base level of the pollutant concentration is due to the urban, regional and continental scale components. The spatial variability can be expected to follow the same sort of pattern, with the microscale and neighbourhood scales sources producing the greatest spatial variability and the urban, regional and continental scales, the lowest. It will also be appreciated that it is not readily possible to determine the precise contributions of pollutant sources at the different scales to

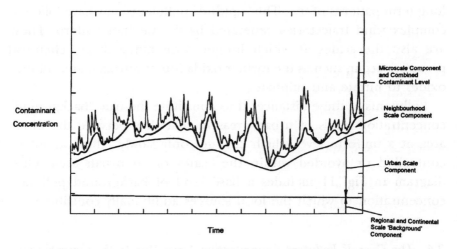

Fig. 13. Hypothetical example of the contributions of pollutants from different source regimes to the combined pollutant level at a point.

the total pollutant concentration level except, within limits, by their different frequencies of fluctuation.

The form of the overall pollutant concentration curve with time in an urban area will depend upon the relative contribution from the different distance regimes. For example, in the UK if there is an anti-cyclonic weather pattern with light easterly winds during a holiday period, then the long range contribution of pollutant sources from Europe will be high and the urban and smaller scale contributions will be low. Thus, the overall curve will show low relative levels of short term fluctuation and low spatial variation. Alternatively, if there are westerly winds carrying relatively uncontaminated air from the Atlantic during a busy working day, the long range contributions will be low but the urban and smaller scale contributions will be relatively high. Thus the overall curve will show a higher level of short term fluctuation over a relatively small "background" concentration.

The question of what should be considered as "background" pollution is not formally defined. It tends to be regarded as a nearly constant, all pervading level of the pollutant over the area which changes only slowly with time. As noted earlier, on this basis it is pollutants dispersing over regional and continental scales that may be regarded the true "background" level of pollutant. However, "background" is also used to describe the longer term average pollutant levels, irrespective of their source. This is a common procedure when, for example, dealing with the additional contribution of a new local source of pollutant. However, in this case it will be appreciated that there will be shorter term excursions in the pollutant level locally which are well above the "background". The definition of a "background" level thus depends to some extent on the purpose to which it is put. From the point of view of, say a single building, any pollutant level which is largely unvarying across the scale of the building might be regarded as a "background" level of pollutant. This would imply spatial scales of the order of a few hundred metres, though the associated temporal scales would be typically some tens of seconds. If, as an alternative, a "background" level was required that indicated the average level of exposure of a building ventilation system over its air exchange rate, a temporal scale of between ten minutes and an hour would be more appropriate, with the implied urban and regional spatial scales.

3. Some Examples of Urban Pollutant Data

3.1. Pollution Monitoring Sites in the West Midlands Area

To provide an illustration of the characteristics of pollutant levels in urban areas, we have used data from the UK National Network sites in the West Midlands to investigate the degree to which pollution levels at a particular site are related to those at different sites within the same urban area. There are three sites normally operating in this large urban area, but for an additional period, recently, BRE carried out measurements of pollutant levels in the same area as part of a building ventilation study (Kukadia and Palmer, 1996 and Kukadia *et al.*, 1996). The ventilation study incorporated both external and internal measurements of various pollutants at a site in the centre of Birmingham. This has in effect provided an additional monitoring site as well as a comparison of indoor/outdoor pollutant levels. The main interest here is in the external measurements. The original report on this study (Hall *et al.*, 1996c) contains additional data to that shown here.

A map of the West Midlands area showing the National Network sites, the site of the building used for the ventilation study and of the meteorological site are shown in Fig. 14.

The buildings used in the ventilation study (site B in Fig. 14) were located in the central commercial district of Birmingham close to a

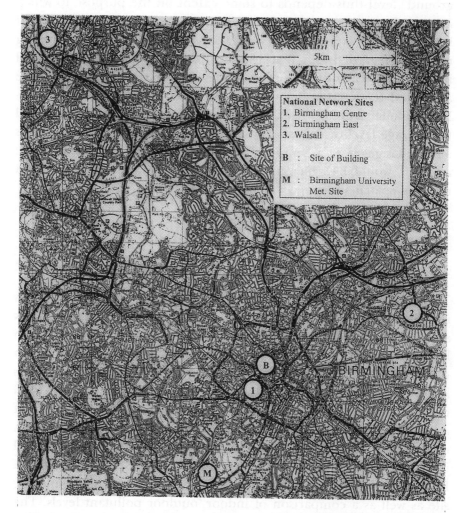

Fig. 14. Map of the West Midlands area showing the various monitoring sites.

busy eight-lane roadway (Great Charles Street). The external measurements were made outside the naturally ventilated building (referred to as the CB Building) about 3 m above the footway at the side of this main road. Of the three network sites (sites 1, 2 and 3 in Fig. 14), the closest to this building is the Birmingham Central site which is about 560 m to the south-west. This site is located in a pedestrianised area (Centenary Square) 100 m from a main road and 10 m from a small carpark. The other two network sites are located further away from the city centre. The Birmingham East site is about 5 km ENE of the building and is sited within the playground of a junior/infant school in a residential area. Finally the third site, at Walsall, about 13.5 km to the north of the city centre, is also a school site in a residential area, but is 200 m west of the M6 motorway (carrying 70,000 vehicles per day) and 500 m north of a smelting plant.

Pollutants monitored at the ventilation study site included sulphur dioxide (SO_2), NO, NO_2 and CO, these are also monitored at the two Birmingham network sites. The Walsall site only monitors NO and NO_2, so we have mainly looked at NO_2 as a common pollutant between the sites. Examples of the latest annual data (taken from AEA, 1995) are reproduced in Fig. 15 for hourly measurements of NO_2 at the three sites. The annual mean levels are broadly similar at 46 ppb (Birmingham Central), 38 ppb (Birmingham East) and 61 ppb (Walsall). However, inspection of the plots in Fig. 15 shows that though there are some observable peaks in common (like that towards the end of December), the shorter term fluctuations show only limited similarities between the sites even on this basis of hourly averages.

Data at the CB Building were available for the period 14–21 February 1996 and so this period was chosen for more detailed study. For the ventilation study, the data were recorded with an averaging time of 5-minutes, however only 15-minute averages were available for the network sites and so 15-minute averages were calculated from the CB Building data. Data for shorter averaging times than 5 minutes is difficult to find; most pollution monitoring instruments need to average over a few minutes to smooth instrument noise from the signal at low levels of concentration. However, there were significant differences between even the 5- and 15-minute averaging times noted

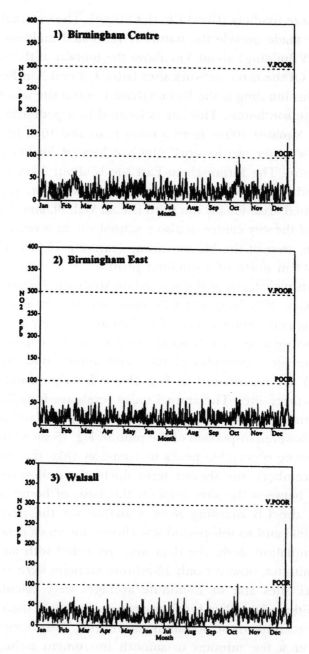

Fig. 15. Annual, hourly mean, measurements of NO_2 for the three National Network Sites in the West Midlands.

above. Figure 16 shows this difference for one example, for SO_2 measurements on 14th February, together with data for a 60-minute averaging time; this latter averaging time matches the hourly mean annual data shown in Fig. 15. The data shows significant additional fluctuations with reducing averaging times, especially for 5-minute averaging time compared with 15. Shorter averaging times than 5 minutes would show much greater fluctuations.

Hourly wind speeds and directions were obtained from Birmingham University as part of the ventilation study. The University is located about 4 km south-west of the city centre, as shown on the map in Fig. 14.

The weather in Birmingham during the week of measurements began with a mobile anti-cyclone crossing the country on the Wednesday and Thursday (14th and 15th of February), resulting in fairly low windspeeds (1–5 m s^{-1}) from the north-west over Birmingham.

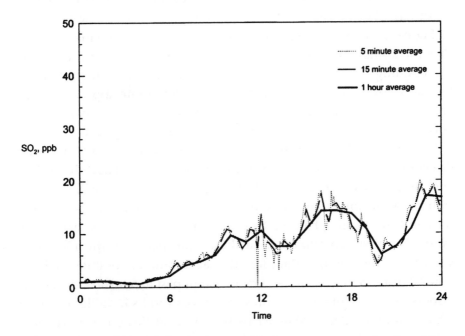

Fig. 16. Effect of 5-, 15- and 60-minute averaging times on concentration fluctuations (SO_2, 14th February).

This period was followed by a return to a changeable west/north-west flow across the country with windspeeds increasing during the second half of Thursday (15th) and during Friday (16th) to 10–14 m per second from the west. There was some intermittent rainfall during Thursday evening and early Friday morning but temperatures were mild. During Saturday (17th), the wind backed towards the south by the end of the day, with windspeeds falling to between 7 and 9 m per second. It was cloudy throughout the day with some rainfall during the evening. On Sunday (18th), a vigorous depression tracked south-eastwards from Scotland to North Germany bringing a very strong northerly airflow across the country. Windspeeds in Birmingham remained around 9 m per second throughout Sunday, Monday and Tuesday (18th, 19th and 20th), apart from a slight lull late on Sunday. The wind direction gradually veered from south-west early on Sunday to north by Monday morning and remained between north and north-west during Monday and Tuesday (19th and 20th). There were frequent rain and snow showers throughout the latter half of the study period (Saturday 17th to Tuesday 20th) and temperatures were low.

Figure 17 shows examples of NO and NO_2 measurements at all the sites for a single day, Wednesday 14th February. Measurements inside the building, when available, are also shown as the solid symbols on the plots. All pollutant levels were relatively low, none approached any limit or guideline value. The mean windspeed for 14th February ranged between 3 and 5 m per second up to about midday, after which it fell steadily to about 1 m per second at around midnight. The wind direction remained consistently in the arc 300°–330°. Thus, the Walsall and Birmingham Central/CB Building sites were nominally aligned with the wind direction, while the Birmingham Central/CB Building sites and the Birmingham east site were nominally aligned across the wind. To a first approximation at these windspeeds, in the 15-minute averaging time of the sampler, the accumulated pollutants from upwind distances of about 1–3 km are swept past it. Similarly, the lateral dispersion of pollutants over these distances results in pollutants over a band of about 200–500 m across the wind (at this greater distance) being averaged across the sampler. Thus, the sampler is averaging pollutant levels over a wedge-shaped area upwind of around

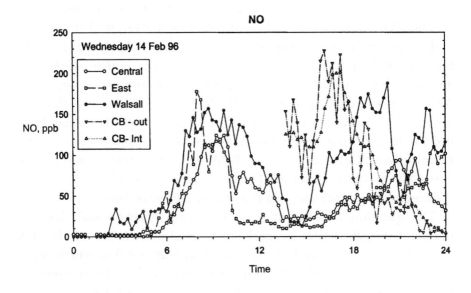

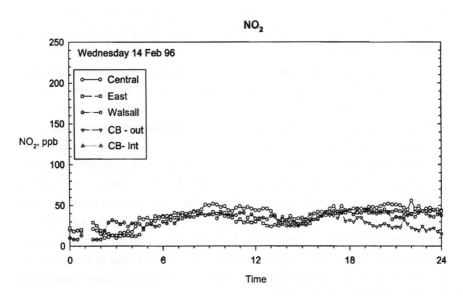

Fig. 17. Measurements of NO and NO_2 for a single day, Wednesday 14th February (15-minute means).

0.25–1 km^2. These distances correspond to the approximate boundary between neighbourhood and urban scales as defined by Munn.

The NO_2 concentrations showed relatively low levels of fluctuation and some similarity in trends between the sites. The NO concentrations showed quite marked variations during the course of the day and high levels of fluctuation even with 15-minute averaging times. Also, on a visual basis, none of the NO concentrations at the sites appeared to show much correlation between one another apart from the broad morning and evening peaks. This applied even to the Birmingham Central and CB Building sites, which were only about 500 m apart. Here, levels of NO_2 were generally higher at the Birmingham Central site, while levels of NO were lower.

The question of what might constitute a "background" level of pollutants for these sites is clearly not a fixed matter. The implication of the term "background" is of some lower limit below which levels do not usually fall, though other definitions have been used for different purposes. In practice, this can only be expressed on some statistical basis which will depend upon the application. For example, a "background" pollutant level for the West Midlands area, encompassing all the sites, might be defined as the common level of pollutant that occurs at all sites, that is the lowest common denominator of the data. In Fig. 17, a lowest common denominator "background" level is effectively the lowest data point at a given time from any site on the plot. All the sites contribute to this minimum at some point. For NO_2, where the level of fluctuation is relatively low, the "background" on this definition usually comprises a significant proportion of the overall concentration at any site. For NO, however, where the level of fluctuation is high, the Birmingham East site produced the lowest concentrations most frequently but not consistently. It will also be appreciated that when there are large fluctuations in pollutant level, such a definition would depend upon the data averaging time chosen. As this increases, the "background" level on this definition would tend towards the long term mean.

Thus for individual sites, it is difficult to define a "background" except on the basis of some pollutant level exceeded at a specific frequency, such as a 10%ile, for some particular averaging time.

3.2. Correlations of Pollutant Levels Between the Sites

In order to quantify the degree to which pollution levels at each site might be related to each other during the week 14–21 February, values of the correlation coefficient (r^2) were calculated between data from pairs of sites for each day and for each pollutant.

This was first investigated with the Birmingham Central and CB Building sites. Since these were the closest, it was thought that they ought to show the highest correlations. Initially, 15-minute averages were used and each 24-hour period was considered as a whole. Correlation coefficients between the data at the two closest sites, Birmingham Central and the CB Building, were calculated and the values obtained are given in Table 1. It can be seen that, with the exception of the last two days of the period, there is little correlation between pollution levels at these two sites. It was thought possible that the correlation might be greater for longer averaging times, since this would have the effect of smoothing out the high frequency fluctuations that are probably due to more local sources and emphasising the common elements in the two data sets. Two longer averaging periods were investigated, 30 minutes and 1 hour. The correlation coefficients obtained are also given in Table 1. In general, as would be expected, the correlations increased with increasing averaging time, but there are also some instances where the converse occurred. The changes in values of r^2 are mostly small and there is still only a limited degree of correlation between the two sites. Two cases, one showing higher values of r^2 (increasing with averaging time) and one lower (decreasing with averaging time), are shown plotted respectively in Figs. 18 and 19. It is of interest that even with such short spacings between sites, the correlations between the measurements can frequently be negligible.

The same approach was used to investigate the relationship between the Birmingham Central and Birmingham East Sites, which have a much greater spacing. We examined the correlation between data at the sites for occasions when the wind was largely blowing from the centre towards the east and compared these results with occasions when the wind was perpendicular to this, across the line of the sites. In the former case, there may be evidence of pollution from sources

Table 1. Correlations between Birmingham Central and CB Building for different averaging times.

Date	CO			NO			NO$_2$			SO$_2$		
	15 min	30 min	1 hour	15 min	30 min	1 hour	15 min	30 min	1 hour	15 min	30 min	1 hour
Wed (14.2.96)	0.13	0.14	0.17	0.36	0.34	0.40	0.10	0.11	0.14	0.07	0.08	0.12
Thurs (15.2.96)	0.01	0.004	0.01	0.12	0.12	0.10	0.01	0.005	4×10^5	0.11	0.13	0.18
Fri (16.2.96)	0.11	0.22	0.29	0.15	0.22	0.27	0.42	0.51	0.64	0.07	0.09	0.12
Sat (17.2.96)	0.001	0.001	0.002	0.001	0.002	0.01	0.10	0.10	0.09	0.05	0.06	0.08
Sun (18.2.96)	0.16	0.21	0.29	0.12	0.23	0.37	0.67	0.69	0.61	0.06	0.07	0.08
Mon (19.2.96)	0.002	0.003	0.004	0.02	0.02	0.03	0.03	0.03	0.03	0.04	0.04	0.05
Tues (20.2.96)	0.52	0.58	0.68	0.47	0.59	0.72	0.58	0.59	0.58	0.13	0.17	0.20
Wed (21.2.96)	0.35	0.52	0.72	0.50	0.57	0.62	0.23	0.29	0.39	0.56	0.52	0.79

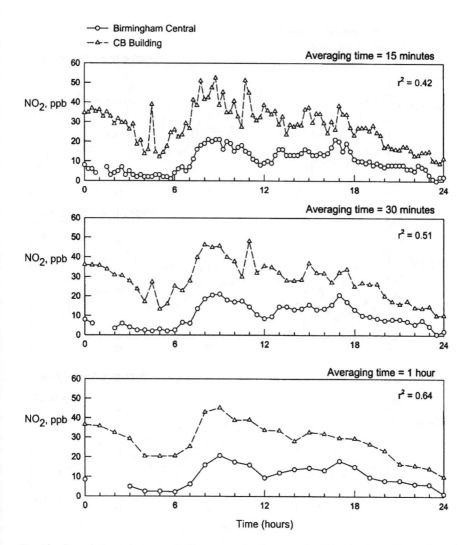

Fig. 18. Correlations between NO_2 concentrations at the Birmingham Central and CB Building sites for different averaging times (data for a period with high correlation, 16th February).

close to the central sites travelling to the eastern site, and so higher correlations may be obtained by allowing for this travel time. Conversely, in the latter case there may be larger correlations since the contributions in the two areas from more distant sources might be more similar as the area over which these pollutants originate is almost the same.

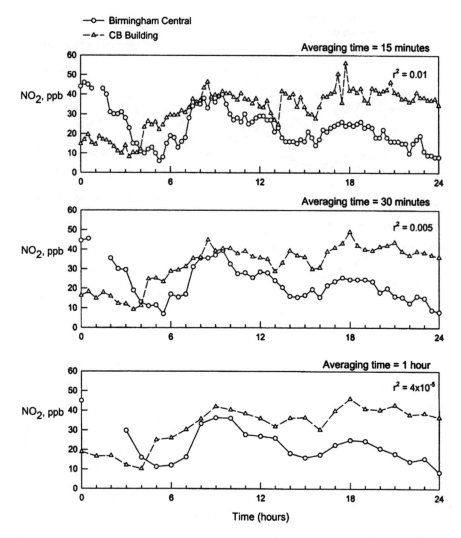

Fig. 19. Correlations between NO_2 concentrations at the Birmingham Central and CB Building sites for different averaging times (data for a period with low correlation, 15th February).

Two days were identified with wind directions largely from the centre to the east sites (i.e. south-westerlies) — Thursday 15th and Friday 16th. These were compared with three days when the wind was mostly from the north-west and thus blowing across a line between the

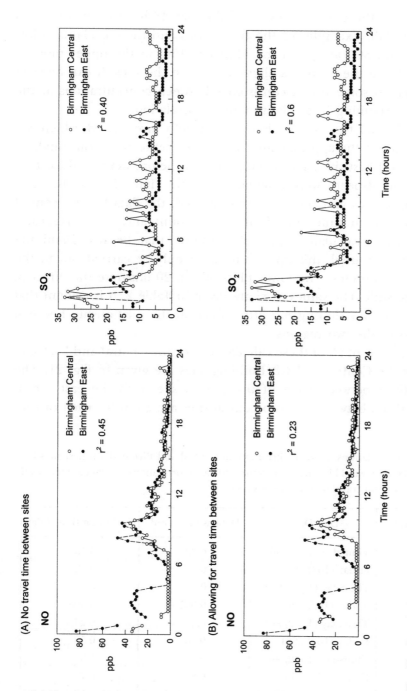

Fig. 20. Correlations for NO and SO₂ concentrations between the Birmingham Central and East sites with an aligned wind (15th February).

central and the eastern sites — Wednesday 14th, Monday 19th and Tuesday 20th. Examples of the time series and the correlation plots for NO and SO_2 are given for Thursday 15th in the upper plots in Fig. 20, when the wind was aligned between the sites. It can be seen that there is some broad consistency between concentrations at the two sites, resulting in a relatively high correlation.

One further type of correlation was briefly investigated, which was to account for the approximate travel times between the Birmingham Central and East sites. The same periods as before were used for this. Average windspeeds were calculated and the time of travel between the central sites and the east site evaluated, the data for the central sites were then time-shifted by this amount and the correlation coefficients re-calculated. It should be noted that, because 15-minute averages were used, this time shift could only be carried out to the nearest 15 minutes. The lower plots in Fig. 20 illustrate the effect of this time shift. The values of r^2 are given in Table 2. It is evident that overall, the time-shifting has little effect on the relationship between the data in the two regions.

The results of all of these calculations for the Central and East sites and for the Central and CB Building sites are given in Fig. 21. The correlations between each of the central sites with the east site as well as those between the two central sites are given for each of four

Table 2. Correlations between Birmingham Central and Birmingham East sites for occasions when wind blowing from central to east (correlations for data as measured and allowing for travel time between sites).

Pollutant	Day/Date	Data as measured	Including time delay
CO	Thursday (15/2/96)	0.09	0.13
	Friday (16/2/96)	0.13	0.02
SO_2	Thursday (15/2/96)	0.4	0.6
	Friday (16/2/96)	0.001	0.008
NO	Thursday (15/2/96)	0.45	0.23
	Friday (16/2/96)	0.46	0.58
NO_2	Thursday (15/2/96)	0.4	0.3
	Friday (16/2/96)	0.8	0.7

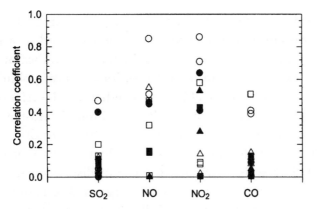

Correlations

● Central and East Birmingham Sites

■ Central Birmingham and CB Building sites

▲ CB Building and East Birmingham sites

Filled symbols are for occasions when wind direction along line of sites (~250°)

Open symbols are for occasions when wind direction across line of sites (~340°)

Fig. 21. Overall correlation coefficients between Birmingham Central, CB Building and Birmingham East sites for all pollutants.

pollutants. There is also a division between wind directions broadly aligned between and across the line of the sites. There is little consistency between the results and the correlation coefficients vary considerably. It can be seen that, overall, the higher correlations tended to be when the wind was across the line of the sites.

4. Discussion

It will be appreciated from both the initial discussion and from the example data analysed above that the prediction of pollutant concentrations at specific sites in urban areas is not a perfectly straightforward matter. However, the way in which different sources, especially at varying distances, contribute to the overall levels at a particular site, and the causes of variability in these levels can be understood. The sample data analysed here comprised only a snapshot of pollution characteristics over a short period. However, this is sufficient to show that wide variations in shorter term pollution levels over even quite short distances in urban areas can be expected as a matter of course. Data from monitoring sites is valuable as an indicator of trends in pollution levels and as an indicator of longer term pollution levels over a wider area, but its ability to predict local variations, both spatially and temporally at other places is limited. Most monitoring equipment has a relatively slow response, of the order of minutes and thus fails to record the quite substantial fluctuations that can occur over shorter time scales. Similarly the use of dispersion modelling to predict the pollution characteristics over large areas is mainly confined to relatively coarse grids, usually of kilometre or greater scale, and to longer term averages (one hour is the minimum period and much longer time averages are more common). Information on these small scale variations is presently quite limited. There is a need for pollution monitoring investigations in urban areas which look at both very short term variations in pollution levels and the spatial variation over short distances.

From the point of view of the needs of a specific site, it is clear that data from monitoring sites even relatively close by can only be certain indicators of broad-based, long term pollution levels and that more detailed data, especially if shorter term variations are of interest, would require site-specific investigations.

5. Conclusions

(1) The paper has outlined the way in which pollutant sources, especially from varying distances, contribute to different features of the total concentration at a point. Both spatial and temporal variations of pollutant concentration can be large and generally increase with reducing averaging time and spatial scale.

(2) An investigation of a sample of data from four sites in the West Midlands confirms these characteristics, with significant levels of temporal fluctuation and spatial variation over relatively short distances. Correlations of data between different sites frequently show negligible correlation, even over short distances or relatively long averaging times.

6. Summary

The paper describes the characteristics of different types of pollutant sources in the way that they are experienced in a fixed locality in an urban area. The locality in this sense can also be a building or part of a building (a ventilation inlet for example). The most important parameter is the distance of the polluting source and therefore the characteristic features of sources at different distances are discussed. The relationship between "local" sources and "background" levels of pollutants, their contributions to the total local pollution levels and the vertical and lateral gradient of pollution are considered. The discussion is illustrated by examples from measurements of dispersing plumes and of urban pollution levels experienced during a specific investigation in a large urban area.

Pollution levels can show large fluctuations, which increase with reducing time scales, and significant spatial variation over short distances. Measured pollution levels on four sites in a large urban area, the West Midlands, showed these characteristics for averaging times down to one minute and spatial separations of 500 m. Data correlations between the sites were generally poor.

Acknowledgements

This investigation was carried out as part of the POLIS project (on integrated urban planning) in the EC JOULE 95 research programme. It was part funded by the Construction Directorate of the UK Dept of the Environment.

Thanks are due to Geoff Broughton and Jon Bower of NETCEN for supplying the shorter time averaged data from the National Network sites in the West Midlands.

References

AEA Technology (1995) *Air Pollution in the UK, 1994.* AEA Technology, Report No. AEA/RAMP/200015001/1 (ISBN 0-85356432-9).
Anon (1996) Road traffic responsible for most air pollution. *Pollution* **26**(11), November, Front Page.

Benarie M.M. (1980) *Urban Air Pollution Modeling.* The MIT Press (ISBN 0-262-02140-4).

Cook N.J. (1985) *The Designers Guide to Wind Loading of Building Structures. Part 1.* Butterworths (ISBN 0-408-00870-9).

Cook N.J. (1990) *The Designers Guide to Wind Loading of Building Structures. Part 2.* Butterworths (ISBN 0-408-00871-7).

Dabbert W.L., Ludwig F.L. and Johnson W.B. Jr. (1973) Validation and application of an urban diffusion model for vehicular pollutants. *Atmos. Env.* **7**, 603–618.

EEC (1985) Council directive 85/203/EEC of 27 March 1985 on air quality standards for nitrogen dioxide. *Offic. J. Europ. Commun.*, No. L87/1.

EPAQS (1995). *Sulphur Dioxide.* Report of the Expert Panel on Air Quality Standards, UK Department of the Environment, HMSO (ISBN 0-11-753135-9).

Georgii H.W., Bush E. and Weber E. (1967) *Investigation of the Temporal and Spatial Distribution of Emission Concentration of Carbon Monoxide in Frankfurt/Main.* University of Frankfurt/Main, Institute for Meteorology and Geophysics, Report No. 11.

Hall D.J. and Kukadia V. (1994) Approaches to the calculation of discharge stack heights for odour control. *Clean Air* **24**(2), 74–92.

Hall D.J., Kukadia V., Spanton A.M., Walker S. and Marsland G.W. (1996a) *Concentration Fluctuations in Plumes Dispersing Around a Building — A Wind Tunnel Model Study.* Building Research Establishment, Report No. CR 16/96.

Hall D.J., Spanton A.M., MacDonald R. and Walker S. (1996b) *A Review of Requirements for Simple Urban Dispersion Models.* Building Research Establishment, Report No. CR 77/96.

Hall D.J., Spanton A.M., Kukadia V. and Walker S. (1996c) *Exposure of Buildings to Pollutants in Urban Areas — A Review of the Contributions from Different Sources.* Building Research Establishment, Report No. CR 209/96.

Hall D.J., Macdonald R.W., Walker S. and Spanton A.M. (1998) *Measurement at Dispersion Within Simulated Urban Arrays — A Small Scale Wind Tunnel Study.* Building Research Establishment, Report No. CR244/98.

Hoydysh W.D. and Dabbert W.F. (1994) Concentration fields at urban intersections: fluid modeling studies. *Atmos. Env.* **28**(11), 1849–1860.

Jones C.D. (1983) On the structure of instantaneous plumes in the atmosphere. *J. Hazard. Mater.* **17**, 87–112.

Jones C.D. (1996) Something in the air. *Chem. Eng.* **28th November**, 21–26.

Kukadia V. and Palmer J. (1996) *The Effects of External Atmospheric Pollution on Indoor Air Quality.* Paper Presented at the 17th AIVC Conference, Gothenberg, Sweden, September 1996.

Kukadia V., Palmer J., Littler J., Woolliscroft R., Watkins R. and Ridley I. (1996) *Air Pollution Levels Inside Buildings in Urban Areas: A Pilot Study.*

Paper Presented at the CIBSE/ASHRAE National Conference, Harrogate, UK, September 1996.

Munn R.E. (1981) *The Design of Air Quality Monitoring Networks*. In "Air Pollution Problems" Series, Macmillan (ISBN 0-333-30460-8).

Oke T.R. (1987) *Boundary Layer Climates*. Routledge (ISBN 0-415-04319-0).

Ott W.R. (1977) Development of criteria for siting monitoring stations. *J. Air Pollut. Control Assoc.* **27**(6) 543–547.

QUARG (1993) *Urban Air Quality in the United Kingdom*. First Report of the Quality of Urban Air Review Group, UK Department of the Environment (ISBN 0-9520771-1-6).

Schwela D. and Zali O. (1999) *Urban Traffic Pollution*. E & FN Spon (ISBN 0-419-23720-8).

Timmis R.J. and Walker C.A. (1989) *Dispersion Modeling of the Effect of Smoke Control on Air Quality in Kirkaldy*. Warren Spring Laboratory, Report No. LR729.

Vignatti E., Bercowicz R. and Hertel O. (1996) Comparison of air quality in streets of Copenhagen and Milan, in view of the climatalogical conditions. *Sci. Total Env.* **189/190**, 467–473.

CHAPTER 13

THE WHOLE BUILDING AND PATTERNS OF DEGRADATION

R. Inkpen

1. Introduction

Degradation of natural building stone occurs at particular locations on different types of buildings that may have undergone different conservation treatments. The buildings may be simple monuments such as standing stones or complex configurations such as Norman cathedrals. In either case, the processes of stone decay cannot be divorced from the built form nor from the socio-economic framework that established and maintains that building. Most studies of stone degradation and suggestions for conservation measures, however, operate as if building stone degradation can be abstracted from the context of its formation. Laboratory-based and small-scale studies have outlined a range of degradation mechanisms for different stone types. From these studies, indices of degradation have been constructed, the effectiveness of different stone types and conservation treatments assessed and dominant agents of decay identified. The underlying assumption has been that degradation processes identified at the small-scale operate in the same manner at the large-scale. Similarly, the environmental system and its variation within which weathering occurs are assumed to be a mirror-image of the conditions constructed and controlled in the laboratory. The building scale is seen as a simple extrapolation of process/form relationships and their associated environmental parameters identified from small-scale studies. Initially, this chapter explores the role and limitations of small-scale studies for identifying and quantifying the processes and relationships causing stone degradation at the scale of the whole building.

Based partly on small scale studies, many workers have used information on process/form relationships to construct classification systems for degradation on buildings. From such systems, the spatial variability of forms thought to be associated with different weathering agents can be identified and mapped across the building (for example in Inkpen *et al.*, 2000; Inkpen *et al.*, 2001). Likewise, the spatial variability of environmental parameters, such as moisture content or chemical composition can be sampled and mapped over a surface. Combining the two spatial distributions a correlation can be established between different weathering agents and forms. Implicitly, such combinations suggest causal relationships between environment, agents and weathering forms. Some of these systems can also be predictive, highlighting where certain types of weathering agents are likely to be dominant on new buildings. Although some of these systems begin to tackle the complexity of relationships causing weathering at the scale of the building, they all tend to look at the forms produced rather than the context of production. Discussion of the limitations of these classification schemes forms the next section of the chapter.

Context is both multi-dimensional and spatially and temporally variable. It can include the general weathering environment which lie beyond the scope of the building itself as well as the composition of building elements. At the large-scale, the nature of weathering environment sets the constraints on the type of weathering agents present and the scope of their actions and so indirectly the range of weathering rates and forms. At the scale of the building itself, weathering context can be studied using the concept of the catchment area as a fundamental unit on the building. This unit is of use in exploring the socio-economic factors that impinge upon the weathering behaviour of the building. Decisions on the financing of construction and repair work as well as what conservation treatments to employ and when, for example, will all influence how weathering agents interact with the stone. Socio-economic factors are not, however, limited to those explicitly concerned with maintainence of the building. There are also wider issues which impinge upon current and future weathering behaviour over which architects, masons and monument authorities have no control. Issues such as sustainability, environmental policy and global climatic change will impact upon the building. As with the larger environment, these large-scale trends form the framework within which conservation is undertaken and set constraints upon the scope of activities and responses available for maintaining the building fabric.

2. Small-Scale Studies of Stone Degradation

Large-scale international and national surveys of stone degradation such as the NATO/CCMS project in Europe and the National Materials Exposure Programme (NMEP) in the United Kingdom have used only small-scale samples to collect data (Butlin *et al.*, 1993; Yates and Butlin, 1996). In both studies, small $50 \times 50 \times 8$ mm tablets of different stone types, Portland Limestone, Monks Park Limestone and White Mansfield in the case of the NMEP, were used to assess weathering. Tablets were exposed at each site, either to rainfall or sheltered from it, and weighed every six months. Other data, such as chemical composition, were collected from specific tablets, but the weight change index was the main measure used in data analysis. Weight change of tablets from each location was averaged to represent the weight change for that site for a particular time period. Data from each site were plotted against indices of the nature of the general weathering environment — such as sulphur dioxide levels and rainfall, averaged or integrated over the same time period. From these data, weathering curves (usually based on linear regression) which express the relationship between the environmental variation and weathering of stone were derived. Using such a relationship, the potential implications of policy decisions that could alter environmental variables (such as pollution levels) could be assessed. For a given decrease in sulphur dioxide, for example, the regression equation predicted how much weathering loss should be reduced by. Such an analysis fitted in with the general move towards the concept of critical load assessment as a policy tool within the European Union (Bull, 1993; Bull, 1995; Cowell and ApSimon, 1995). Studies using runoff slabs to derive chemical mass balance equations for stone loss (Livingston, 1986; Reddy, 1988; Moses, 1996) have likewise tried to link general environmental variables and their variations to changes in stone loss over a given time period.

Laboratory-based studies concerned with single or, at most simple combinations of, weathering agents (Goudie, 1974 and 1999; Robinson and Williams, 1982; Smith and McGreevy, 1983) have tended to concentrate on illustrating the power of process or processes to cause degradation. The quasi-standard test for durability of limestone building stones in the UK, the sodium sulphate crystallisation test (Ross and Butlin, 1989; also prEN 12370), for example, uses a saturated salt solution and a heating and cooling cycle to induce "unnatural" weathering cycles. Weight loss of standard cubes (40 × 40 × 40 mm) under this regime are then used to assess the general durability of the stone to weathering relative to a standard set of limestone cubes. The test was originally designed to simulate frost weathering rather than general durability, but even within these terms the samples are exposed to rates of temperature change and a supply of salt that are unrealistic to say the least. Use of the test results as an indicator of durability is based upon the assumption that weathering behaviour in the building bears a direct relationship to the behaviour of the stone in the test.

More complex weathering simulations that try to model reality in greater detail by either matching salt supply to that found in reality or by combining more variables of appropriate magnitudes (Goudie, 1974; Smith et al., 1987; Smith and McGreevy, 1988) are still limited in the direct relationship of these results to the weathering of stone on a real building. As laboratory-based simulations, they represent simplifications of the real world. The variables controlled are those thought by theoretical analysis, other laboratory studies and field observations to be those important for weathering. The weathering system is, however, defined by the worker and so only those variables identified, probably those capable of measurement, and which are capable of simulation at some of their properties will be analysed. This can severely restrict the type of weathering system simulated. Salt supply, for example, can vary in both the mode of delivery, its spatial and temporal characteristics, in composition and in quantity (McGreevy, 1982). Altering any of these, either singularly or in combination, could alter the amount of degradation induced in experimental samples. Variables that are difficult to measure or model may be excluded from study as will those not identified or considered inappropriate, such as the second to second variability of sulphur dioxide levels. Likewise, the experimental sample is analysed in isolation from its position on the building. Position could dramatically influence the types of weathering agents affecting the stone and specific micro-environment within which the stone weathers. Similarly, position can be an indication of the past weathering history of the surface. Past weathering will mean that, unlike most experimental work, the stone which weathers will already by altered and not a pristine, fresh stone sample. Past weathering will alter the physical and chemical characteristics of the stone surface at least as well as resulting in the storage of weathering products in the stone, the so-called memory effect (Smith *et al.*, 1988; Cooke, 1989; Inkpen, 1991).

The above discussion highlights the limitations of small-scale field trials and laboratory-based studies in explaining weathering at the scale of the real building and even real building surface. This is not to imply, however, that such studies do not serve a useful role. Controlled examination of specific weathering agents and associated mechanisms under limited, simulated conditions help to identify the potential power of agents and combinations of agents of particular characteristics to cause degradation. From such an analysis, a deeper understanding of the role of specific mechanisms or group of mechanisms in causing decay can be gained. It cannot, however, be assumed that the results of such studies can then be translated directly without further consideration and thought to explain patterns and forms of degradation observed in the real environment on building surfaces and over a building as a whole. Such studies can, however, point to particular forms of weathering that may be associated with particular processes or combination of processes. This can then be used to devise classification systems based on assumed processes of formation as well as suggesting which other environmental parameters to measure the variation of.

3. Classification of Building Degradation

Classification systems usually try to link a process or agent or combinations of processes and agents to particular types of weathering features. Links may be established from laboratory studies or from the common association of particular features with the presence of specific weathering agents. Sometimes, however, this tends to result in correlation being equated with causation. The presence of high concentrations of salts and a particular set of weathering forms, for example, may be taken to indicate degradation by salt weathering mechanisms. Such a conclusion may only identify one weathering agent whilst there are many other contingent factors such as the state of the stone, the continued supply of salt, the removal of weathered products and the position of the stone block or weathering feature that are also important determinants of weathering.

One of the simplest classification systems by Camuffo et al. (1982) divides weathered surfaces on the basis of colour. White surfaces are those that are rainwashed and are areas of surface loss. Black surfaces are areas of deposition where black crusts, such as sulphation crusts, form. Grey areas are areas of dust deposition rather than transitional zones between black and white areas. Location is central to this classification system as the degree of exposure has an important bearing on whether a surface is rainwashed or not. Other classification systems have concentrated on identifying and mapping a selected range of features. Robinson and Williams (1996), for example, identified and mapped areas of active and inactive cavernous weathering on sandstone blocks on churches of different ages in West Sussex. From these data, they suggested that a great deal of cavernous weathering occurred during the 18th and 19th century. This would suggest that this form of weathering is associated with pollution from industrial and urban activity which began in this time period. Antill and Viles (1998) used a classification system based on colour — a visual surface disruption to classify over 700 individual stone blocks in an external wall of Worcester College, Oxford. The maps produced were constructed by sketching the wall and then colouring each block individually according to its class. The patchy and distinct patterns of specific decay forms could then be related to the location of the patches on the wall and their relationship to other decay forms. Inkpen *et al.* (2001) carried out a similar study, but integrated the data into a geographical information system (GIS). In the latter study, the use of a pre-defined unit of classification, the stone block, was largely avoided by using the GIS maps of stone blocks as the base map upon which decay forms were mapped. This meant that the observer was not restricted to either percentage covers per block, nor to allocating a block solely to one decay class.

Zezza (1996) identified and mapped different types of weathering on four buildings in the Mediterranean: Cadiz Cathedral, Spain, Bari Cathedral, the Church of Sta. Marija Ta., Cwerra, Malta and the Sanctuary of Demeter Eleusis, Greece. Zezza used a methodology identified as Integrated Computerized Analysis (ICA) for Weathering to relate different types of information about weathering in a common framework. This method combines identification and mapping of building lithology with information on the distribution of weathering forms and other indices of alteration such as ultrasonic identification of sub-surface weathering forms.

Although the above integrated systems are in the early stages of development, the potential for deriving useful information for both understanding weathering behaviour and developing appropriate conservation strategies cannot be underestimated. Central to these schemes, however, is developing an appropriate classification system for weathering features. Fitzner (1990) developed a system based on a hierarchy of feature classification (Fitzner *et al.*, 1992; Fitzner *et al.*, 1996). The classification system has four groups of weathering forms at the highest level in the hierarchy: loss of stone material, deposits, stone material detachment and fissures/deformation. Within each of these groupings, there are further sub-divisions into 29 weathering forms with finer divisions down to individual weathering features. Further differentiation is possible by adding an intensity parameter to each individual feature. The relative success of these systems for aiding conservation work is in the early stage of evaluation.

The Building Research Establishment (BRE) in the UK uses the durability index in combination with a knowledge of the structure of a building to define damage zones (BERG, 1989). For each area of the building, the mix of factors that produce its weathering environment are reduced to a potential damage indicator. The classification is then used to identify which durability class would be appropriate for particular zones. Approaches such as these tend to ignore the problem of equating damage with a laboratory-derived measure of durability. The damage potential of any zone is the result of a combination of factors which vary both spatially and temporally. The same damage potential may be present on different parts of the building, but result from the combination of different sets of factors. The durability index is a general index, whilst weathering environments will be specific combinations of different factors.

Despite a range of classification systems for weathering forms, there is a tendency to imply a simple causal relationship between process and form. This may result from the focus on specific areas of damage on a single surface or façade. The BRE damage index and the damage index derived from Fitzner's work point to spatially, and presumably temporally, distinct zones of degradation. Zonation of forms may be explained by reference to solely process or agent differences. Consideration of the position or rather context within which forms occur on the building may be useful in identifying potential damage zones and also in providing a full explanation of such forms.

Context, like durability, is an elusive, but vital term in stone weathering. At the large scale, context can include factors such as the pollution environment which determine the types and quantities of pollutants available for reaction with the stone. A simple first approximation, pollution environment could be divided into urban and rural (Cooke, 1989). This division assumes that predominantly local pollution sources contribute to the creation of the pollution environment. Urban, residential and industrial areas with large local producers of sulphur dioxide are the areas with aggressive pollution environments. Rural areas have lower densities of low use and, it is assumed, less polluters, and so have lower levels. Recent trends, in Western Europe at least, suggest that this simple division of the environment and weathering rates may have be valid in the recent past (Attewell and Taylor, 1988; Kupper and Pissart, 1976; Cooke *et al.*, 1995), but there is a narrowing and, in some areas, parity in pollutant levels (Inkpen, 1989). Despite this, a difference in weathering between urban and rural locations persists (Honeybourne and Price, 1977; Jaynes, 1985; Jaynes and Cooke, 1987; Inkpen, 1989).

Although the larger scale context may set the limits to the types of interactions possible, characterisation of it does not fix the absolute nature and values of pollutants that react with the stone on different parts of the building. As the scale changes, so does the context and hence the factors that become influential in limiting and defining stone degradation. The building itself is not a passive element, it creates its own context rather than simply reflecting the larger scale environmental conditions. In this manner, it is analogous to the relationship between an organism and the environment in evolution. A simple view would be that the organism responds to the environment as it changes (Levins and Lewontin, 1985). The organism, however, only obtains information about the environment mediated by its senses. This means that environmental fluctuations may be reduced or enhanced in magnitude by the senses, and the organism responds accordingly. Similarly, the organism will create its own environment to alter its context. Fur and feathers, for example, could be seen as mechanisms by which the organism reduces the impact of the general environmental conditions by the creation of a particular, beneficial micro-environment. The building can be viewed in the same manner. Fluctuations in variables such as sulphur dioxide can be measured over small temporal and spatial scales, but the effect of these scales of change on the weathering of stone is unclear. These changes are mediated throughout the surface of the stone, which may have differential responses to different scales of change and to different types of pollutants. There is little in the literature to suggest that appropriate scales of change have been determined for the response of a building or even a single surface.

There is evidence to suggest that the impact of environmental changes is also mediated through an environment of the buildings own making. The structure of the building itself results in the creation of numerous micro-environments that either enhance or reduce both the magnitude and frequency of environmental conditions as measured away from the building. These smaller scale micro-environments are the context within which weathering takes place, and again study of the appropriate scale on which to monitor these micro-environments is relatively poorly developed. Creation of these micro-environments is not, however, a random process. Any building was designed for a purpose and that dictates the general structure and layout of the building. Conservation of the building adds another layer of design. Combining these human inputs into the physical system results in a structure with unintentional, but designed micro-environments within which weathering occurs. This means that it should be possible to identify and map different types of micro-environments on a building and, to a certain extent, predict the conditions in each.

Identification of micro-environments does not mean that distinct units of study have been identified. Micro-environments may merge and interact with each other across a building surface. Analysis of weathering would be aided if it could focus on a distinct, functional unit of study. Defining sections of the building as a catchment area could be of help in establishing such a framework. Catchments have become a fundamental unit for analysis in hydrology (Chorley, 1969) and the particular characteristics of the concept make them particularly relevant for studying buildings, i.e. ease of demarcation, nested, ordered and interconnected nature of catchments, the topological and geometric integrity of the catchment, the catchment as the basic erosional and depositional unit in the landscape and as the basic unit of input/output analysis. A building catchment represents the area contributing to a runoff outlet or set of outlets, which may be linear or point features, e.g. a drainage channel or a number of drainage points across an area. Definition of the catchment area and characterisation of it using a few simple parameters indicates how water and material is transported about and between catchment systems. In particular, such descriptions could help to clarify the hierarchical and interconnected nature of the drainage system and the location and effectiveness of storage and removal networks across the building. Inputs and outputs are now tied to a defined, functional unit and any surface studied can be positioned within the context of this unit.

Building catchments, unlike their "natural" counterparts, have not developed in response to tectonic and erosional processes, rather they may have been designed with that specific purpose in mind. Within each catchment, there will be elements that can be regarded as either unconnected micro-catchments or flow barriers depending on the scale of study. These are usually isolated features which tend to deflect or concentrate water flow within the catchment. Their presence can, therefore, be important in determining patterns over material movement within the catchment and so influence the context of any surface under study. Seemingly similar surface in terms of properties such as slope, material and conservation histories may differ significantly in terms of their connectivity to other areas of the building, and hence in the material flows likely to influence their weathering behaviour. This framework also provides a conveniently scale-independent basis for incorporating and interpreting how the human dimension is likely to impact upon stone degradation.

The importance of context can be illustrated by the current environmental concern with climatic change. Various scenarios of climatic change have been suggested and are continually being refined and recalculated (Intergovernmental Panel on Climatic Change; Impact of Climatic Change on UK). Within the UK, data implies that there will be a general rise in temperature and rainfall, but that this change will be highly variable spatially. It is also likely that extremes of weather will increase in frequency as climatic change occurs over the next 50 or so years. Modelling of climatic change using General Circulation Models tends to model atmospheric movement and change on the scale of large cells of a resolution (e.g. 20×20 km) that makes prediction of conditions for particular locations difficult. If only general trends for the UK are considered, then three key contextual factors can be highlighted; i) a rise in carbon dioxide levels, ii) an increase in temperature (between 1.5°C and 4.5°C over 50 years) and rainfall amount and variability with global rainfall increasing by between 3% and 15%, and iii) a rise in sea level of between 20 and 50 cm over 50 years.

Local details of the impact of climatic change are sketchy. The global models used to predict large-scale climatic change usually use grid cells of a resolution greater than 20×20 km. Change below this scale is not modelled and the accuracy of prediction for a few cells of this size is relatively low. For the UK as a whole, only broad potential regional trends in climate are possible to predict. For Southern Britain, for example, climatic change is likely to result in increased winter precipitation and lower summer precipitation. Additionally, increased storminess of weather patterns coupled with prolonged summer heat will probably produce a more Mediterranean-style climatic regime. In these circumstances, assessment of current degradation processes and conservation practices in existing Mediterranean areas would help to predict the likely response of buildings to this type of change in climate. At a more local scale, it is virtaully impossible to predict the exact impact of climatic change at the scale of a single urban area, let alone a single building. Only general trends can be put forward and limits set to the potential range of scenarios for any local area. This may not be a major problem in determining the impact of such climatic changes on buildings for two reasons. Firstly, the magnitude of the changes outlined is not as significant as some urban to rural changes in environmental conditions that a number of historic monuments have already experienced over their life as urbanisation has progressed. Under these changing conditions, it is only when such climatic changes have been coupled with pollutants that severe damage appears to have resulted. Secondly, as well as the urban location of most monuments ameriolating the impact of climate change, the response of building materials will not be immediate nor necessarily dramatic. The level of temperature that is likely, 2–3°C, is of a similar or less magnitude than current differences in temperature resulting from aspect around the same building.

The rise in carbon dioxide will contribute to the "greenhouse" effect, but will only minimally alter rainfall acidity (Trudgill and Inkpen, 1993). Potentially of greater significance is the change in climate and sea level. The impact of a changing climate on a particular site will be a function of the precise combination of changes, e.g. rise in temperature, increase in rainfall or rise in temperature and decrease in rainfall, and the nature of the building under consideration. Buildings with short service lives (30–50 years) may be relatively unaffected by such changes. For such materials already in service, their expected service lives are too short for economically feasible alterations to the fabric to be important. For materials with short expected service lives yet to be used, a knowledge of climatic change can be incorporated into their design or specification.

In relation to both types of building materials, there are a limited number of possible alterations climatic change will induce. The first possibility is that the mix of weathering agents and the intensity of process operation will change as climate changes. Within geomorphology, the work of Peltier (1950) and climatic geomorphology in general is of relevance as he outlined the dominant forms of weathering associated with different climate zones. A transition from, say, a temperate to a more Mediterranean climate will be associated with a change in the suite of weathering processes. A second possibility as suggested by Brimblecombe (1997) is that a gradual change in environmental conditions can result in a change in the rates of operation of different weathering processes and so in a change in the dominant set of weathering processes. For both of these alternatives, any change in climate will produce changes in both the range of conditions experienced as well as the cycles of climatic variables, such as freeze-thaw activity, which will influence the rate and nature of degradation. A third possibility is that the change in conditions will alter the relationship between variables in the weathering system. In this case, the whole structure of the system will be re-defined by a change in climate. In this alternative, there is no gradual transition. Change is abrupt and dramatic with weathering forms previously adjusted to environmental conditions weathering rapidly as the relationships that define the weathering system alter. A final alternative is that the change in climate may not alter the dominant set of weathering processes, but may alter the extremes experienced by the building. In this case, the development of forms associated wtih extreme events will increase. These forms are likely to become more common and persistent as the operation of weathering processes during "normal" periods will not be sufficient to remove them from the building.

The impact of climate change does not depend solely on changes in climatic variables however. Any impact will be mediated by the building environment itself. An increase in external air temperature, for example, could result in a decrease in the amount of heating required in some buildings. This will alter the thermal gradient through the building materials and so reduce their thermal stress. Decreasing the steepness of the thermal gradient can also influence particle deposition onto the building surface by altering the flow of air close to the surface and so altering the settling velocities of particles. Likewise, an increase in the frequency of extreme conditions can increase the number of times high heating temperatures are required. Thus, although overall heating requirements may be decreased, the frequency of episodes of high thermal stress in building materials may increase. A similar complex-mediated sequence of alternatives can also be put forward for changes in rainfall amount. A decrease in the amount of rainfall can result in increased periods of sulphation for limestone surfaces. Increases in the frequency of extreme events, however, will result in the generation of runoff of higher velocity and potentially greater erosivity. The paths of such flows will still, however, be dependent upon the configuration of catchments on any individual building. Each building will also have a unique history of conservation work and building practices and decisions. This will mean that each building will have a unique set of weathering forms associated with its past environmental conditions and past building practices. Context is therefore, important not only for determining the mix of changes associated with climatic change in any area, but also for determining the mediation of these changes at the scale of the individual building.

4. The Building as the Physical Representation of Socio-Economic and Cultural Factors

A building stone is not merely a physical part of the building. Its weathered state and treatment also represents the underlying socio-economic and cultural currents of the society the building is within. What is an acceptable level of degradation, who decides this and what should be done are all questions that require a human agent to make a decision. Any decision will involve an individual or group of individuals in some particular historical and social setting (Inkpen, 1999). Decisions cannot, therefore, be divorced from the circumstances or context of the questions posed. In this sense, any building stone by its presence and subsequent weathering and treatment history represents the socio-economic and cultural values of society over time.

The importance of social and cultural values for what appears to be a purely physical phenomenon can be illustrated by looking at how durability and durability criteria have been used in stone degradation. Durability itself is recognised as a complex, vague and ambiguous term (Turkington, 1996). The BRE (1984) view of durability is based on the physical characteristics of the stone, resulting from a combination of the internal structure of the stone and the nature of the cementing material. The most recent digest from the BRE (1997) whilst still viewing the rate and form of weathering as being a function of the internal characteristics of the stone explicitly recognises the complex nature of durability and its dependence upon the location in which the material is placed and the relationships between the material and the rest of the building environment.

> *"Durability only needs to be assessed if stone intended for longer life or harsher exposures ... or severe environments. Because such assessment must take account both of the stone's position in the building and the weathering to which it will be exposed, a single durability test cannot be of universal application"* (BRE Digest 420, 1997; p. 3).

The digest regards the normal service life of a building of 60 years as too short a time for stone durability to be an issue, as seen in the particular cases cited above. This illustrates the impact of defining durability relative to economic concerns rather than as a purely physically determined entity.

Porosity and, in particular, microporosity have, however, continually been put forward as the key characteristic in controlling durability (Knöfel, 1991), despite studies showing that stones with similar structural and mineralogical properties exhibiting differing weathering behaviours. Garden (1980) views durability not as a fixed property of the stone, but as an outcome of the interaction of material and environment, including the in-service expectations. Nireki (1980) similarly, viewed performance over time as a vital element in durability. Performance implies that the stone is continually being assessed in relation to the fulfilment of some designed function. On this basis, the expected performance forms the criteria for assessing durability and not an absolute property of the stone. Duffy and O'Brien (1996) again highlight the importance of expected future performance for defining durability. They, however, point out that performance itself is different for different interest groups, such as architects and engineers. They settle on evaluating stone durability as:

" ... *should be concerned with assessing how resistant a stone is to changes of its physical, mechanical and aesthetic properties over time*" (Duffy and O'Brien, 1996; p. 254).

Central to their interpretation is the view that the surface environment is the important determinant of durability. They view the surface environment by determining the supply of weathering agents and the range of weathering processes. Using this definition, they view durability not as an absolute measure, but as being dependent upon the surface environment upon which the stone is exposed.

Despite the increasing complexity of the views of durability outlined above, there is still a tendency to view it as a physical phenomenon only (Cramer, 1994; Drever, 1994). The terms used, such as "performance" and "aesthetic", imply an important input from socio-economic and cultural concerns into the definition of durability. A central question which none of the above authors explicitly address, but which will implicitly be incorporated into the agendas of each interest group is, why is a concept such as durability required in the first place? Durability is defined by the architects, testing agencies and legislators in relation to a product, be it stone or any other building material. The drive to define durability via some standardised and repeatable measure is part of the view of building material as a commodity. In the case of natural building stone, a highly variable natural material is being characterised by a set of measurable parameters defined as important within a scientific and technological framework. It is the behaviour of these parameters in controlled conditions that become the descriptors of the material and the means of representing the material to other parties interested in purchasing stone. In this manner, the socio-economic framework is constraining the type of parameters considered to be valid for characterising building materials.

Putting a value onto stone also defines the temporal dimension of durability. Expectations of durability vary with the type of construction for which the stone is used, an issue recognised by Lewry and Crewdson (1994). They divide durability testing in four types: benchmark tests, reference or comparative tests, environmental or stress tests and site testing. Within this framework, they suggest that the construction product directive in the European Union (1989) and BRE digest (1994 and 1995) will promote the harmonisation of standards. As part of this scheme, they view the development of design standards that simulate the response of materials in every environment as central. They suggest that accelerated testing in this framework becomes significant as it provides data on the most important factors and mechanisms causing degradation. From these tests, models and damage functions for the end use environment can be constructed. The context of a building and the material within it is replaced by a set of experimentally derived damage functions developed and assessed over a time scale significantly less than the lifetime of the material. The imperative of standardisation within a market and the confidence in a product influence the time scale over which durability is determined. In this case, the time scale of assessment is reduced and formalised as a damage function applicable to a given material in a given environment.

In what could loosely be described as cultural monuments, value lies in the continuance of original material or its replacement by identical material. Durability in this context has a long time frame with material expected to survive over hundreds of years. The value of the stone lies not only in itself, but in the context into which it is placed. Stone used for conservation work has value imparted to it by the cultural significance of the monument and also takes on expectations associated with such a context. Within a modern building, stone may be used as a facing material. In this situation, durability is defined by the expected life-time of the building for its current use by its current occupants, a shorter time frame. The value of the stone is defined by its context as one element in a "disposable" building. This not only determines the manner in which the stone is used, as a thin skin (4–5 mm) designed for decorative purposes, but also the type of properties that affect its durability. Valuation of stone in each context uses different time frames and highlights different stone/environment relationships as important in determining the definition of durability.

The design of the building and conservation of the building fabric are immediately obvious human elements important for the degradation of stone. As with the physical environment, however, there are larger scale socio-economic factors that will impinge upon these human factors and affect both the types of conservation treatments applied and the evaluation of degradation itself. Current concerns over sustainability provide another useful illustration of the wider socio-economic context and its potential impact upon stone degradation. Sustainability itself is a multi-layered and complex term (Redclift, 1987; Redclift, 1991; Simmons, 1990) which has meant that it has been open to a range of interpretations ranging from maintaining current practices, the *status quo* view, to radical social and economic re-structuring as the means to achieve sustainability. Redclift (1987 and 1991) notes that sustainabiltiy can be interpreted in varied and often contradictory ways by different interest groups. Different definitions of the term become accepted in different locations partly as a result of the power relationships between these interest groups giving sustainability a political dimension. The Brundtland Commission (1987) highlights the importance of the political dimension for defining and acting to ensure sustainable development which is seen as:

> "*A process in which the exploitation of resources, the direction of investments, the orientation of technological development and institutional change are all in harmony, and enhance both current and future potential to meet human needs and aspirations*" (Brundtland Commission, 1987; p. 46).

Often, however, sustainable development has been legitimated as a purely economic entity. Pearce *et al.* (1991), for example, interprets sustainable development as non-declining human welfare over time defined in terms of standard of living and in retaining a stock of capital assets. As with durability, economic value is used to re-define the concept and in so doing the concept becomes open to manipulation via the market. Within the quarrying industry, for example, there is evidence of an explicit acceptance of the economic definition of sustainable and hence on the need to define the environment and environmental impacts in similar terms.

"Better environmental performance requires better profitability, not less. There is a need to slow down the rate at which environmental demands are being made upon us, to enable us to adjust to all these changes." (Jackson, 1994; p. 33).

Using an economic value for measuring all impacts and remedial measures means that sustainable practices are often assessed using only this criteria. Attaching a monetary value to the impact of a set of practices therefore becomes a vital element in assessing the sustainability of any conservation or building practice.

How these diverse views of sustainability are mediated by those involved in conservation will determine the impact of such diffuse social concepts on stone degradation. An interpretation of sustainability could be mediated via legislation in, for example, the need for an environmental impact assessment of conservation practices undertaken. This may involve determination, in monetary terms, of the ecological costs of undertaking a particular practice for local or regional scale ecological systems. Such an analysis, however, highlights another problem with sustainability. Sustainable practices at the scale of the building may involve costs that are unsustainable at a different scale and in different locations. Use of replacement stone from original quarry areas may limit the environmental cost, however, defined in a historic monument, but the transportation of that material and its removal will involve environmental costs at two scales. The area from where the stone is quarried will suffer environmental alteration as the stone is removed as well as local environmental problems from the traffic involved in transporting the stone. At the scale of the region and globally the removal and transportation of the stone will produce atmsopheric pollution both from the quarry functions and from the power required for the quarry to operate. Sustainability is, therefore, potentially open to different interpretations at different scales.

Likewise, sustainability also needs to be viewed as a temporary complex concept. As economic growth rates change and as society itself changes, then the valuation assigned to different types of buildings will change. Their worth as items of cultural heritage will alter and their status as objects worthy of continued conservation alter. This perception may change over time if the degradation caused by pollutants or other factors increases to a point where the costs of conservation outweigh the benefits of sustaining the monument.

It should also be remembered that the buildings themselves are not isolated individual entities. Brimblecombe (1997) noted that cultural heritage should be viewed as embedded, as architectural assembles rather than just as objects. In this manner, an analogy could be drawn with ecosystems and species preservation. Viewed in this manner, any single building would need to be valued in the context of both its local "ecosystem", city or cultural tradition, as well as valued for its contribution to the cultural diversity of an area. In this way, similar arguments to those employed, concerning conservation of biodiversity for future generations, can be made in relation to cultural diversity of a historic building stock. This adds a long-term dimension to the temporal considerations of sustainability in relation to cultural heritage.

The potential requirement for assessing the life cycle impact of the processes and techniques used is another area where legislated definitions of sustainability will impinge upon building degradation. A policy such as replacement of degraded stone with equivalent stone may have to consider the environmental impacts of quarrying of the replacement stone, the methods used to insert the replacement and remove the old stone and its disposal. In other words, the whole production and consumption system for any product will need to be evaluated in terms of its environmental impact and its sustainability compared to alternative conservation strategies. This means that not only does the physical environment become an important explicit determinant of defining an effective treatment, but the impact of that treatment as defined by its sustainability becomes a factor requiring consideration. Once you extend this form of analysis to conservation strategies and techniques, such as lime washing or epoxy resin emplacement, a range of issues concerning appropriate space and time scales of impact become clear.

Historic Scotland's policy in relation to the preservation of carved stone (Maxwell, 1996), for example, can be examined in the light of a board definition of sustainability. The percepts set out governing relevant action to protect stones (Table 1); can be re-interpreted as statements concerning the relative importance of sustainability issues. The percepts outlined can be taken as a general statement of

Table 1. Historic Scotland's percepts governing relevant action to protect stones.

✔ Retain stone *in situ* wherever possible particularly if it believed they have not been removed from their original locations.

✔ Where stones were retained *in situ* and protective enclosures were needed, each protective enclosure should be purpose-designed to create the correct environment for the stone with due regard for its surrounding conditions and location.

✔ Stones that needed to be removed should be kept locally.

✔ Where stones had to be removed consideration should be given to their relocation within an existing local structure. The long-term stability and public access to such a building should also be taken into account.

✔ Movement of stone to remote locations such as museums should be considered only a last resort.

✔ Consideration should be given to replacing stone with casts where movement of the original was essential.

Based on Maxwell, 1996, p. 936.

preferred options in conservation treatment as suggested by D'Avino (1996), where the preservation of the original material is seen as integral and fundamental to the conservation of the form. In this view, the material is the "soul" of the form and should only be removed or altered as the last resort to maintain the built form. Built into this view is a particular view of the material as a repository of cultural information and by implication conservation is both a "*critical-historical judgement*" and "*scientific-technical knowledge*". Leaving the original material wherever possible implies a policy of non-intervention with the degradation system. Such a policy can be interpreted as one of long-term sustainability of the current relationships within the system. Maintaining the material may, however, imply altering the current relationships between the stone and its local environmental conditions (constant or altered) as suggested by the use of protective enclosures. Such enclosures would have associated with them an environmental cost, both in terms of their production and in terms of their impact upon the environment once in place. Comparing the different types of impact using a common measure or formula is difficult if not impossible, and each factor such as visual impact, alteration of runoff characteristics will be measured and assessed differently by different people.

Where intervention is necessary based on the "scientific-technical" side of the equation, it is usually only on the basis of a consideration of the small surface studied. The type of treatment employed will be a function of the general conservation policies of the appropriate management and the available treatments known to the consultant and management. Each treatment can also be considered in terms of its sustainability at different scales. Limewashing, for example, appears to be a relatively environmentally harmless treatment. (The following is not meant to imply that limewash is a hazardous treatment, it is only an illustration of the factors that need to be considered when making a full analysis of the impact of using limewash. A similar argument could be put forward, for example, when using new, quarried stone for restoration work.) The application of a limewash to a surface involves the use of no hazardous chemicals and is not viewed as hazardous to the stone itself. At the local level of the building surface this maybe the case, but to get the limewash a whole set of production processes need to occur. These would need to be considered in a full life cycle analysis and they may not have such a benign environmental impact. Likewise, applying the limewash is not the end point for the limewash within the environment. Over a period of time the lime is removed from the surface and transported initially within the building catchment of which the surface is a part and then out of the building system altogether. The interaction of the lime with the rest of the catchment and the potential interactions of the lime with the extralocal environment would need to be considered in a full life cycle analysis. Similarly, the frequency and magnitude of repeated treatments could influence the magnitude and frequency of effects associated with lime production and lime waste in the environment.

An important consideration is the economic aspect of conservation of whole buildings. For management of resources at the level of the building, it is important to know which materials are most sensitive to degradation. Using this knowledge, appropriate scheduling of repairs and conservation work in line with resources can be drawn up. This requires a good inventory of both the types of material in a building and their location. A combination of these factors can be used to plan where requirements for conservation may outstrip available resources in the future.

The individual building also needs to be considered in the wider context of economic priorities of funders, such as central government, as well as the planning context which may constrain both the flow of resources and the type of work undertaken. Despite the prevalence of cost-benefit analysis in this context as a means of comparative evaluation of competing demands, it is recognised that the valuation of cultural heritage is difficult (Lowenthal, 1994). In particular, the importance of the purpose of the valuation for its outcome is a key point for the valuation of cultural heritage. Mohr and Schmidt (1997) argue that the introduction of the concept of non-use valuation may be an appropriate means of valuing cultural heritage. Non-use valuation has been used in relation to natural resources and can be built into a complex set of different types of non-use. Non-use valuation assumes that when confronted with a change in the use or characteristics of an object, a person can be compensated for these changes. The individual decline in well-being experienced can be translated into an acceptable range of alternatives with monetary value for that individual. In economic terms, there is an assumption of substitutability in the valuation of cultural heritage, i.e. individuals are willing to make trade-offs (Mohr and Schmidt, 1997; p. 337). This requires that the non-use value of cultural heritage usually be related to some measurable, market-valued goods or services. As cultural heritage is viewed as a social good, its valuation requires a consideration of trade-offs at the social level in competition with other social goods. The valuation of self-selected sets of individual experts carries equal weight with that of any other individual in society, although the potential of different individuals to influence the valuation of heritage for others may differ.

Valuation of cultural heritage may be made by techniques such as contingent valuation. This method uses different hypothetical situations, such as the current state of decay of a building and its characteristics after conservation, and by various means, drawn into providing a monetary valuation. The monetary valuation provided is concerned their with willingness to pay for or accept different types and states of repair of a building. The valuations can be in pure monetary terms of willingness to forgo some other resource. Despite problems with the operationalisation of the technique, it does provide a comparative means of determining the "general" value placed on cultural heritage and their conservation in different contexts.

Importantly, economic valuation cannot be divorced from the context provided by sustainability mentioned above. Viewing culturally important buildings as part of a wider "architectural ecosystem" means that valuation cannot be made for one building in isolation. Maintaining the diversity of this architectural ecosystem may not only be important for maintaining the economics of tourism to the urban area. Snickars (1997) also suggested that the general attractiveness of an urban area will be affected by the amount of maintained cultural heritage it contains. If a high quality cultural buildings or set of building are allowed to degrade beyond a certain threshold, this can spark a reaction in the type of resident attracted to the area. Once began the exodus and image of the urban area is difficult to reconstruct and a cumulative economic impact upon the whole urban area is felt. This illustration highlights the need to value buildings not only in terms of their components, but also in terms of their relationship to their milieux.

5. Conclusion

Studies of degradation have tended to concentrate on extrapolating small-scale field and laboratory studies of process/form relationships to the scale of the whole building. Extrapolation in this fashion assumes that the different spatial and temporal scales of processes and forms associated with the building as a whole unit have no impact upon the operation of agents and processes of degradation. Whilst these small-scale studies do help in developing a detailed understanding of the mechanisms of degradation, they do not help in understanding why degradation occurs with a particular magnitude, frequency and form at particular locations on a building. Mapping patterns of degradation is an important element in recognising that order and pattern are present at the scale of the building itself. Such patterns cannot, however, be simply reduced to explanation by reference to the presence of particular weathering agents. The question of "why there and now"? still remains. Contextualising stone degradation by consideration of the wider environment and how its influence is mediated to a particular surface is essential to understanding how and why degradation occurs. Using a catchment framework within which to analyse and contextualise, such degradation provides a physical, functional unit for simple systems analysis.

From the above examples of durability and sustainability, however, it is clear that the role of socio-economic factors in stone degradation should not be undervalued or regarded as a mere extension of the physical environment of the stone. Focusing on a specific problem of degradation tends to limit consideration of appropriate and effective treatments and policies to those of effects identifiable and measurable only on the surface of stone itself. Using a catchment framework for contextualising the surface widens the scope of potential effects, including a larger portion of the physical environment. Recognition of the socio-economic basis of conservation work and the general trends which influence this also means that the appropriateness treatments need to be viewed within a wider and more socially and culturally sensitive context. As with the physical environment, however, general trends are translated to the stone in a variety of ways, and manner of this mediation will dramatically influence how concepts such as sustainability are used in any particular project on any particular building. Importantly, analysis and understanding cannot be solely focused on the scale of the surface. Such a scale of study limits the spatial and temporal horizons of analysis and therefore of explanation and understanding of stone degradation.

Acknowledgements

The author would like to thank Dr. Peter Collier and Professor Mike Taylor for useful discussion of ideas in this paper.

References

Antill S.J. and Viles H.A. (1998) Deciphering the impacts of traffic on stone decay in Oxford: some preliminary observations from old limestone walls. In *Aspects of Stone Weathering, Decay and Conservation*, eds. Jones M.S. and Wakefield R.D. Imperial College Press, London, pp. 28–42.

Attewell P.B. and Taylor D. (1988) Time dependent atmospheric degradation of building stone in a polluting environment. In *Engineering Geology of Ancient Works and Historical Sites*, eds. Marinos G. and Koukis G. Balkema, Rotterdam, pp. 739–753.

BERG (1989) *The Effects of Acid Deposition on Buildings and Building Materials in the United Kingdom.* Building Effects Review Group Report, HMSO, London.

BRE (1984) *Decay and Conservation of Stone Masonry.* BRE Digest 177.

BRE (1994) *Standardisation in Support of European Legislation: What Does It Mean For the UK Construction Industry.* BRE Digest 397, September 1994, CI/SfB(A).

BRE (1995) *A Guide to Attestation of Conformity Under the Construction Products Directive.* BRE Digest 408, August 1995 Ci/SfB(A).

BRE (1997) *Selecting Natural Building Stone.* BRE Digest 420, January 1997, CI/SfBe.

Brimblecombe P. (1994). The balance of environmental factors attacking artifacts. In *Durability and Change: The Science, Responsibility, and Cost of Sustaining Cultural Heritage*, eds. Krumbien W.E., Brimblecombe P., Cosgrove D.E. and Staniforth S. Wiley and Sons, Chichester, pp. 67–79.

Brimblecombe P. (1997) Group report: heritage and monuments. In *Saving Our Architectural Heritage: The Conservation of Historic Stone Structures*, eds. Baer N.S. and Snethlage R. John Wiley and Sons Ltd., Chichester, pp. 389–403.

Brundtland Commission (1987) *Our Common Future.* Oxford University Press, Oxford.

Bull K.R. (1993) Development of the critical loads concept and the UN-ECE mapping programme. In *Critical Loads: Concepts and Applications, ITE Symposium, No. 28*, eds. Harnung M. and Skeffington R.A. pp. 8–10.

Bull K.R. (1995) Critical loads and levels — a policy tool. In *The Effects of Air Pollutants on Materials, Proceedings of a Workshop Held in London, 27–28 July 1994*, eds. Lee D.S. and McMullen T.A. AEA Technology, pp. 19–22.

Butlin R.N., Yates T.J.S., Coote A.T., Lloyd G.O. and Massey S.W. (1993) *The First Phase of the National Materials Exposure Programme 1987–1991.* BRE Report CR253/93.

Camuffo D., Del Monte M., Sabbioni C. and Vittori O. (1982) Wetting, deterioration and visual features of stone surfaces in an urban area. *Atmos. Env.* **16**, 2253–2259.

Chorley R.J. (1969) The drainage basin as the fundamental geomorphic unit. In *Water, Earth and Man*, ed. Chorley, R.J. Methuen, London, pp. 77–100.

Cooke R.U. (1989) II. Geomorphological contributions to acid rain research: studies of stone weathering. *Geograph. J.* **155**, 361–366.

Cooke R.U., Inkpen R.J. and Wiggs G.F.S. (1995) Using gravestones to assess changing rates of weathering in the United Kingdom. *Earth Surf. Proc. Landforms* **20**, 531–546.

Cowell D.A. and ApSimon H.M. (1995) Estimating the cost of damage to buildings by acidifying atmospheric pollution in Europe. In *The Effects of Air Pollutants on Materials, Proceedings of a Workshop Held in London, 27–28 July 1994*, eds. Lee D.S. and McMullen T.A. AEA Technology, pp. 53–62.

Cramer F. (1994) Durability and change: a biochemist's view. In *Durability and Change: The Science, Responsibility, and Cost of Sustaining Cultural Heritage*, eds. Krumbien W.E., Brimblecombe P., Cosgrove D.E. and Staniforth S. Wiley and Sons, Chichester, pp. 19–25.

D'Avino S.T. (1996) Chemical and physical techniques for the conservation of stone materials: interdisciplinary approach and specific nature of the discipline. In *Proceedings of the 18th International Congress on Deterioration and Conservation of Stone, Berlin, 30 September–4 October 1996*, ed. Riederer J., pp. 923–928.

Drever J.I. (1994) Durability of stone: mineralogical and textural perspectives. In *Durability and Change: The Science, Responsibility, and Cost of Sustaining Cultural Heritage*, eds. Krumbien W.E., Brimblecombe P., Cosgrove D.E. and Staniforth S. Wiley and Sons, Chichester, pp. 27–37.

Duffy A.P. and O'Brien P.F. (1996) A basis for evaluating the durabiity of new building stone. In *Processes of Urban Stone Decay*, eds. Smith B.J. and Warke P.A. Donhead, London, pp. 253–260.

Fitzner B. (1990) Mapping of natural stone monuments — documentation of lithotypes and weathering forms. *Analytical Methodologies for the Investigation of Damaged Stone*, eds. Veniale F. and Zezza U. (organisers). Proceedings of the Advanced Workshop, Pavia, 14–21 October 1990, La Goliardica Pavese.

Fitzner B., Heinrichs K. and Kownatzki R. (1992) Classificatino and mapping of weathering forms. In *Proceedings of the 7th International Congress on Deterioration and Conservation of Stone, Lisbon, 15–18 October 1992*. Laboratorio Nacional de Engenharia Civil, pp. 957–968.

Fitzner B., Heinrichs K. and Volker M. (1996) Monument mapping — a contribution to monument preservation. In *Origins, Mechanisms and Effects of Salts on Degradation of Monuments in Marine and Continental Environments*, ed. Zezza F. Proceedings European Commission Research Workshop, 25–27 March 1996, Bari, Italy, Protection and Conservation of the European Cultural Heritage Research Report No. 4, pp. 345–355.

Garden G.K. (1980) Design determines durability. In *Durabiltiy of Building Materials and Components*, eds. Sereda P.J. and Litvan G.G. ASTM STP 691, pp. 31–37.

Goudie A.S. (1974) Further experimental investigation of rock weathering by salt and mechanical processes. *Zeitschrift Geomorph.* **Suppl. Bd. 21**, 1–12.

Goudie A.S. (1999) Experimental salt weathering of limestones in relation to rock properties *Earth Surf. Proc. Landforms* **24**, 715–724.

Honeybourne D.B. and Price C.A. (1977) *Air Pollution and the Decay of Limetones.* HMSO internal note, BRE N117/77, cited in Jaynes S.M. (1985).

Impact of Climatic Change on UK. *http://www.doc.mmu.ac.uk/aric/gcc/ukimpact.html*

Inkpen R.J. (1989) *Stone Decay and Atmsopheric Pollution in a Transect Across Southern Britain.* Unpublished Ph.D. thesis, University of London.

Inkpen R.J. (1991) *Stone Weathering and the "Memory" Effect.* Department of Geography, Portsmouth Polytechnic, Working Paper, No. 17.

Inkpen R.J. (1999) Atmospheric pollution and stone degradation in nineteenth century Exeter. *Env. Hist.* **5**, 209–220.

Inkpen R.J., Collier P. and Fontana D. (2000) Close-range photogrammetric analysis of rock surfaces. *Zeitschrift fur Geomorphologie* **N.F. Suppl. Bd. 120**, 67–81.

Inkpen R.J., Fontana D. and Collier P. (2001) Mapping decay: integrating scales of weathering within a GIS. *Earth Surf. Proc. Landforms* **26**, 885–900.

Intergovernmental Panel on Climatic Change. *http://www.ipcc.ch/*

Jackson N. (1994) MPG6 — the hangover worsens. *Quarry Manag.* **21**, 33.

Jaynes S.M. (1985) *Studies of Building Stone Weathering in South-East England.* Unpublished Ph.D. thesis, University of London.

Jaynes S.M. and Cooke R.U. (1987) Stone weathering in southeast England. *Atmos. Env.* **21**, 1601–1622.

Knöfel D. (1991). Causes, mechanisms and measurement of damage in cultural heritage materials: the state of the art: concrete structures. In *Science, Technology and European Cultural Heritage: Proceedings of the European Symposium, Bologna, Italy, 13–16 June 1989,* eds. Baer N.S., Sabboni C. and Sors A.I. Butterworth-Heinemann, pp. 138–147.

Kupper M. and Pissart A. (1976) Vitesse d'erosion en Belique de calcaires d'age primarie exposes á l'air libre ou soumis á location de l'eau courante. *Report of the Commission on Present-Day Geomorphological Processes (International Geographical Union),* pp. 39–50.

Levins R. and Lewontin R. (1985) *The Dialectical Biologist.* Harvard University Press, Cambridge, Massachusetts.

Lewry A.J. and Crewdson F.E. (1994) Approaches to testing the durability of materials used in the construction and maintenance of buildings. *Construct. Build. Mater.* **8**, 211–222.

Livingston R.A. (1986) Evaluation of building deterioration by water runoff. In *Building Performance: Function, Preservation and Rehabilitation, ASTM STP 901,* ed. Davis, G. American Society for Testing and Materials, pp. 181–188.

Lowenthal D. (1994) The value of age and decay. In *Durability and Change: The Science Responsibility, and Cost of Sustaining Cultural Heritage,* eds. Krumbien W.E., Brimblecombe P., Cosgrove D.E. and Staniforth S. Wiley and Sons, Chichester, pp. 39–66.

Maxwell I. (1996) Historic Scotland's work in the preservation of carved stone. In *Proceedings of the 8th International Congress on Deterioration and Conservation of Stone, Berlin, 30 September–4 October 1996*, ed. Riederer J., pp. 935–954.

McGreevy J.P. (1982) Frost and salt weathering: further, experimental results. *Earth Surf. Proc. Landforms* **7**, 475–488.

Mohr E. and Schmidt J. (1997) Aspects of economic valuation of cultural heritage. In *Saving Our Architectural Heritage. The Conservation of Historic Stone Structures*, eds. Baer N.S. and Snethlage R. John Wiley and Sons Ltd., Chichester, pp. 333–348.

Moses C.A. (1996) Methods for investigating stone decay mechanisms in polluted and "clean" environments, Northern Ireland. In *Processes of Urban Stone Decay*, eds. Smith B.J. and Warke P.A. Donhead, London, pp. 212–227.

Nireki T. (1980) Examination of durability test methods for building materials based on performance evaluation. In *Durability of Building Materials and Components*, eds. Sereda P.J. and Litvan G.G. ASTM STP 691, pp. 119–130.

Pearce D., Barbier E., Markandya A., Barrett S., Turner R.K. and Swnason T. (1991) *Blueprint 2: Greening the World Economy*. Earthscan Publications Ltd., London.

Peltier L.C. (1950) The geographic cycle in periglacial regions as it is related to climatic geomorphology. *Ann. Assoc. Am. Geograph.* **40**, 416–436.

prEN 12370. *Tests On Natural Stone — Determination of Resistance by Crystallisation Cycles.*

Redclift M.R. (1987) *Sustainable Development: Exploring the Contradictions.* Methuen, London.

Redcilft M.R. (1991) The multiple dimensions of sustainable development. *Geography* **76**, 36–42.

Reddy M. (1988) Acid rain damage to carbonate stone: a quantitative assessment based on the aqueous geochemistry of rainfall runoff from stone. *Earth Surf. Proc. Landforms* **13**, 335–354.

Robinson D.A. and Williams R.B.G. (1982) Salt weathering of rock specimens of varying shape. *Area* **14**, 293–299.

Robinson D.A. and Williams R.B.G. (1996) An analysis of the weathering of Weladen sandstone churches. In *Processes of Urban Stone Decay*, eds. Smith B.J. and Warke P.A. Donhead, London, pp. 133–149.

Ross K.D. and Butlin R.N. (1989). *Durability Tests for Building Stone.* BRE Report 141.

Simmons I.G. (1990) Ingredients of a green geography. *Geography* **75**, 98–105.

Smith B.J. and McGreevy J.P. (1983) A simulation study of salt weathering in hot deserts. *Geograf. Ann.* **65A**, 127–133.

Smith B.J., McGreevy J.P. and Whalley W.B. (1987) Silt production by weathering of a sandstone under hot arid conditions: an experimental study. *J. Arid Env.* **12**, 199–214.

Smith B.J. and McGreevy J.P. (1988) Contour scaling of a sandstone by salt weathering under simulated hot desert conditions. *Earth Surf. Proc. Landforms* **13**, 697–705.

Smith B.J., Whalley B. and Fassina V. (1988) Elusive solution to monumental decay. *New Scientist* (2nd June), 49–53.

Snickars F. (1997) How to assess and assert the value of cultural heritage in planning negotiations. In *Saving Our Architectural Heritage: The Conservation of Historic Stone Structures*, eds. Baer N.S. and Snethlage R. John Wiley and Sons Ltd., Chichester, pp. 349–370.

Trudgill S.T. and Inkpen R.J. (1993) Impact of acid rain on karst environments. In *Karst Terrains, Environmental Changes, Human Impact*, ed. Williams P.W. Catena Supplement 25, pp. 199–218.

Turkington A.V. (1996) Stone durability. In *Processes of Urban Stone Decay*, eds. Smith B.J. and Warke P.A. Donhead, London, pp. 19–31.

Yates T.J.S. and Butlin R. (1996) Predicting the weathering of Portlands limestone buildings. In *Processes of Urban Stone Decay*, eds. Smith B.J. and Warke P.A. Donhead, London, pp. 194–204.

Zezza F. (1996) Decay patterns of weathered stones in marine environments. In *Origins, Mechanisms and Effects of Salts on Degradation of Monuments in Marine and Continental Environments*, ed. Zezza F. Proceedings of the European Commission Research Workshop, 25–27 March 1996, Bari, Italy, Protection and Conservation of the European Cultural Heritage Research Report No.4, pp. 99–130.

INDEX

The Natchez Trace followed the old Chickasaw trail and was not yet a good road. The going was bad for men, worse for wagons. Trees had to be cut down, canebrake hacked away, bridges contrived. Cold rain washed out fires and turned the trail to mud.

Day after day the general strode along, now stepping up his pace, now dropping far behind, to be with different companies. The men watched him with growing affection.

"He's a tough one, our general," a private said.

"I never knew him till now. He's sound as hickory." The man held up a hickory cane he had cut as an aid on the march.

"Sound as hickory!" passed down the line. Before the march ended, the word "Old" was added in affection.

"Old Hickory's lookin' after us best he kin."

Andrew Jackson

Frontier Statesman

by Clara (Ingram) Judson

ILLUSTRATIONS BY LORENCE F. BJORKLUND

Follett Publishing Company

CHICAGO

Author's Foreword

Looking back across the pages of United States history, five men stand out uniquely as great Americans — George Washington, Thomas Jefferson, Andrew Jackson, Abraham Lincoln, and Theodore Roosevelt. Their lives span the years from our beginnings to the first World War, and each man in his own way made a distinguished contribution in the American struggle toward a land for the free.

These men well knew that the ideal of freedom was a hazardous thing. They knew, too, that it never could be casually passed on to the next generation; that it must be earned by each citizen in his own day. By spoken and written word and by vital leadership, each of the five influenced events and drew mankind nearer the goal of equality under the law.

In recent books I have tried to portray significant developments in our country's history as revealed through the lives of our great men. The work was incomplete without Andrew Jackson, the man who lived midway — not at the beginning, though he initiated many new ideas; not near our time, for the frontier wilderness he knew has long since vanished. He was of heroic mien, and has been the subject of many books. But in the scores I read, I found his personality blurred by partisanship. What manner of man was he, really?

I went to Nashville, Tennessee; to his beautiful home, The Hermitage; to the state library and archives where newspapers of his day, rare volumes, and letters are faithfully gathered and treasured. And there I came to know Andrew Jackson.

It was a thrill to hold in my hand the map he drew of the battle of Horseshoe Bend; to see the field writing kit, with its ink and sand bottles, its candle and sealing wax; to read his letters to Rachel, his reports to his government, letters to friends and kinfolk. Only in this intimate way can one understand the vigorous mind, loyal heart, and integrity of our frontier president.

Many people aided me in my study. My grand-daughter, Jane Canning, was with me in Nashville as we explored the Jackson country. Then, as I stayed on alone, Mr. Robert Quarles, Director of Archives, State of Tennessee, allowed me to read Jackson letters and documents now housed in a beautiful new building. Mr. Fred Estes, Director of the State Museum, showed me Jackson relics, including the writing kit which Mr. Quarles later had photographed for me so it could be accurately pictured in our illustrations. Miss Isabel Howell, Director of the State Library, searched out precious volumes and later read my manuscript and made valuable suggestions.

I am also indebted to Mrs. Douglas Wright, corresponding secretary of the Ladies' Hermitage Association, who helped me in countless ways; to Mrs. W. P. Cooper, vice-regent, and Miss Fermine Pride, second vice-regent of the association, who personally showed me treasures of The Hermitage, and to Mrs. Anne Orr Callahan, who lent me material. And as always I am grateful to Miss Louise Borchelt and Miss Florence Davison of the Evanston Public Library for constant help in unearthing facts.

C.I.J.

Evanston, Illinois
June 9, 1954

Andrew Jackson,
Frontier Statesman

Andrew Jackson, Reader

A FRECKLED REDHEAD sat on a rail fence, ready to jump.

"You git down off that fence, Andy Jackson," a cattleman shouted. "You know well as I kin tell you that the cattle pen's no place fer a seven-year-old this morning. Git down!"

"You let me in the pen when you want work done," Andrew retorted glibly. "Maybe you'll want me in a minute."

Several cattlemen grinned. A man couldn't help but be amused at that boy's quick talk, though James Crawford, the owner, said they spoiled the lad.

"We'll not be asking you anything this day, Andy. Not with the beeves in fine fettle ready to start to Charles Town."

"I want to go along," Andrew said, glad to have the heart of the matter come up. " 'nd I'm going."

"Sure enough you shall — someday," James Crawford called. "I've told you, Nephew, when you're old enough to take one hundred sixty miles on the road you're to go. It's no picnic."

Before Andrew had an answer a new voice called from the big house, a little distance away.

"Andy! Andy Jackson!" one of the Crawford boys shouted. "Your mother says to come. It's time for school."

"Aw, school," Andrew complained scornfully. "*School!*"

Reluctantly he slid from the fence, tearing his faded home-spun pants on the way down. He strode to the house in sullen protest.

His mother was busy at the great table in the center of the kitchen. A glance at her son showed his state of mind.

"It's no use for you to act up, Andy," she said kindly. "When you are old enough, Uncle James will take you on the drive. Till then, my son's going on with his schooling." She smiled at the angry boy. His hair, so like her own, was uncombed; the deep blue of his eyes showed he was in a temper. Then she saw the tear in his pants.

"Wash your face and hands, Son, and comb your hair. I'll get needle and thread. You're not fitten for school as you are."

Andrew stalked to the pump and washed. He combed his hair with the comb that hung near. As he returned to the kit-

chen he heard one of the Crawford girls talking as she washed the dishes.

"It's no wonder Andy hates field school, Aunt Betty. He's learned all the schoolmaster kin teach. Andy's read the three books over'n over, and he kin beat the master figuring."

"Andy's teachin' me to figure." The smallest cousin looked up from the hearth where she polished steel knives with sand. "He helps all the young-uns, Andy does."

" 'nd I'm sick of school!" Andrew burst out, unappeased by praise. "I kin read and write. I want to go with the men."

Without a word, Mrs. Jackson knelt down and sewed up the rent in his pants. Andrew stood rigid, an attitude more expressive of protest than any words. As she sewed, Mrs. Jackson thought of the field school where the children went daily.

It wasn't really in a field. South Carolinians wouldn't waste cleared land on a schoolhouse. This log cabin stood in a grove of scrub pine on soil hardly fertile enough to be worth clearing. The building was about twenty feet square with an opening on one side where a door hung by leather hinges. There were no windows.

Inside a long slab was fastened to one wall for a desk; its surface was smooth as an ax could make it. Bark-free logs rolled near were the seats. School equipment included three books, a stout switch, and a lazy schoolmaster. Rarely would a competent man bother to teach school in this year, 1775. It was a fill-in job at best.

School bored Andrew. He had taught himself to read by the time he was five; no one knew just how. Likely from newspapers left by travelers from Charles Town. But Mrs. Jackson was never daunted by her son's moods. She cut the thread with a kitchen knife, tested the mend, and rose to her feet.

"There's this to remember, Andy," she said. "With your father dead, you and your brothers have to make your own way. You take to readin' better'n Hugh or Robbie. You're goin' to get an education. My mind is set on that. Maybe some day you'll be a preacher. That's my dream, Andy. I'd have a pride if you were. For the now, be thankful for a home and a school nearby —"

"But Mother!" Andy interrupted. Then he paused.

"But what?" His mother was annoyed. "If you have something to say, say it out. Don't stand there glowering 'but.' What worries you, Son?"

"It's stupid to stay inside on a fine day, Mother," Andrew burst out. "I want to go on the cattle train. Truly I do."

"Of course you do! Any boy of spirit would!" Her smile was understanding, her voice warm. "And someday you shall. Your uncle has promised. When you are older, his boys likely will have farms of their own; he'll need you. But he will not want an ignoramus. Now run on — maybe you kin catch the others."

Andrew's shoulders drooped, but he turned toward the door. Something in his look made her stop him.

"Remember when the traveler came last week? The men were away, and you were the only one who could read the Charles Town paper to us?" Andrew nodded indifferently. That had happened before. He thought little of it.

"I was proud, Andy. Remind me some evening, and I'll tell you about your father and his father and why you are here in the Waxhaws of South Carolina. I think you ought to know, now that you are getting older."

Andrew grinned. That last phrase pleased him. With a whoop he ran swiftly along the road to school.

A couple of hours later the cattle drive passed the log schoolhouse. The master let his pupils go out and watch this big annual event. Beeves, plump from fattening in the pens, were already covered with red dust that hung like a copper cloud over the outfit. Days later, men and cattle would soak themselves clean in a creek before going into the city. Andrew eyed them with longing. He'd be glad for the discomforts on the way if he could share the excitements.

"Bet I kin throw you," a voice behind taunted him.

"Bet you kin't!" Andrew whirled, got a grip and threw his challenger before the boy had got a hold.

"You will, will you?" he said, grinning.

"Boys! Boys! No fighting! Git to your lessons!" The schoolmaster, switch in hand, called from the doorway.

"That man doesn't know the difference between fightin' n' wrestlin'," Andrew said in a scornful whisper. But it was useless for pupils to protest. They went inside and to work.

A few evenings later members of the Crawford and Jackson families gathered before the great fireplace in the kitchen, each busy with an evening task. The girls knitted or spun. Boys whittled wooden pegs or spoons. The house seemed quiet with the men and older boys gone.

Mrs. James Crawford — Aunt Jane to the Jackson boys — lay on a long-seated chair with her feet on a stool, well covered with a blanket. All of her eight children were strong and well, but the mother grew weaker daily. It was fortunate that her sturdy sister, Elizabeth Jackson, could take the many cares of the household from her frail shoulders.

"Aunt Betty," one of the Crawford girls remarked, "the other morning you said you'd tell Andy about his father. A tale would be nice this night."

Firelight played over eager faces turned expectantly toward Mrs. Jackson. A good tale made evening work pleasant.

"Please, Aunt Betty," another girl said.

Mrs. Jackson smiled quietly, and for a few minutes she knitted in silence as her thoughts went back to long ago.

"This story is for Andy," she began. "Hugh, fetch wood handy. The tale is long. The fire may die low.

"The Jacksons lived in Carrickfergus — that's a word that's fun to say." She rolled it off her tongue with a round Scotch-Irish brogue as Hugh set down an armful of wood. "We all lived around near, in Northern Ireland where our forefathers had come from Scotland many years before.

"My family, the Hutchinsons, were weavers. Your grandfather Jackson, Andy, had a store. He did right well but none of his sons liked store-keeping. That worried the old man. Sam was a sailor. Hugh went soldiering, and after the French were licked in Canada his company was sent to fight the Cherokees in the Carolinas. Andrew — I mean your father, Andy — tilled a farm though why he cared for the land no one could see. A man wasn't allowed to buy even a half acre for himself. And rents were so high a man could hardly make a living for all his hard work."

"Farmin's work here," a Crawford lad said resentfully.

"And why not?" Mrs. Jackson's eyes flashed. "Work is work. But look at what your father has done! He'd not been here five years before he owned many acres. Now he has great tracts of land and cattle for Charles Town each year. He could never have done that in Ireland. Never, I tell you!"

She eyed the children closely, then went on quietly. "Well, then, your Uncle Hugh came home from America. He told tales of this country around here — the Garden of the Waxhaws he called it — a neat phrase that sounded wonderful in Ireland.

My Andrew set his mind on coming right over. I wanted to go, too. My four sisters were in America. We sailed to Philadelphia, bought a wagon, and journeyed down to the Waxhaws."

"That's a funny word, Waxhaws," Robert Jackson remarked.

"We thought it sounded nice," Mrs. Jackson said. "It's the name of a tribe of Indians that used to live around here where now North and South Carolina meet. But my sakes! The place was no garden when we got here! Soil red. Trees thick. But Andrew could buy land. He felt like a king! We still own the two hundred acres he got by Twelve Mile Creek. It's your inheritance from him, boys," she spoke proudly. "With luck he'd soon have had more; he worked hard. You were about three, Hugh, and you, Robbie, were little more than a baby when —"

"What about me?" Andrew interrupted.

"Bide your time, Andy. You'll hear soon." Mrs. Jackson picked up a ball of yarn that had rolled off and went on.

"We cleared land and put in a small crop. We started a cabin, too. Next year he cleared and planted more and we made the cabin cosy. But late winter of that second year he strained himself trying to move a tree he'd felled. He lived only a few days. Before he died, he told me to keep that land for you boys — he was Scot as well as Irish and thrifty.

"Seemed like I had come to the end of the world when he died. I knew he'd want to be buried in a churchyard, so I walked to the nearest neighbors. We made a coffin. There was a heavy snow, right then. Somehow we got the coffin home, fixed him for burying, and dragged it by sled the many miles to the churchyard. Nothing else mattered to me, then."

The children were silent. Even those who knew the story were awed at the thought of the stricken woman and her babies.

"But in the worst of it, God provided," Mrs. Jackson went on in a lighter voice. "A woman with little children couldn't work a farm. But my sisters lived near; their husbands were prosperous and kind. We went first to your Aunt Margaret Mc-Kemey's, where you were born, Andy, a few days later, March 15th, 1767. See, you are in the story now, as I promised. Then, soon, we came to this house."

"A lucky day for us, that was," Aunt Jane Crawford added. "With me so ailing, I don't know what we'd have done without you."

"What happened next?" Andrew prodded as his mother sat silent. The tale could not end now. "That isn't all?"

"No, Andy. That isn't all. You boys have to live the rest of the story yourselves." She looked at her sons lovingly. "Hugh, Robert, Andy — your father chose America as the place where you are to live your story. I've been thinking that you might live it better if you know how and why you got here."

The Crawford children looked enviously at their cousins. Aunt Betty seldom spoke so seriously; her tone gave importance to the three boys. They watched her for more. But the evening was over. She rose to help her sister to bed.

Boys and girls put work away and did their final chores. Then they ran up real stairs to bed. The Crawford house had two rooms on the second floor, one for boys, one for girls. Few families in the Waxhaws were this comfortable.

When the men came back from Charles Town they brought startling news. Far to the north, at a place called Boston, colonists had become so angry about taxes that they tossed boxes of good newly arrived tea into the harbor rather than pay the duty on it. The event was news here, though it had happened months before. England had closed the port, it was said, and Boston people were getting hungry. Neighbors gathered at the Crawfords to discuss all this.

"We kin send 'em corn and cattle," one man suggested.

"I've got barley I could send along," said another.

"What do those British think we came to America *for?* to make money for *them?*"

"We'd not take such things laying down — not we Carolinians. If Boston folks fight, we'll fight, too."

"They's nobody around these parts to fight," a neighbor drawled. "Food's more needed than bullets, today."

So food was started on its way, by the long overland route. It could not go by boat; the port of Boston was closed.

War talk fascinated Andrew. Though he was the youngest brother, he was leader in games when the chores were done. They drilled with sticks for guns. They wrestled and raced, the winner thus proving his fitness for soldiering. Andrew devised war games with corn of different colors for different companies. It was so exciting that days fairly flew by.

Andrew was eight when Uncle James sent him to a better school. William Humphries won Andrew's interest by showing him a map — the first the lad had ever seen. The paper seemed like magic. Rivers and towns, so far away one might never have known of them, were there — plain as day. Andrew was so intrigued that he studied hard. He improved in reading and figuring and learned some Latin.

The reading, especially, was useful when he was at home. For people were more than ever interested to hear newspapers read. They wanted to know about Boston. Andrew read about Bunker Hill, about George Washington being made commander in chief and taking charge of the army at Cambridge. Some of the neighbors actually knew Washington because they had lived in Pennsylvania when the French and Indian War was fought.

In June of 1776 war came to Charles Town. Robert Crawford, Uncle James' brother, raised a company of volunteers and marched to defend the town. They did not fight, for the fort repelled the invaders and saved the city. But they brought home tales of the battle they had seen from afar. After that boys added forts and harbors to their war games.

On a hot August day in the summer of 1776, Andrew was sent to Uncle Robert's house on an errand. Just as he arrived a rider dashed up and flung himself off his horse.

"News!" he cried. "Hear ye all! I have news!" People ran from hither and yon. They saw him wave a paper he'd taken from his pocket. "Read the news!"

"Fetch Andy," Uncle Robert said, but Andrew was close by. "Andy, read it to us." The rider held out the paper.

"You the reader?" he asked before he let it go.

"Always am," Andrew replied. "Is it long?"

"Good 'n' long. And big words! Bet you don't know 'em."

Andrew glanced at the paper — and at the gathering audience. Big words indeed! Declaration. Independence. Political. Unanimous. Lucky the schoolmaster had taught him some Latin. It helped. He stepped away and gave a quick, prudent reading, saying the strange words aloud so that when he read he would not stumble, they would roll off easily. But people were eager.

"Andy! Come and read! We're waiting."

So he took his place on the horse block and began.

The War
Moves South

For a full minute after Andrew finished reading the Declaration of Independence, a hush lay over the listeners. People in this isolated community were not prepared for the ideas in a document like this one. Then a babble of questions began.

"States — that paper says 'states.' We're a colony."

"Are we on our own now, rid of England?"

"Ye think England will take this laying down?" Men looked to James Crawford to give answers.

"The paper Andy read makes it plain that we are not colonies of England any more," he told them. "We're states, united together in freedom from England. England won't accept it. General Washington is already meeting them, remember? This paper will help him. It makes clear what they are fighting about."

"Everyone ought to hear this paper," Robert Crawford said. "Andy better read it on the church steps, come Sunday."

Men approved that suggestion. As the group broke up people talked vigorously: men and women about the Declaration; boys, regretting that fighting was so far away.

Post days, after that, people came quickly to hear the latest news. Washington and his army were on Long Island. Congress was in session in Philadelphia. It seemed as though the Continental Army had more defeats than victories. But General Washington did not lose his courage. More travelers passed through the Waxhaws than before, and long herds of cattle and wagon trains of food went north.

Mrs. Jackson insisted that Andrew stay in school. Perhaps he could not have endured it but for Mr. Humphries' maps. With these, Andrew could figure out war news that he read in the Charles Town newspapers.

So months went by while Andrew went to school and did chores.

One morning in the spring of 1778 Uncle James Crawford tilted back his chair to show that he had finished breakfast. Then he turned and looked thoughtfully at Andrew.

"Well, Andy," he remarked, "I hear you're eleven now. Lookin' pretty tall and healthy, too. You've less fat than my boys, but I'll wager you eat as much. What you aim to do this day?"

The question amazed Andrew. When Uncle James joked, something was up. Did he have a twinkle in his blue eyes?

"Why, I'm doin' whatever 'tis you want done, Uncle," he replied meekly. The children around the table snickered.

"Reckon you're right at that," Crawford admitted. " 'Cause what I want done is somethin' you've long been teasing for. You're to make ready to join the cattle drive tomorrow."

"Uncle James!" Andrew could not trust his voice to speak.

"Betty, see that his clothes are clean. He'll need an extra shirt. My older boys and your Hugh are goin' — it's a good time to break Andy in. We start at sun-up."

"Yes, sir." Andrew had found his voice. "Kin I take m' dog 'nd m' gun? I don't need another shirt. I'll wash mine evenings."

"You do that, naturally. But you need an extra for town. No dogs. I carry a gun. I'll do any shootin' needed." He grinned at Andrew's crestfallen expression.

Andrew began his chores in a happy daze. He swaggered a bit before the children but was meek as a dove with the men. They had him help sort beeves for the journey — man's work. At noon he dashed to the house and found his mother.

"Maybe I better ride over and tell Mr. Humphries I'm not comin' to school for a while," he suggested.

"Yes, you should," she agreed. "And, Andy, be polite. It ain't nice to be too happy, leavin'." Andrew grinned and was off.

He minded his manners while talking with the master. But afterward, he boasted to his mates about going on the drive.

"You think you're so smart," one retorted, annoyed. "I kin lick you easy." This seemed likely; he was a big boy.

"Try it!" Andrew taunted.

The big boy rushed at Andrew and all too soon rubbed his face in the red dust. He turned to gloat — and Andrew was up and at him, only to be downed again. But not to be licked. A crowd gathered. Bets — marbles, knives — were made. On the fourth round Andrew got the big boy down and sat on him, hard. The crowd yelled as the challenger admitted defeat.

"Trouble with Andy," he said annoyed, "he don't know when he's licked."

Andrew grinned, brushed his pants, mounted his pony and went home. There, his mother scolded about his rumpled clothes.

"Fightin' agin?" his uncle noted. "Any more and you stay home tomorrow — hear me?"

At sun-up, next morning, the cattle drive left for Charles

Town. The experienced men rode along the string of beeves watching strays. Andrew had the beginner's job, the end of the line where dust enveloped him as he kept stragglers in line. Even this task excited him. He quite fancied himself as he galloped after a bull who wandered off into an unfenced wood.

Men used to the drive had little pleasure in keeping cattle in line, letting travelers pass on the narrow trail, and feeding men and animals. But they all liked the evenings in camp.

After the cattle were settled to grazing, the men washed themselves and their shirts in a creek. Then they told tales around a campfire until sleep overcame them.

Andrew could hardly believe his eyes when, a day or two later, he saw Charles Town. Houses, not cabins; streets, not roads or trails — it was bewildering. And beyond the town of a thousand people was the harbor with ships along the wharf and the blue sea in the distance. The astonished boy from "up-country" was so amazed that Hugh and Jimmy Crawford laughed.

"You wait till we git the cattle settled in the market pens, and we'll show you the town," Jimmy promised. "Father'll let us off; I'll ask him myself."

At twilight, the cattle neatly bedded down, the boys put on clean shirts and set out. The houses were more beautiful than any Andrew had ever imagined. On the sidewalks — imagine wooden walks for people — elegantly dressed ladies and gentlemen strolled in the cool of the evening. Carriages drawn by fine horses drove by with the people bowing right and left to friends. The ease, the elegance, astonished Andrew.

"Someday I'm coming here and be as grand as any of 'em," he said.

"Truck on all that," Jimmy was not impressed. "You'd be a silly to pleasure yourself that way, I say. Me, I'd ruther see a cock fight. Come on, I'll show you."

Cock fights were not news to Andrew. Passing travelers had told of the battles in the cockpits of Charles Town. Andrew had even tried to match two of his mother's best roosters — and gained nothing but a scolding. But what he had pictured was nothing like what he now saw: men crowding around the pits, betting, shouting; handsome cocks battling for life in bloody combat. The boys had a struggle to drag themselves away when time for their turn at watch in the pens drew near.

"We'll be here a week," Jimmy promised; "we'll come often."

But alas for the boys. The market was good. Crawford cattle were sold by the second day — and at big profit, too. War was good for farmers, men said as they started home. Andy's most precious souvenir was a scrap of paper on which he had copied directions for feeding and training fighting cocks. It seemed important.

Some of the profit from the cattle sale belonged to Mrs. Jackson. To Andrew's dismay she promptly arranged for his further education. An academy some miles away had been recommended to her. She determined that now her son was to start on his classical education to be a preacher.

Andrew set out boldly, hiding his misgivings. But this school pleased him. The students liked sports better than Latin; as he did. He made friends and even did well in his studies. He got a reputation as a fighter and a runner and a stickler for fair play. *He* would fight a larger boy. But let a big boy pick on a smaller one, and Andrew soon made him bite the dirt.

On his brief visits home his mother noted his growth and his improved manners. His Latin book fascinated her.

"You make me proud, Andy," she said shyly.

This year when Andrew was at school was proving to be a hard time for the young republic. The war seemed at a stalemate — except for George Washington. He held on with dogged determination that decided the British to turn southward. They took Savannah and headed for Charles Town.

Robert Crawford again raised a company for defense. Jimmy Crawford and Hugh Jackson joined up, with other sixteen-year-olds in the Waxhaws. They helped to save the city, but at frightful cost. Hugh was one of the many killed and buried in an unmarked grave.

But Americans could not hold the town. The British brought more forces, and in May of 1780 Charles Town was in their hands.

The British began a march of terror into the Carolinas that lasted for months. Andrew's school closed. He got home just as the British Colonel Tarleton and a hundred horsemen descended on the Waxhaws. They killed a hundred and thirteen people; a hundred and fifteen lay wounded as the cavalry galloped off, searching for Captain Crawford's company that chanced to be somewhere else at the moment.

Andrew helped to bury the dead and carry wounded to the church. Mrs. Jackson and other women nursed their men tenderly, but many died. The lad who had liked drills and war games saw a new side of war now, and it sickened him. Hugh was dead. The smell of blood was horrible. Wounded men cried in misery and pain.

As he worked, helping, Andrew felt a growing hate of the British. Why should they come to a peaceful settlement — no

soldiers were there. Why should Waxhaw people have to bow
to a faraway king — or be killed. His rage became a bitter
need for revenge. It absorbed him. He could hardly eat or
sleep.

The dead were hardly buried before the raiders were back,
killing again. A third time they came; the whole summer of
1780 was a nightmare, a terror. Men were away, fighting.
Women and boys did their best, but stock was stolen, crops
taken, people killed.

When at last the British seemed to have gone north, Robert
and Andrew walked miles to Captain Crawford's camp and
begged to enlist. Robert was accepted as a private. Andrew
was made a messenger for Major Davis. The officer welcomed
a boy who was a good rider and knew the trails. He did not
guess that the tall lad was only thirteen years old. He equipped
his messenger with a horse and a pistol — and ever after was
Andrew's ideal of what an officer should be.

To Andrew's dismay, this choice service was short. After
the battle of Hanging Rock, Major Davis sent the wounded
home under the care of the Jackson boys. They knew the way.

Once home they had a new problem. Mrs. Jackson and
several neighbors had decided to go to Charlottesville, North
Carolina, and wait out the war. She was determined to take
her sons with her.

"But Mother," Robert exclaimed, "we *have* to fight!"

"How can I let you go?" she cried. "Hugh dead . . ."

"Someone's got to lick the British!" Andrew said angrily.

"You're right, Son." Mrs. Jackson squared her shoulders.

"I'd fight in a minute — if I was a man. Go. And God go with you." Her eyes were bright as she watched them leave the next day.

They had many miles to walk. And travel was dangerous because of Tories, people who had long lived around there and continued loyal to the English king. Children as well as adults who walked on the road or worked near in the fields might suddenly be asked by a traveler, "Who you for?"

Life might depend on whether the answer was, "The King," or "The Congress."

The Jacksons and their kin would promptly say, "The

Congress." But the two boys knew that their chances of getting to Uncle Robert would be better if they were not seen. So they hid in the woods when they guessed someone was coming near.

October sunshine was warm and cheerful when the brothers left their mother. Maples and dogwood flamed red in the woods. Nuts spread on the ground tempted the boys to fill their pockets.

"We'll find more later on," Robert decided. "My pockets are full of food Mother packed — that's better'n nuts, even these."

The third day the weather changed. Rain lashed the red leaves and winds tossed them, sodden, onto the ground. The boys were drenched and cold when night came. Yet they hesitated to make a fire and betray their whereabouts.

"What's that on ahead," Andrew whispered at twilight, "fire?"

Peering from behind tree trunks, slipping from bush to bush, they came near a small campfire and two travelers.

"I wonder they got it going," Robert said.

"Tain't any fire to boast on," Andrew whispered. "Listen. Maybe we can tell if they're friend or foe." The boys strained to hear.

"Nolichucky Jack — ever hear of him?" Andrew whispered.

"Never," Rob said, "but he must be something to hear them tell it, brave, smart — listen, Andy! Hear that? 'Licked the British.' These men are for the Congress, I'll bet. I'm going to ask 'em." Together the boys slipped from the tree's shelter and crept close to the strangers.

"I heard you talk about licking the British — who you for?" Andrew asked them boldly.

The men jumped up, hands on guns, but when they saw the boys, one said, "For the Congress, of course. And don't be so touchy. Ain't you heard that Nolichucky Jack and his men have licked the British at King's Mountain and the British are on the run?"

The boys stared. "Who's Nolichucky Jack?" Andrew asked.

"John Sevier. Called mostly by the river near his home. Sit down, lads. Warm yourselves and hear a tale." The boys came near.

"Jack and his men are independent folk. They like plenty of room. When they heard the British were coming west of the mountains, the news didn't set well. Jack decided to teach 'em a lesson that they'd remember a while. When war is made, Nolichucky Jack wants to make it himself — see?"

The boys didn't, exactly, so they listened carefully.

"Jack's got a thousand men with him. Some call 'em outlaws. They wear buckskin, carry tomahawks and muskets, and heed no politics. They're for freedom and fair play. They didn't like what they heard about Tarleton's raids. So when a scout brought word that those British had come west and fixed themselves up on a mountain, Jack put it up to his men.

" 'Do we go?' Not a man lagged. They went. They packed provisions and ammunitions and were on their way that fast.

"As they drew near the mountain a woman slipped in to

camp at night. 'I'm for Congress,' she told Jack. 'You goin' for to lick those British?' Jack nodded, wondering at her.

"'Then go straight ahead. They're three miles yet. They're on a high place they call King's Mountain. I've been there. I sold 'em chickens. You-all kin lick 'em if you work fast.'

"Quick like he is, Jack ordered 'em to march. Soon they could see breastworks on the ridge and tents white as snow — and the flag of Britain flying so fine.

"'Dismount and tie your horses!' Jack ordered. 'Yonder is the enemy. You know your duty, men. Don't fire till you kin see their eyes.' Those men climbed up right under cannon that could not be fired *down* to kill 'em. They cleaned up the place right well — more prisoners than victors."

Andrew leaped to his feet. "They got 'em all?" he cried.

The men grinned. "Jack'd go fer you, son. Yes, they got 'em all. The British ran up the white flag."

"But they hain't gone from around here," Robert remembered.

"No. But they will be."

Andrew got up. The men looked at him in surprise.

"Ain't you staying by the fire?"

"No. We're on our way. But thank you for the news."

Robert followed his brother. But when they had gone a ways he remarked, "That fire felt fine, Andy. Didn't you want to stay?"

"Sure did," Andy said. "But the rain stopped. I'm thinking Uncle Robert will like to hear the news about King's Mountain. A victory will encourage him."

"Clean My Boots!"

THE NEXT DAY the boys found Uncle Robert, the Waxhaw company, and other soldiers. Uncle Robert was a major now. The boys told the story of victory at King's Mountain.

"Is it true?" Andrew was eager to believe. "Is the war won?"

"It's true I reckon, Nephew. We've had rumors of Nolichucky Jack from others. But the war's not won — though this will discourage 'em plenty, I reckon."

"May we enlist?" Andrew asked anxiously.

"Sure may. Go to the quartermaster for your muskets."

Over campfires men talked of Nolichucky Jack, and Andrew got his first knowledge of the land beyond the mountain. "Up-West," men called it. Tales told were all the more exciting because he was hungry, wet and cold most of the time as they ranged up and down their assigned area, hunting the enemy.

Andrew was doing sentinel duty on a May evening in 1781 when he heard the dread cry, "Tories are coming!" Peering from behind a tree, he dimly saw a party ahead and fired. Then he ran for a nearby house. Six others were there. They took stations and fired. They thought they saw some Tories fall.

At that moment a bugle behind the house sounded loudly the Americans' cavalry charge. The Tories fled. For minutes no one made a sound. Then a sergeant spoke.

"Who blew that bugle?"

"I did, Sir." The men whirled and saw a meek little man come in at the back door. "I didn't have m' musket handy, so I grabbed the bugle. Guess they must-a thought we got more men 'n we really have." He stared uneasily as the men rolled on the floor laughing in hysterical relief.

But the next time Andrew and Robert were not so lucky. Word of Tory skirmishing got to the British, and they sent trained troops to help the country people. Major Crawford heard this and moved his men. They took a firm stand in the Waxhaw church, just as a company of British dragoons dashed up.

They set the church on fire and shot or captured Americans who attempted to escape. Andrew dashed from a back door and made off for the swamp — he had seen Robert go in that direction. A dragoon followed, but local boys could easily shake pursuit in that maze.

The brothers met and spent a cold, exhausting night. In the morning they crept to the edge of the swamp.

"See any one?" Robert whispered.

" 'Nary a soul. We're near Cousin John's cabin. Let's go there. I'm hungry." They crept out cautiously and ran to the house. They were just starting to eat steaming bowlfuls of hot mush when a clatter in the yard alerted them.

"It's the British," one of the children cried. Andrew

and Robert hid behind the big loom as an officer and three
men came in and began wrecking the room. One thrust his
saber through the fabric in the loom. Andrew darted out to
save himself.

"You there!" the officer yelled at him. "Clean my boots!"

Andrew drew himself up haughtily.

"I may be your prisoner, Sir. But I am not your boot-
black! I claim treatment due a prisoner of war."

The British soldiers stared. The room was still. The face,
neck, ears of the officer grew red, and his throat worked. He
was so angry that words would not come out. Then he lifted
his sword and swung it, hard.

Andrew raised his left arm and dodged. That arm saved
his life, but the sword cut his wrist to the bone and gashed
his head. Andrew staggered, and blood spurted, quick and
red. Children screamed. Cousin John's wife sprang to steady
the boy.

"Tie him up!" the officer ordered.

"But, Sir!" The housewife faced him bravely. "He's
bleeding —"

"Let him bleed! Serves him right!" Then his eyes gleamed.
"I have use for him. He surely knows the way to Thompson's.
Have him lead you to that rebel." This was an order.

As the officer spoke the housewife grabbed cloth and tore
it into strips. She bound Andrew's wounds the best she could
so hurriedly — and checked the cruel bleeding.

"Enough! Make him run! Now get out!" the officer cried.

A soldier prodded Andrew with a musket. The lad ran

out of the house, out of the dooryard. The soldier mounted and rode at his heels. As he ran, hatred of the British, begun at the massacre in the Waxhaws, mounted. Andrew all but choked with the evil of it, with longing for revenge.

"Kill him if he misleads you!" the officer shouted after them. The threat helped to clear the fog in Andrew's brain. This was not the time for hate. He must think. Neighbor Thompson lived near. Andrew might run the short way through the woods or a roundabout route, across a meadow, in full sight of Thompson's house. Running was misery, but Andrew chose this long way.

At the house the soldiers searched and found no man, no papers. As Andrew hoped, Thompson had seen them coming.

Angrily the soldier, and Andrew, returned to report. The officer glowered at Andrew — had he tricked them?

"You walk all the way to Camden, to prison," he ordered. "Give him no food or water and make him run!"

Cousin John was heartsick at the boy's plight. Bravely he protested, but in vain. Robert was already on the way to prison, and John's house was wrecked.

Somehow Andrew lived through that forty-mile journey. He was hungry, thirsty, and sick from throbbing wounds. If he could only find Robert; that thought helped sustain him. In Camden, he was thrown into a loathsome prison room. Robert was not there. A smirking Tory pulled off Andrew's coat and boots; he never saw them again. Later someone set a jug of water and a hunk of dirty bread inside the door.

Soon more prisoners came and Andrew heard dread words. "They've got the pox!"

"I'm *glad* Robert isn't here," Andrew told himself. "Glad!"

Youth, a sturdy constitution, and his determination for revenge kept Andrew alive. He learned to chew nasty bread. He made himself lift his head, roll over, move about the room. Slowly the wrist and head throbbed less. Days passed. He had no idea of time — none of them had. New arrivals were too ill to talk even if they had news, which was doubtful.

One day a stir outside sent Andrew to the tiny window.

"I see American soldiers," he reported tensely. "Somethin' goin' on." His words roused listless men.

"You sure, Andy?"

"Certain sure. I kin see it plain."

"Maybe General Greene's come. I heard General Washington put him in place of Gates."

"But nothin' happened," another complained.

"It's happening now!" Andrew's voice broke with excitement as he watched. Men crowded close, hoping for a glimpse.

At that moment the keeper dashed in with boards and nailed up the window. Men dropped to the floor, too weak to endure this added disappointment. The room stank from the smallpox sick. Two died that day. Andrew noticed he felt feverish.

The next morning Andrew had an idea. He managed to force out a knot in one of the boards. With this for a peephole he could see artillery and soldiers, but they were far off. His head ached too much to make sense of the distant action.

"Are we winning?" a prisoner asked.

"Looks like it," Andrew tried to sound assured. Then fever overcame him, and he fell to the floor.

Hours, or was it days later, he never knew, he heard a word that caught his attention.

"Andrew Jackson! Come out!" It must be part of his dream.

"Jackson! Stir yourself!" That sounded real. "Stir yourself!"

Andrew rolled over, raised up on his elbow and staggered to his feet. The door was open. The sunshine blinded him as he walked out. Before him was a vision.

"Mother! Robert!" He ran toward them. Robert leaned weakly against Mrs. Jackson, but she had an arm for Andrew, too. A quick look showed her the boy's red swollen face — the pox. Both boys! But her voice was steady.

"It's all right, Andy. We're going home."

Andrew couldn't believe it. "Has Robert been *here?*"

"Yes, all the time. They guessed you were brothers so they kept you apart."

"But you, Mother? You're here!"

"And you're free, Andy. I heard my boys were prisoners. I heard some British were to be exchanged. So I came to Camden to get you two on the exchange list. Robbie's awful sick. Help him to mount, Andy. I could git only two nags — the rest are stolen or with our army. You ride the other one yourself. I'm well. I like to walk.

Andrew helped get Robert onto the horse; then he put out his hand for his mother to mount.

"You're ridin'." She got on. They set off, Andrew steadying his brother.

When they had gone several miles, Mrs. Jackson felt safer. She stopped by a creek and had Robert lie on the grass while she washed him and put on clean linen. They ate homecooked food and drank the clean water from the brook. As they went on, Robert grew worse. He did not speak sense, but prattled foolishly.

"Don't try to talk, Robbie," she told him. "Soon you'll be home in your own bed." The next day a cold rain poured down. They pushed on as fast as they could.

Late the second night they were at home. Robert was hardly conscious, and Andrew had a raging fever. The cold rain, just as the pox sores were breaking out, made him desperately ill. He did not know that his brother died two days later.

After the burying, neighbors tried to comfort Mrs. Jackson.

"A rider says General Greene won at Camden. He's got Cornwallis on the go at last. We're safe.

"Virginia's getting her share, now. I heard that Colonel Tarleton's up there. He nearly arrested Governor Jefferson in his own house. That Tarleton is a terror. We know."

Days were peaceful enough while Andrew made a slow recovery. The drying sores itched, and he was tired and listless. That summer of 1781 was hard in the Waxhaws, despite the absence of fighting. Many neighbors were dead. Men and boys were away with the army. Food was scarce; even seed corn had been sent to the soldiers. Every household nursed sick and wounded.

Weeks dragged by. Gradually Andrew could move around the house. Then he helped with a little garden. He noticed one day that his mother studied his face carefully. Then she went at a big washing and a round of baking. The next day she wakened Andrew early.

"Andy, I've not told you because I wanted you to get well as fast as you could. But I had news a while back of your cousins, Joseph and William. Those Crawford children I helped raise are dear to me. Now they're sick, in a prison ship in Charles Town harbor."

Andrew stared at her, shocked. She had known this and had not told him?

"Two neighbors are goin' with me and we're leavin' now, for Charles Town. Maybe their boys are in prison, too. We may not be allowed to help, but we've got to try."

"I'm goin' with you," Andrew said quickly.

"I was sure you'd say that, Andy. But you're not strong enough yet for the journey. Anyway, nursin's women's work. We'll manage."

"What shall I do here, Mother?" Andrew was dismayed. His uncle was still away; Aunt Jane was long since dead. The older girls were married, and the younger ones lived around with them to help with the babies. Cousins and neighbors were dead, or in prison — maybe some still alive in the army. One didn't know.

"You know well enough how to manage around the place, Andy. Keep it best you can against the day your uncle and the boys come home. They've always been good to us." She paused, thinking.

"I'll be coming back," she went on confidently, "but just in case I shouldn't, sit down, Andy, and listen. Remember words I say now. Be kind, and don't hurt the feelings of others. If you get into a dispute, settle it in fairness; don't take personal fights to law. Never slander or let yourself be slandered. Bide by the truth. You'll need friends. Make them by being honest; keep them by being steadfast. To have a friend, you must be a friend. And remember, let your wrath cool before you act." She hesitated, twisting the corner of her apron, making sure she had said what she intended. Then she rose quickly.

"Take care of yourself, Andy."

The two neighbors had ridden up while she was speaking. Elizabeth Jackson went to the horse block where her mare was pawing at the ground, ready. She mounted and rode off.

Andrew stared after her, his throat working painfully. Breakfast was waiting for him indoors, but he could not eat. He carried the bowl outside and scraped the mush into the chicken yard. Then he looked around. He was alone. No use to speak. No one was there to listen.

Weeks later a traveler told Andrew that he had heard

Mrs. Jackson found her nephews and other youths who needed nursing. Later a rider said she was ill with a fever.

Andrew was fourteen and a half when he learned that his mother was dead. He went into the house and sat down quietly. It seemed that he could see her, bending over the fire, cooking. See her knitting, weaving, tending her sister or the children. And now she was gone. Buried in an unmarked grave near Charles Town, the rider said.

One of the cousins came home in a few days and brought a new husband; they would stay, they said. Then a man from town brought a package for Andrew Jackson. He opened it and saw a dress of his mother's and a few trifling possessions.

"Her Bible's not here," Andrew said in a dull voice. "I'd like right well to have her Bible." Then he turned and walked off into the woods.

"Shall we go fetch him?" the cousin said anxiously.

"No. He wants his own," her husband answered. "He's suffered too much for his years. But Andy's got spunk. He'll come back. We'll be here with him then."

Rich Man,
Poor Man

WHEN ANDREW JACKSON came back to his kinfolk, they found him changed. He was serious, willing to listen when his uncles talked. Life in the Waxhaws had changed, too.

Lord Cornwallis had been defeated at Yorktown in October, that year, 1781, though peace papers were not signed until September of 1783. But people in the Carolinas could not wait for that. When the fighting ended, those left alive moved back to their homes and farms. Tories who had fought against the new republic or had been spies and informers, found themselves hated. They moved to Canada or to England.

During the British stay in Charles Town officers had taken most of the nice homes — a usual wartime custom. The evicted owners fled north into the "up-country" for refuge. These city folk now moved back. A few had lived in the Waxhaws region and would be missed. Andrew had several friends he was sorry to see depart.

"When you come to Charles Town, our house is your home," they told him. Andrew promised. But he did not expect to go to the city. There were no cattle to sell.

Days were filled with work. Schools opened. Andrew attended first one and then others as he lived with various rela-

tives. His Uncle Robert, James Crawford, and George Mc-
Kemey were glad to have an extra hand. But Andrew was
restless. The days of adjustment were hard for everyone.

Waxhaw men, like many patriots over the country, were
impoverished. In the Carolinas houses had been damaged or
burned to the ground. Furniture, cooking utensils, tools, had
been stolen or wantonly destroyed. Countless things were
needed — horses, cattle, chickens, grain, new buildings. There
was no money for buying and nothing to give in trade.

"They say if you make out an account the state will pay
for what you lost," a neighbor reported.

"So I've heard," Uncle James said. "I'm making a record
now. So is Robert."

Major Robert Crawford had served his country with distinction for seven years. Now he had nothing left but his land.
He turned in a modest claim for part of his losses. The state
treasurer fussed and trimmed it down. The major was not
surprised; he knew the state had very little money. Finally he
collected a small fraction of what the treasurer had agreed was
a fair claim.

Other Carolinians, hearing of such things, packed up and
moved West. Beyond the mountains they would find Indians,
toil, and hardship, but they could make a new start. Americans
paid a big price for this new thing called freedom.

Andrew Jackson heard talk about these things, and it
occurred to him that he too, was paying a price. Not in money,
for he had none to lose, but in something dearer, something
that could never be replaced. His mother and two brothers

were dead. Indeed, his father's death, though long before the war, was part of the price paid for Andrew to be born in a land where men now were to have equality under the law.

As he began to work on farms, Andrew saw, too, that his health was not as good as before because of army life, prison, and smallpox. All his life he would carry the scars from that officer's sword and from smallpox. His heart was scarred, too, with hate which smoldered in him. It was not surprising that he found it hard to settle to work at old tasks.

Uncle Robert apprenticed him to a saddler who had a good leather business and needed a helper.

"Now why did you do *that?*" Mrs. Crawford demanded.

"Well," Crawford tried to think why. "I guess because Andy likes horses and horses need saddles. He might as well learn to make good ones." Andrew's aunt was not impressed.

Saddle making did not interest Andrew. Even school was better than sitting all day on a bench. After trying it for six months he surprised his old school teacher by appearing unannounced.

"Is there any room for me?" he asked and was welcomed. A few weeks later Andrew surprised his teacher a second time by leaving, without saying good-by.

Andrew tried several schools and liked none. His uncles were worried. One youth alone could not farm that land up by Twelve Mile Creek, and it was too far away for them to help. They agreed that Andrew was bright — but what should they do about him? They didn't know.

The month that Andrew was sixteen, 1783, a letter came

for him from Ireland. He chanced to be at Uncle Robert's, and family and neighbors gathered quickly. Andrew felt a thrill of importance as he broke the seal and glanced at the page.

"Grandfather Hugh Jackson is dead!" he exclaimed.

"He was a good man, Andy. You never saw him; but you kin be proud of him."

"He was a respected merchant in Carrickfergus," Uncle Robert added. Then he turned to a boy near. "Run to the cattle pen and fetch Uncle James and any others you see. Tell 'em Andy has news. Go ahead, Andy. You kin read again when folks come."

Andrew had scanned the letter to himself. "It's from a

lawyer," he said in an awed tone. "Grandfather left more'n three hundred pounds for mother 'n Hugh 'n Robbie 'n me — but they're dead!"

"Then the money is all yours, Andy," Uncle Robert took the paper Andrew handed him. "Yes, that's plain."

"Andy's got an inheritance!" Mrs. Crawford called to kin who were approaching. They listened eagerly as Andrew read.

"Three hundred pounds sterling! That's a lot of money."

"He kin buy stock and set up farming with that." Uncle Robert figured quickly.

"You could take half and go to college," Uncle James said. "You'd have plenty left to set up in a profession. I al'ays heard your grandfather was thrifty."

Andrew's blue eyes sparkled; his face flushed. This was the best moment he'd had in years. But he could not speak. A lump in his throat was too big for talking.

"Andy better go right to Charles Town and git the money," Uncle Robert said to his brother. Then he turned to Andrew. "That colt you've been training is shod 'n ready. You're welcome to it. If I was you, I'd set off this very day."

Mrs. Crawford roused from thought. " 'Melia, stir up the fire. Jimmy, fetch water and fill the wash kettle. Andy, fetch your clothes. Lucky for you it's a quick drying day. I'll have clean things ready time your uncles git through talking."

While Andrew and his uncles conferred, the women and children scurried about. This tall youth of sixteen was no longer a problem; he was a rich relation. Even Andrew, absorbed in listening, could feel a difference. He found it pleasant.

Three hours later he rode away. His homespun coat and breeches were well brushed; his saddlebags bulged with food and clothing; his head buzzed with advice. The world seemed good — he was going to town and soon his pockets would be full of money.

The next morning he reread the lawyer's letter. Amazingly his good fortune was still true, not a dream in the night. He fed his horse, washed and ate, taking his time; the day was fine. Soon he rode on, thinking all the while about the money.

What should he do? Buy stock and build a cabin? Sensible, but not intriguing. Or study law? Was it Uncle James who had talked about law?

"Most of the lawyers in the Carolinas were Tories," one of the uncles had said. "Now they are disbarred from the courts. This makes a fine opening for a loyal American, like you, Andy." I ought to think about that, Andy told himself.

The next day he met some north-bound travelers.

"Where you bound fer?" one asked Andrew.

"For Charles Town," Andrew replied. The traveler nearly fell off his horse, laughing.

"It's plain you're from up-country. You don't know the news." Andrew was vexed so the man added, "There ain't no such place any more."

"Charles Town! Where's it gone?"

"Hain't gone. It's in-cor-por-ated. That's somethin' important. And the name is Charleston now — less like namin' after a British king," he added.

"Where you stayin'? Not that it's m'business."

"You're right. It hain't," Andrew agreed cheerfully. "But I'd thought of Quarter House Tavern." He named a place he remembered from the cattle drive.

"Um-m, that's all right," the traveler conceded. "Some inns are fancy. You might not feel at home. Luck to you."

As Andrew rode on he pondered on that remark. He was glad to know about the change of name and pleased with its un-British sound.

The encounter stirred memories. He thought of the fine houses he had seen; the elegantly dressed women, the men, dandies in satin coats and breeches, lace ruffles and jeweled snuff boxes. *Those* people would feel at home in Charleston. He remembered the races and the sleek, handsome horses; the fighting cocks and the excitement of those contests. He was surprised how much he could remember that he had not thought of in years.

He got to the city at noon, stabled his horse and settled at the Quarter House Tavern — it was still a good hotel. When he had washed and put on a clean shirt, Andrew walked to the lawyer's office and introduced himself.

"We were expecting you, Mr. Jackson," the lawyer said. In an hour the business was done. Andrew Jackson walked out of the place with a tidy fortune in his pocket.

On the way over, Andrew had noticed the sign of a tailor shop. Now he walked directly there as though that had been his plan all the time. The tailor measured him. Andrew then ordered three suits, one to be ready the next day. The tailor would have to work all night, but he did not mind. Mr. Jackson had paid the extra cost without a murmur.

After dinner the new heir walked four miles on the Broad Path to the military wall where he could see the fort. On the way he saw gangs of slaves and free workmen, building fences, repairing houses, making gardens. How quickly repairs were made here — not slowly as in the Waxhaws.

Andrew did not try to find friends. He did not speak to several who looked at him inquiringly. People could wait until he had his new suit, tomorrow. He was not ashamed of the clothes he wore, but he liked the drama of change. It would be the better for waiting.

The new coat and breeches fitted him well, but made his heavy boots seem shabby. The tailor knew a cobbler who could serve a new customer at once. The cobbler suggested a hatter and a barber. Andrew hardly knew himself when, that evening, he strolled along Broad Path swinging a cane like other gentlemen.

Within a week he had met several friends. They greeted Andrew cordially, entertained him and introduced him to other young people. Soon he was attending races and cock fights, betting freely as others did; many an evening ended with a ball. None of this was planned; he drifted — and found it very agreeable.

Now and then as they strolled along the street or sat at dinner talk turned to the future.

"What are you planning to be, Mr. Jackson?" he was asked.

"Oh, I don't know," Andrew always pushed the question aside. Why must a person have to *be* something? The present

was so delightful, why make plans?

Andrew was usually unlucky with his bets. But no matter. He had plenty of money.

Then, quite suddenly, his money was gone. Gone! Grandfather's earnings, gone! Andrew couldn't believe it. Frantically he turned out pockets, hunted through clothes, dashed to the stable to search saddlebags. No money, anywhere.

On the way back to his room the innkeeper stopped him.

"I'm callin' to your mind, Mr. Jackson, that your bill has not been paid for three weeks." The tone was polite, but very firm. "There are charges on it from the tailor and the hatter, too. You plan to pay today, no doubt?"

Andrew muttered something and pushed by the man. He shut the door of his room and leaned against it. Then he saw, on the table, that unpaid bill. He picked it up and scanned it — the sum was about forty pounds.

An hour's hard thinking brought no answer to this problem. The room was suddenly intolerable. He went to the stable and got his horse, hoping that a ride might clear his thoughts.

In the space before the stable a group of men were throwing dice. One looked up at Andrew and his horse.

"That's a beautiful animal you have there, Sir," he remarked. "I could use him. I'll throw you for him. How much?"

"Forty pounds," Andrew heard himself say.

"Done! Clear off, men. This is going to be good."

Andrew handed the bridle to the nearest man and squatted where the challenger pointed. He shook the dice with steady hand — then tossed.

"You've won!" The shout seemed to come from far away. "You keep the horse and get the money!" Men eyed him enviously as the challenger peeled bills from a roll and paid.

"I'll throw another day — with better luck," the man said casually. "I like that horse."

Andrew took his winnings, the first of any amount since he had come to the city. Then he turned and took the bridle.

"Thank you, mister," he said politely. He went to the hitchrack and tied his horse. The animal turned and looked at Andrew inquiringly — weren't they going someplace?"

Suddenly a strange something that had possessed Andrew during these gay days was gone. This beautiful horse — why, he had nursed the tiny colt when it was a lanky newborn creature. He loved it. His uncle, who needed horses, had given it to Andrew out of love for his orphaned nephew.

"And, I, miserable wretch," he whispered, leaning against the horse's silky neck, "squandered my grandfather's money and wagered you on the turn of the dice!" Andrew's lips curled in loathing for this youth, himself. He threw his arm around the horse and rubbed his stricken face in the smooth neck. The horse neighed softly, content.

Over where the dice had been thrown men still talked loudly about Jackson's luck. Andrew hardly heard them. He put his lips close to the horse's ear and whispered a vow.

"Never again, never, I tell you, shall I throw dice." He patted the soft hide and added, "And fool though I am, I keep my word."

Within an hour he had paid the hotel bill and two smaller ones he suddenly recalled. Then he packed his bag, mounted his horse, and rode north.

The thought of what his uncles would say was not pleasant. But they could say nothing as bad as his opinion of himself.

"I reckon I've learned my lesson," he remarked to the horse as they stopped at a spring. "It came high. I hope it lasts. It ought'er."

Across
the Mountains

ANDREW HAD a miserable time after his return to the Waxhaws. But he did not blame his relatives when they scolded him. He had, indeed, been a wastrel and a fool. When he got tired hearing of his faults, he simply rode away.

A rumor drifted back that he was going to school. "Bone-carriers," as gossips were called in the Waxhaws, did not say how his tuition was paid. Later the uncles heard that he was *teaching* school. That might be true; he knew more than many schoolmasters.

"That lad has had a time of it," Uncle James said when he heard. "I'd hoped to do right by Betty's boy."

"You tried," Robert Crawford comforted.

"Sometimes tryin' is not enough. He's smart, Andy is. I'd like to see him amount to somethin'. Wonder if he's sold that farm of his. It ought to bring good money now that folks are moving more."

"He might sell and we not know," Robert said. "Land's his."

Around Christmas of the next year the Crawfords heard that Andrew was in Salisbury, North Carolina. Then a traveler told that the youth was studying law with Spruce Macay, a lawyer of standing. The uncles were relieved and pleased.

Andrew was pleased about himself, too. He had come to Salisbury a stranger; got himself work and a chance to study. These facts helped to wipe out his chagrin about losing his money. This town was not a city, but it was an agreeable place to live. Two wide shaded streets crossed at the center and public wells on these streets provided water and sociability all day. A score of good houses and many cabins were set in pretty gardens — Andrew liked it much better than a farm in the Waxhaws.

Salisbury people liked Andrew Jackson, too. Now about eighteen, he was tall, slender, and intelligent-looking. He was an easy talker, and his quick smile made his long face, pockmarked and scarred with a sword, very intriguing to the girls. His skill in riding and games earned the respect of the men.

The Rowan House, where he got room and board, was a

pleasant home and a center of gay social life. For two years, Andrew worked and studied law with two other young men in the small house by Spruce Macay's garden that was his office. His special friend was John McNairy, one of the students. Together they copied long reports, ran errands, and read law books. Together they mixed fun with this hard work.

After those years, Andrew studied law with Colonel John Stokes, one of the best lawyers in North Carolina. Stokes was a favorite with Andrew. He was not only a fine lawyer, but he had served nobly in the war. The colonel had lost a hand in battle and he wore a silver ball, attached to that wrist. Often at a critical moment in court Stokes pounded on the table to enforce a point. The sound, like a bell ringing, was impressive. Andrew noticed that it was effective, too.

The summer of 1787 Andrew Jackson got his certificate to practice law in the courts. The paper said that he was "a person of unblemished moral character and competent in the knowledge of law."

The new lawyer, now a few months past twenty years old, was more than six feet tall, and he moved with the grace of a natural athlete. His dark auburn hair fell over his forehead, hiding some of the pockmarks. His long face was not handsome, but it had a look of distinction. And he had a knowledge of law equal, at least, to that of most lawyers in that time and place.

Courts in North Carolina, as in some other states, were held for one week, twice each year. Then judge and lawyers moved on to another town, making a "circuit." The first court

held in Salisbury after Jackson was certified was in November
of 1787. This was two months after the writing of the Constitution of the United States was finished in Philadelphia. Andrew
hardly noticed that event in a far-away city because his own
first day in court absorbed his mind.

On that morning, he strode down the street wearing a
new broadcloth coat and a ruffled shirt. His eyes were deep
blue, intent and eager. His hair was plastered smooth with
bear grease, and he carried himself with an air that was noticed
by townfolk and strangers in town for the court.

At the close of the second day the judge spoke to Andrew.

"You ought to go Up-West, young man," he said. "They
need lawyers. I hear you're as good a rider as General Washington — none could be better than *him*. You could take the
journey."

"Thank you, Sir." Jackson was pleased with such approval. "Maybe I should follow the court here awhile, though?"

"You're right. Get experience. You'll do well, Up-West."

The judge was not the first to suggest that Andrew Jackson should move west. People talked constantly of such a move.
By "west" they meant the long, oblong strip beyond the Alleghenies known as the Western District of North Carolina.
Fur traders had been the first to tell of wealth out there. A few
brave souls had settled in the rich valley beyond the Blue Ridge
— they called their village Jonesborough. Later French Lick
was a settlement on the Cumberland River.

North Carolina found it hard to govern such a far-off region
so they decided to give the western district to the United States

in payment of war debts. Settlers were not consulted and when they heard, they were angry.

"Pushin' us off, are they? They daren't!"

"We kin make our own state, if we're not wanted." They organized what they named the State of Freeland with John Sevier — Nolichucky Jack — as governor. The federal government might have seen by this that western settlers wanted government. Instead they paid no attention at all; the new state had many troubles. The people tactfully changed the name to State of Franklin and asked Benjamin Franklin to help them. Even he could not interest government officials in a place so far away. So in three years, North Carolina took the district back. This put Sevier out of his job and made him angry, which bothered no one in North Carolina.

Meanwhile Andrew Jackson rode the circuit. He stayed in the last town and took a job at a store.

"What you doin' that for?" John McNairy asked, annoyed.

"I earn a better living clerking than lawyering," Andrew told him and bade John good-by. He might have stayed there but for the fact that John turned up a few months later.

"Look at this," John waved an important-looking paper.

"Judge! You a judge?" Andrew exclaimed as he read.

"Yeh, me," John grinned. "Appointed by the governor of North Carolina for the Western District. Want to go along?"

"You're the one appointed," Andrew reminded him.

"Yes. That bothered me at first. I wanted you to go, too. Then I discovered that I, the judge, can appoint. I make you public prosecutor. Want it?" Jackson knew that this job, often called prosecuting attorney, was one he would like.

"Want it?" he exclaimed gaily. "I've *got* it, John." They began at once to make plans.

The five-hundred-mile journey to French Lick was over mountains, through Indian lands and vast unknowns. They calculated the dangers carefully — and courageously decided to face them. They went to Morgantown to make ready.

Horses were, of course, the only means of transportation. Jackson had two good ones, well shod. He always had one or more good horses; he selected well and cared for them wisely. Now he packed saddle bags with law books, clothing, two pistols and food. He carried legal papers in his wallet and his gun cradled in his arm. McNairy made the same preparations.

The two came to Jonesborough in the late summer of 1788. They saw fifty or more cabins and a log court house — a cabin about twenty-four feet square. They had not expected this; court was often held in a house. They met townsmen and talked business.

"There's a lot of legal work here," McNairy said to Jackson as they conferred later. "Cases of boundary lines and unpaid debts."

"We better stay awhile," Andrew agreed. "They say a train of settlers is going through to French Lick this fall. We might join up with them." This seemed good sense so they got board and room at Kit Taylor's cabin.

Jonesborough men liked Andrew Jackson's way of working.

"That Andy Jackson goes right at it," a client told around the village. "He listened to my say. Then he heard Jake —

that's fair enough. Then he says to us, 'You two ain't so far apart but you could git together.' And fust thing we knew, we had! Out of court, too."

The visiting lawyers made themselves useful in many ways. They helped fell trees, build cabins, and husk corn. Andrew liked it all — but for one man.

A Colonel Love had a horse and boasted of his speed. Andrew thought his own horse as good or better. He stood the boasting for a while, then challenged Love to race their horses on a track near town. With horses so important, many a village had a race track before a schoolhouse.

"It's your risk," Love said pompously. "Mine'll win."

Jackson laughed. "We'll see," he said, confident too.

The men hired jockeys, trained daily, careful to use the track, one in a morning, the other in an afternoon. Townspeople were excited; bets were made. Love was the favorite, but Jackson had backers too.

Jackson's jockey got sick the day before the race. A pity, for not another one could be found.

"Will you call off the race?" Jackson was asked.

"Not me! I'll ride him myself!" And he did. But the weight was a handicap even to a good horse. Jackson lost the race and more money than he could afford. He lost more, too; self-esteem. But he blamed only himself.

"I just got carried away with my own notions," he said, ruefully as men gathered around him. "I got a good horse."

Later he remarked to McNairy, "Mark it up as another lesson. Looks like I always have to learn the hard way."

In the fall the forty migrating families came to Jonesborough. They were an amazing group. Some had a wagon, some only one horse for a big family. All were going to take up government land given in payment for war service.

Jackson and McNairy joined up, and they set out over the mountains, a cruelly hard trail. Nights found men so exhausted they would hardly take watch duty.

"We hain't seen an Indian," one complained. "You're just feared at night."

"We got plenty o' work without staying up nights!" That fact was plain enough.

Before twilight, the train halted. Some made fires; some hunted game. Women spread quilts over bushes for shelter;

others cooked supper. Parents crawled under cover with the little ones and slept heavily. Men at watch dozed.

One night when Jackson had the watch some sixth sense made him uneasy. He sat against a tree, long legs stretched out. The camp was silent — then an owl hooted, far away.

"Who-o? Who-o?" The answer was near — and too prompt. Jackson listened intently.

"Who-o?" This was from the other side of camp.

"Who-o? Who-o?" Jackson's scalp pricked. This sound was behind him and very close. Slowly he crawled to Mc-Nairy.

"We're surrounded by Indians," he whispered in his friend's ear. "I'll bet they'll attack at dawn. We've got to wake the camp and get out of here fast."

It was difficult to waken weary sleepers with whispered

words. Still more difficult to persuade them to pack and move. But the two lawyers persisted. By dawn they were far away. They left bright, warm fires. "To fool the Indians," Jackson said, as people grumbled.

"All that fuss — for what?" men asked when at last Jackson let them stop for breakfast. "I ain't seen any Indians!"

Later an exhausted traveler overtook them. He panted out a sorry tale.

"Fifteen of us were traveling along, late last night," he said. "We came on camp fires, warm and cosy. Why folks left such a good place we couldn't figure. We turned in, warm and happy. Indians got us at daybreak. I'm the only one left." No one doubted where that massacre had been. Jackson and McNairy were suddenly heroes.

Maples were crimson; nuts were sweet eating when the party came to French Lick. They found that this settlement, once a fur trading post, was now called Nashborough in honor of a war hero. Villagers welcomed the newcomers.

"Two lawyers are along," one local man told another. "Good ones, I hear. Now maybe we kin get justice by the law."

"Have you got a place to stay?" someone asked Jackson. "My name's Overton, John Overton from Kentucky."

"No place yet," Jackson answered, shaking the outstretched hand. "I need one. Can you recommend me?"

"Glad to. My place suits me fine. It's out a ways, across the river. I'll take you with me, come evening."

Men crowded in to talk with Jackson about legal business. He was going to like this place; he saw that already.

Frontier Lawyer

Tʜᴇ ʟᴀsᴛ of the travelers were straggling into town when Jackson and McNairy with a couple of local men went over to Boyd's Tavern for dinner. On the way, Jackson's keen eyes observed many details of the settlement.

He saw that, in addition to the fort and cabins in the stockade on the river bluff, there were several stores and two inns, called taverns. These were all built of logs. Further along he saw homes, many of them log cabins, some were bark tents, easy to erect. Beyond were several wagon camps such as his own party was setting up.

"That shabby cabin you see over there," a townsman pointed out, "is the court house. Its run-down condition shows the way law has fallen into disrespect here. I'll wager you'll want it cleaned up first thing, Judge McNairy."

John McNairy's face flushed, and Jackson glanced away to hide a twinkle in his eyes. That was the first time they had heard John's new title used. It sounded fine. Also the respectful way another man stepped ahead to open the door was pleasant. The conversation continued as they ate.

"Collecting honest debts is more important than the state of the court house. I hope you will get right after debtors,

Judge McNairy. Nowadays, a man thinks he can agree to pay for what he wants — and then skip out or defy the law and stay. It's not right."

"You'll have your hands full of business, too, Mr. Jackson," another said earnestly. "I'd like to engage you to take a case for me at once."

The men from the East seemed to be readily accepted. No one mentioned that they were young, Jackson still several months under twenty-two and McNairy not much older. Jackson, the taller, carried himself with special dignity that was reassuring to a client who needed legal help. Suddenly John Overton noticed the time.

"We must go, Jackson. We've a ride ahead of us," he said. "Will you come with us, McNairy? I live at Widow Donelson's; she has several cabins." He had already explained his invitation to Jackson.

McNairy hesitated. Then he said, "Thank you, but I'll stay in town for now. See you in the morning, Andy."

"You may change your mind," the man left with McNairy remarked. "Inns in town are fair, but Mrs. Donelson is a real cook. She raised eleven children, and she likes people around."

McNairy smiled. "The place sounds good. But I want to see about cleaning up the court house. I want the law respected."

Jackson and Overton mounted, rode to the river, and ferried across before they talked much. Looking back, Jackson saw gleams of candlelight in cabin windows and a ruddy glow from campers' fires here and there. Then they galloped on.

"Donelson," Jackson remarked, as they breathed their horses. "I've heard that name somewhere."

"Of course you have," Overton assured him. "John Donelson was a prominent man. He came from Pittsylvania County in Virginia. He was a partner of James Robertson in the expedition of 1779 — all fine folks. Donelson was mysteriously killed three years ago. They were living in Kentucky then, but Mrs. Donelson moved back here. Most of her children are settled around Nashborough. Her old place has several cabins, and she likes to have company. Perhaps you didn't realize that we still have Indian troubles — two or three settlers killed a month ago around-about here. She feels safer with a man near, come night."

Overton talked so interestingly that the journey seemed short. Mrs. Donelson welcomed them, and the men went on to Overton's cabin close by.

At breakfast the next morning Jackson met a married daughter, Mrs. Rachel Robards, who helped to serve them. She was a beautiful young woman, with serious brown eyes, an intelligent face and kindly manner. As the men rode off Jackson asked about her.

"Is her husband dead, John?" That was a likely reason for a married daughter staying with her mother.

"Not dead. Just too insanely jealous and quarrelsome to live with. I'm a relative of his; I know. It's a sad situation."

"Then we'll not discuss it," Jackson said firmly. He had been raised to honor women, and he was no gossip. He turned the talk to business.

But in one way or another, as days passed, the Donelson family turned up in conversation. Gradually Andrew Jackson pieced together new information with facts he recalled hearing in Salisbury, till he had the story.

John Donelson, Rachel Robard's father, was born in Maryland and as a young man moved to southwest Virginia, then new country. His thousand-acre plantation on the Roanoke River was due north of Salisbury. Donelson was a skilled surveyor, like George Washington; and he had an iron works, like George Washington's father. Donelson was elected by his county to represent him in the House of Burgesses at Williamsburg. He served while Washington, Jefferson, Patrick Henry, and other famous men were also members. Governor Botetourt trusted Donelson and sent him west to do some surveying about the same time that Washington explored down the Ohio selecting land for soldiers after the French and Indian war.

As for Mrs. Donelson, she was a Stockley of Tidewater Virginia. She made a charming home on the Virginia frontier.

It was in the year 1779 that Donelson and his friend James Robertson decided to leave comfortable homes and stake everything they had on a move west. They knew the journey would be hard. They decided to divide up. Robertson with a few of the men would brave Indians and an unknown mountain trail, and select a site somewhere on the Cumberland River. Donelson was to gather the families who wished to migrate and build boats on the Holston River. Travel by water seemed a safe way to go.

Donelson built the flagship, named the *Adventure*, and then helped others as they waited for the autumn rains to swell the rivers. When they set out the 22nd of December, there were more than thirty boats of all sorts; forty men and a hundred and twenty women and children.

This migration proved to be one of the most dramatic and daring pioneer journeys in American history. Music with flute and guitar, singing and dancing on the larger flatboats kept people in good spirits. Several of Donelson's eleven children were married and had children of their own. His youngest daughter, Rachel, made herself useful tending babies, amusing children, and nursing the ailing.

Donelson guided them down the Holston to the Tennessee, up the Tennessee to the Ohio, up the Ohio to the Cumberland and on that river till they found Robertson. He thought it would be about a thousand miles. With the winding of the rivers the voyage proved to be twice that long. Indians they'd expected to avoid pestered the flotilla constantly. Smallpox, cold, death, haunted the voyage. Only the sturdiest survived to meet Robertson at French Lick in May.

Such courage and daring thrilled Andrew Jackson.

"Robertson surely chose his location well," he remarked one day as the pioneer's name came up. "Land around here is fertile, I hear."

"Yes, it is," Overton agreed. "What Robertson didn't figure on was Indians. They almost wiped out the settlement the first year.

"It got so that a man working his field had to be guarded. Children weren't safe picking berries folks needed for food.

"That's why Donelson moved to Kentucky," Overton added. "But most of his married children stayed around here." When the others had gone, Overton added another fact.

"That move to Kentucky's what led to Rachel's marriage up there. She was a beautiful girl, and Lewis Robards seemed both charming and devoted. When Widow Donelson moved back here she thought Rachel was happy. Later it turned out that Robard's temper made her life wretched. He actually sent for Sam Donelson to fetch his sister home. Rachel felt that was a disgrace; she had tried to be a good wife."

Jackson dropped the subject. But it occurred to him that

this recent affront was what had stirred up the talk about the family that he had been hearing. Nashborough people respected the Donelsons and disliked Robards.

However, law business was so absorbing that Jackson saw very little of the Donelsons. Settlers wanted land deeds recorded; assault and battery cases were common, and suits for debts. Jackson had seventy suits for debt in his first month at Nashborough.

"I kin pay you in land," a client often said. "Cash is scarce."

"Land suits me fine," Jackson agreed. He was so busy he had no time to inspect what he got or figure how many acres he had earned.

After the term of court ended, early in 1789, Jackson and McNairy made ready to ride on. Judge McNairy's circuit covered some fifty miles with many tiny settlements. Men gathered by the court house — now clean and tidy — to bid them a safe journey.

"You'll be back soon," someone called.

"We need both of you here!"

"Well," McNairy adjusted the lead rein on his spare horse. It was hard to leave such friendly people. Jackson spoke for him.

"The judge has to hold court twice a year, all around. We're not really leaving. Just doin' a job. We'll be back when we've been the circuit."

For the next two years the lawyers traveled continually. They came back to Nashborough, in turn, to hold court and do private law business.

During that time the village grew rapidly. Its name was changed to Nash Ville — and soon changed again to one word, Nashville.

The United States of America went through many changes in these same two years. The Constitution was ratified by a majority of the states, and the union was assured. George Washington was elected president the first winter that Jackson was in Nashville. While Washington was being inaugurated in New York, Andrew Jackson was traveling a mountain trail.

The country was very beautiful that spring. Arbutus, bluebells, Dutchman's breeches, made a gay carpet in the forest. Jackson led a spare horse, and his saddle bags held the usual books, papers, clothes, and food. When he could, he stopped for a meal at a settler's cabin. A traveler was always welcome to share simple food and tell his news. When he felt reasonably safe from Indians, Jackson shot a wild turkey or some prairie

chickens and broiled the meat over a fire. When he dared not risk that he munched food from his bag.

McNairy, and another lawyer or two, sometimes traveled with him. But Andrew Jackson's work was not all for the court. He was alone with only his horses for company, much of the time.

Jackson gained a reputation for wise adjustments of disputes out of court. He enjoyed that. He was willing to take long journeys, to hold tedious interviews to get facts that would help his cases. He made the two-hundred-mile journey to Jonesborough more than once and went to Natchez, five hundred miles through Indian country on wilderness trails.

Thomas Green of Kentucky had moved to the Spanish city of Natchez. He had business with a brother in Nashville and in September of 1789, Jackson agreed to go and attend to the matter. He found the Greens living in a handsome house. They became good friends as he worked on their legal problems. They were very pleased with his work and paid him well.

"We wish you could stay here," Mr. Green said. "Then taking land in part payment would have some meaning."

"I like the land and house," Jackson assured him. "Maybe I'll come back and stay in it someday."

"That would please us. We enjoy a gentleman of your attainments. Do come."

Jackson was happy to be wanted. And too tactful to tell his kind hosts that he had no desire to move to a Spanish country. He was proud of being an American. Perhaps some Nashville person would be moving, though, and he could sell that nice place to advantage.

On the long Chickasaw trail home Indians threatened him repeatedly. He slept only lightly, sitting up against a tree. His musket lay across his knees, his loaded pistol at hand to give him a second chance for life if attacked.

In all these experiences Andrew Jackson was gaining in knowledge of law and the ways of men. Nashville people held him in respect and friendship.

Each time he came back Jackson was saddened to hear more talk about Lewis Robards and his hateful treatment of his wife. He saw that Mrs. Donelson was worried. Robards had come down to persuade his wife to return with him. She hesitated. She had done that once and failed in spite of the efforts of Lewis's mother, a devoted friend of Rachel.

Once when Jackson was at the Donelson's, Robards turned on him in jealous rage. Andrew was shocked. He packed his bags and moved in to town immediately despite Mrs. Donelson's protests. He got a room near the court house and ate at the inn.

Soon after, Overton told Jackson that Robards accused Rachel of fondness for him.

"But John!" Jackson was horrified. "That's a lie!"

"We know that," Overton shrugged. "You're so strict about women that folks call you straight-laced! Truth doesn't matter to Lewis. I think he's crazy."

Time moved along, with Jackson away weeks and months. Near the beginning of 1791, he chanced to be in Nashville and to ride out to attend to some legal matter for Mrs. Donelson.

He knocked on the door.

"Come!" Mrs. Donelson called.

But on entering he saw that Mrs. Donelson and Rachel were standing tensely by the fireplace, not looking toward him.

"Oh, excuse me!" he exclaimed and turned to withdraw.

"No. Come in, Andrew," the older woman said. "You might as well know now — everyone will soon. We have just had word from Lewis. He threatens to divorce my Rachel."

"Divorce her — why — why — he can't!" Jackson was shocked. The word divorce was seldom mentioned on the frontier; divorce was hardly possible in many places. Always it was associated with immoral behavior. "He *can't!*" Jackson exclaimed.

"Oh, Lewis can if he sets his mind on it," Rachel said in a tired voice. Then she added, "He says he will haunt me! I am sure he plans to kill me. We must all die — but I cannot bear to die by his cruel hand, in anger!"

"You will not die! Put that out of your mind!" Jackson's eyes grew deep blue, muscles of his face tightened. "He dare not hurt you!"

"Lewis will dare anything when in a rage," Rachel said sadly. She turned and saw Jackson's white, tense face. "But I should not trouble you. Forgive me!"

"It is nothing! I —" Then he remembered Robards' talk. Others could help now, better than he could.

"If you will sign this paper, Mrs. Donelson," he said, "I'll ride back quickly and send Overton out. John will know how to help you." In two minutes he was gone.

Rachel
and Andrew

JOHN OVERTON did have a suggestion that pleased Mrs. Donelson and her daughter.

"I heard today that old Colonel Stark is going to Natchez on his flatboat," he told them. "Perhaps you could go with him, Rachel. It's a mean trip in winter. His wife is going with him though — maybe . . ."

"Oh, John, it would be wonderful!" Rachel interrupted. She flushed happily at the thought of going far away.

"People from Kentucky have moved there." Mrs. Donelson hesitated, trying to think of any she might know.

"Likely you have in mind the Green brothers," Overton reminded her. "Andy and I do business for them. Remember that Andy went to Natchez a while back? That was for the Greens. They're nice folks — hospitable, too. I'm sure they'd be glad to have you stay with them, Rachel."

"When does Colonel Stark plan to leave?" Mrs. Donelson asked.

"Right soon. Tomorrow maybe."

"Do see him quickly, John," Rachel begged. "I could be ready in an hour." So John galloped back to town and found Stark.

"I'd be mighty glad to help any Donelson," Stark said when Rachel's need was explained to him. "John Donelson did me more'n one good turn in his day. My wife's goin', and she'll look after Mrs. Robards." Then he scratched his head thoughtfully. "But a passenger's a worry, and I'm not as young as I once was, John. I ought to have an extra man in case of Indian trouble — thar's been some. Kin you go along?"

"I just could not go, Colonel. Not right off." Overton was dismayed at this new problem.

"How about her brother Sam?"

"I'll ask him." Overton was off in a hurry.

Sam could not go at once, either. Overton asked several brothers and brothers-in-law, but none could leave the next day.

"I wonder about Andy," Overton said to the last one. "He'd be a help on the way — and he's been to Natchez."

"He's the one to go! Tell him how we're fixed and ask him."

"Of course I'll go," Jackson answered when the need was explained. "Mrs. Donelson's been like a mother to me. I'm glad to help out. You can take over my work, John. I'll buy a horse down there and get back as fast as I can. The Greens will welcome Mrs. Robards. They've got a nice place." Even as he spoke Jackson had begun to sort papers to leave with Overton.

At dawn the next morning Stark's boat shoved off, and the long voyage began. Some days were mild, some cold and stormy. The water was high, and they drifted safely north on the Cumberland to the Ohio River, then down to the Mississippi and on south. Only usual difficulties broke the tedium of their days — great floating trees and a couple of Indian scares. The voyage seemed easy to a woman who had journeyed on the *Adventure* when she was twelve — old enough to remember it.

The Greens of Natchez were friendly and cordial. Mrs. Robards was enchanted with their handsome house, delightful manners and her freedom from daily worries. The gay life in an old Spanish town was a change from hardier customs of Kentucky and Tennessee.

Don Estéban Miró, Governor of Louisiana, lived in Natchez. He was a popular gentleman and a skilled diplomat. It was Miró's tact and statesmanship that had made friends for Spain in quarrels with Westerners about river trade.

As soon as Mrs. Robards was settled, Jackson bought two horses and joined a party of men traveling to Nashville. He got there in time for the April session of the court. Legal business had piled up in his absence, and he worked hard to catch up.

A few weeks later John Overton chanced to go to Kentucky on a case. While there, he called on Mrs. Robards, senior and told her that Rachel had gone to Natchez.

"It's a pity she had to leave her home," Mrs. Robards said sadly. "My son had not heard of it, I am sure. He appealed to the legislature of Virginia for a divorce from Rachel. I regret this deeply. She has had a hard time."

"A divorce!" Overton was shocked. The dissolving of a marriage was a most unusual act in this time and place. Lawyers had little reason to understand such laws in various states.

"Yes, that's what Lewis told me," Mrs. Robards said. "He

found that because he was a native of Virginia he must apply in that state. His brother-in-law had charge of it and sent word that the bill had passed the House of Burgesses."

"The Donelsons will want to know of this," Overton said. "My work is about ended here; I'll ride home tomorrow." On the way he wondered whether the family would feel relieved that Rachel was free or disgraced that she was divorced.

He was pleased to see that Mrs. Donelson took the news well. "Now Rachel can come home!" was her only comment. "Someone must take the word to her — maybe Sam can go."

Overton left that problem to her and rode back to town. There, he stopped by Jackson's office to tell his news.

"Rachel free?" Jackson was incredulous. "Are you *sure?*"

"Oh, yes." Overton was positive. "I had it from Mrs. Robards herself. Moreover, several townsmen spoke to me about it. No one likes Lewis — the sympathy is all for Rachel. Mrs. Donelson is pleased. She wants one of the boys to take the word to Rachel."

"If Mrs. Donelson will let me," Jackson said quickly, "I'll go. I'll start for Natchez today." Suddenly he felt free to admit to himself a more than casual interest in the attractive Rachel.

He hurried off; had a brief talk with Mrs. Donelson and was on his way to Natchez. The good wishes of the Donelson family went with him.

When Andrew Jackson told Rachel that her husband had divorced her, she covered her face with her hands, shocked and shamed. But gradually, as she listened to his words, her hands dropped, relaxed, and color returned to her face. Andrew was asking her to marry him.

"Your mother knows what I am saying," he added. "She gives her approval." This pleased Rachel.

So it turned out that on an August afternoon a wedding took place in Thomas Green's house. The pretty bride looked small and dainty as she stood by her tall, slender groom and they spoke their marriage vows. Her brown eyes were tender, her bearing as proudly erect as his. Tall candles burned in crystal sconces in the high-ceilinged parlor and friends who gathered for the ceremony wished them well.

Andrew Jackson was now twenty-four years old and his bride only a little younger. They went to live in the small log house on land he had earned as a fee for legal work. The cabin was by Bayou Pierre; it overlooked the wide Mississippi. There Rachel and Andrew spent happy days.

But the prosecuting attorney for the Western District had duties. Much as he regretted it, Jackson felt that he must be in Nashville for the autumn term of court. He heard of a group of travelers going north, and made plans to join them.

Rachel Jackson was surprised at her reluctance to leave Natchez. She had never been so happy, and the location of their small home was beautiful. But she knew they must go.

The long journey, though tiresome, was made safely, and they came to Nashville in October. Jackson's desk was piled with work, and the court opened soon. But he took time to find a suitable home for his bride. Fortunately her brother John wanted to sell Poplar Grove, his place on the south side of the Cumberland River. Andrew bought it promptly. The property was a small farm with a comfortable little house.

"Fix it any way you like," Andrew told his wife. "Don't skimp. I want you to have the best."

"I have that now, Mr. Jackson," Rachel said proudly. Like most women of her time, she used her husband's formal title, even when they were alone. She began at once to make their home attractive and comfortable. Her husband had his days full of court work and other legal business.

Nashville lawyers — there were several now — were kept busy. The flow of settlers from the east was astonishing. Daily,

families arrived; a man and his wife, small children and a baby,
usually with one horse and a rifle. Children carried a bundle
of clothing, bedding, an ax, a cooking kettle. Their dream was
to own land — often they had bought fifty or seventy-five acres
back home. For this land they had left kin and friends and
endured a hard journey.

When they found their land, the father and sons built a
cabin and cleared a field while the mother and daughters gathered wild food, made garden, dressed skins, made soap and
candles. Those lucky enough to own a cow had milk. All wore
homespun. Women braided long hair and wound it around
their heads. Children chinked the log cabin with mud to keep
out winter storms and poked out the chinks in summer to catch
a breeze.

These settlers wanted a clear title to land. Eastern speculators were not always to be trusted, they had heard. President
Washington was distressed by land speculators who sold land
in violation of Indian treaties and then begged the government
for new treaties to help innocent victims. All this made work
for lawyers in the new region.

Courtroom scenes were often dramatic.

One day Jackson and Judge McNairy came to a cabin far
from Nashville. They were to hold court here. The small room
was crowded, and a fight was going on. No man paid the
slightest attention to the officers of the law. Jackson quietly
edged his way in, unnoticed. Then, suddenly he whirled
around, a pistol in each hand, pointed at the crowd. There
was a sudden, awed silence.

"If you gentlemen will kindly let the court come to order," he said calmly, "we'll save time. My bulldogs make quite a noise if they get to barking." He looked around agreeably for a full minute. Then he lowered his arms.

"The court is ready to convene, Judge," he announced.

In a corner one man was heard to whisper, "They better not stir up anything while that Jackson is here. I've seen him put two bullets in the same hole at thirty paces." A reputation like that had to be earned, on the frontier.

After the fall term of court, Jackson took long journeys on his private cases. Settler families came to know him.

"Paw!" a child who saw Jackson ride up would run into

the field. "Paw! Mr. Jackson's come. He's got your papers." Their business done, Jackson was often persuaded to stay for supper and the night. If the host had enough quilts or skins he hung a curtain across the cabin to make a private corner for their distinguished guest.

Along about this time an honor came to Andrew Jackson that he valued because it showed he was esteemed by his fellow townsmen. He was elected a trustee of Davidson Academy, Nashville's best school. Like George Washington and other men who had had little formal education, Jackson cared very much about schooling, and he knew that the Davidson trustees were the most honored men in the community. He accepted humbly. The election meant much to both Mr. and Mrs. Jackson; they valued the place it gave them in the town, even more than the good living Jackson was making.

In the autumn of 1793, John Overton joined Jackson in Jonesborough where they were to try a case before the fall court. Overton had been to Kentucky to get records needed on this case. He spread them out on a table for study. Suddenly he rose with a dismayed cry.

"Andy! Look at this! Lewis Robards is filing suit for divorce for adultery — filing now, in Kentucky!"

"He can't!" Jackson exclaimed. "A man can't get *two* divorces." Blood drained from his face; the scar was livid.

Overton clutched the paper, reading rapidly.

"This says that the Virginia legislature gave him permission to sue in Kentucky, where all the facts could be presented to the court."

"But you said — you were there — two years ago!"

"And it was quite clear. Old Mrs. Robards told me herself. I talked with several men. They all said the same."

"You didn't actually see the papers?" Jackson was now astonished that they had overlooked this point.

"No, I couldn't. Robards had them and he was away. An there was no question about it to give me concern."

"You'll have to go to Harrodsburg and see those papers, John. I'll finish here. I've got to know the truth."

"However it turns out, you'd better have another ceremony, Andy. Just to protect children if you have them."

"We're married now; everyone knows that." Overton knew it, too.

"I'll ride fast. I'll meet you in Nashville."

When he got home and told his wife, Jackson was not surprised to hear her say his own words.

"We're married now. Everyone knows that. We've been married more than two years, Mr. Jackson."

"I know that. But John says . . ."

"Wait till he comes," she suggested.

John's report confirmed the record he had chanced upon in Jonesborough. The Virginia legislature had granted Robards only the right to sue in Kentucky, a most unusual procedure. The brother-in-law who had presented the bill had faith that Rachel was innocent of adultery. He thought he was doing her a kindness to require suit brought where she could defend herself. Robards had kept secret the real terms of the bill; had merely told that it had been passed.

"He's kept it to himself all this time!" Jackson cried.

"I wonder, Andy — of course this is just a hunch — I wonder if he waited till he had collected Rachel's share of her father's estate. You know that held over a long time because his death, alone in the wilderness, was a mystery. The estate wasn't yet settled when I was in Kentucky two years ago."

"No use dreaming up reasons to explain a man like Robards," Jackson said. "The problem is, what to do now."

"Talk with Rachel's brothers. The divorce is clear now. I made sure of it this time. And of course everyone knows you were married. But another ceremony would be advisable, I'd say."

It was not easy to convince Rachel of this. But when she saw that her family and the lawyers agreed, she consented. Early in 1794 a simple civil ceremony was performed and duly registered at Nashville. As Mr. and Mrs. Jackson left the court house and mounted to ride home her face was sad; her brown eyes troubled.

"I'd give a good deal if this hadn't happened," Overton said as they rode away. "Rachel is good and true. Now she's hurt, down inside. That Robards! It just isn't right!"

"He said he'd haunt her," Sam Donelson remembered.

The two walked off silently, moved by the unease they could not explain.

A Man
of Business

During Andrew Jackson's early years in Tennessee, people in the west felt far away from the United States government. Philadelphia, the capital, was a long, long journey over the Appalachian Mountains. The "West" seemed like another country. Rivers were highways. It was natural that Spain and her port at New Orleans seemed to westerners more important than Philadelphia.

Perhaps easterners sensed this when the first attempt to bring the west into the new nation was made in 1790. The wilderness parts of Virginia and North Carolina were organized and given an odd, long name — The Western Territory of the United States South of the Ohio River. Andrew Jackson was one of several officials in this territory who took the oath of allegiance to the United States. He was then appointed prosecuting attorney, the office he had held for North Carolina.

Settlers felt proud of the new organization. Many had served in the Revolutionary War and were pleased to be noticed. Talk about an alliance with Spain died down. In 1792, Kentucky became a state.

But as months passed, Westerners saw that federal government did little or nothing that helped in daily problems. Nashville men talked heatedly as they met in town.

"President Washington ought to let us settle the Indians," someone said. "Only one way to do that — fight 'em!"

"Washington thinks the Indians have been badly treated."

"Well, they have! But whose fault is it? Land speculators', not settlers'."

"Philadelphia folks don't know what Indians are like," a bystander muttered shrewdly. "Just let 'em hear *one* war whoop . . ." Men grinned, agreeing. Jackson paused to add his word.

"Two men were killed and scalped not a mile from my place last week. A man hates to leave his wife and come to town for business."

"Washington worries about broken treaties," Overton added. "*Our* worry is homes burned and people slaughtered. If it keeps on, no one will dare to farm around here."

Indians got so bold that in 1793 they defied the United States government. John Sevier raised his own army and pushed the Indians back from the settlements. The next year James Robertson did a similar brave thing. His chance came in an unusual way.

Joseph Brown, a white youth who had been taken prisoner by the Cherokees, managed to escape. He came to Robertson.

"I can guide you to a secret place," Brown said. "You hunt Indians where I show you 'n you can lick 'em."

Robertson believed him. Like Sevier, he raised his own army, made a surprise attack and killed many Indians. The rest moved far to the west.

After these raids the whole area around Nashville prospered. New settlers came. The town grew rapidly. Jackson was liked, and he had all the business he could take. Personal problems of his marriage, never much noticed, were all but forgotten while his success was plain to see.

"I wonder if Andy knows how much land he's got," a neighbor remarked. "I heard a man offer him six-forty to take a case."

The figure meant acres, not dollars. Much of early Nashville business was done by barter. Goods and services were paid for with land, a horse, a cow, or a metal object. One man bought a farm with a few metal bells. Settlers needed many things they had not been able to carry west — tools, knives, cloth, cooking utensils. After harvest they bought what they could with corn or cotton.

One evening as the Jacksons sat by their fireplace a new idea came to Andrew's mind.

"There ought to be a store out here. Takes time to ride to Nashville. A smart merchant could make money and help folks out here at the same time."

"Who could do it, Mr. Jackson?" Rachel wondered. She saw that his eyes were deep blue, a sign that he was thinking hard.

"*I* could do it!" he said.

"You'd have to go to Philadelphia to buy goods!" Rachel was amazed at him. "That's eight hundred miles of wilderness."

Jackson laughed. "Anybody know wilderness better'n I do? But don't you fret yourself, Wife. I'll ride fast, work hard and come home soon as ever I can." His mind was already made up.

"Yes, the idea is sound," he went on. "I'll sell land to city folks and with that money buy goods to stock a store. Overton might like to sell some acreage we own together."

In a few days Jackson rode away. He was prepared to sell thirty thousand acres of his own, fifty thousand that he owned with Overton, and some for a neighbor. He traveled fast. At

the journey's end his face was gaunt, his clothes shabby and his hair so long he tied it with an eelskin.

After that speed, he was annoyed to find that city men dallied with business. They wanted land; a frontier man would have said yes or no to Jackson's price; city men delayed. Jackson's patience, never his best quality, was worn out by the time the land was sold.

Then he was shocked to learn that the purchaser expected to pay with notes, written promises to pay later, not in cash as he had expected. But he had to conform to custom.

David Allison paid for a vast amount of land with four notes, one due each year following. Using these notes instead of money, Jackson bought merchandise for his store. These goods must be hauled to Pittsburgh, loaded on a boat for Nashville. Finally his business was done. Gifts for Rachel filled his saddlebags, and he rode off for home.

Talk in town had changed in Jackson's three months' absence. People were excited about Tennessee being a state. A census showed that the territory had more than the 60,000 needed for statehood. So a convention to write a constitution was set for January of 1796; Andrew Jackson was elected one of the delegates. Though he was not yet twenty-nine years old, he was one of its important members.

The convention worked hard for twenty-five days. The document they adopted was what Thomas Jefferson later called one of the best state constitutions. Among other points it gave a man the right to vote after six months' stay in the state, and it demanded free navigation of the Mississippi.

Statehood was granted in June, 1796, and John Sevier was elected governor. Andrew Jackson was elected the first representative to the Congress of the United States.

Through these busy months Jackson had been thinking about getting a new home. His marriage was happy, and they hoped to have a big family. He was successful and he wanted Rachel to have every comfort. Soon after the convention he bought a fine tract of 640 acres about three miles from Poplar Grove.

The land, known as Hunter's Hill, lay high along the Cumberland River. On it Jackson built a good frame house — a real luxury, then. The building was well planned and was the largest house around Nashville. The view from the veranda over river and forest was unusually beautiful.

"Like it, Wife?" Jackson asked on one of their frequent inspections.

"It's wonderful — like a Virginia house!" Rachel could hardly believe her good fortune.

"You are to make a list of furnishings you want," Jackson told her. "I'll order them from Philadelphia."

Rachel Jackson had a gift for hospitality. Poplar Grove had made a kinsman, a well-to-do-neighbor, or a peddler welcome. Now she showed she had good taste in furnishing the new home. The list that Jackson sent off included mirrors, china, chairs, and settees as well as books, music, and a harpsichord such as Mrs. George Washington owned.

That whole year of 1796 was filled with activity for Andrew Jackson — the new home, the new store, the new state. Reluctantly he left, in the fall, to take his seat in Congress. As always, he rode a fine horse and led another. His bags held books and papers. He planned to buy new clothes in the city.

Mrs. Jackson stood by the doorstep as he mounted.

"Be sure you get the right things to wear, Mr. Jackson."

"I promise!" He grunted down at her, "but no wig!"

Her eyes twinkled with mirth. Andrew Jackson in a wig! "Then don't get one," she laughed. He never did.

"This trip is silly," Jackson said, lingering. "I'm not the one for Congress — just sittin' and talkin' and votin'."

"But people want you, Mr. Jackson."

"Or so they think. I'd rather do a job I can work at and know when it's through I've done something."

Rachel smiled. She had not wanted him to go — but she did not say so. She watched him ride away; watched until he was quite out of sight. His leaving meant heavy work for her. He had hired an overseer for the farm, but no competent men could be found for such jobs. The mistress would have many extra responsibilities.

Rachel filled the house with nieces and nephews. She made small toys for them; a gourd doll with painted face and a pretty dress was a favorite. She found bits of broken dishes for playing house, strings of Indian beads and a tiny cooking pot she had brought from Virginia.

"My father made this for me in his iron works," she told the delighted children. "Once, when we were on the *Adventure* I thought it was lost — but I found it. You may play with it." These treasures were kept in a rush basket near the fireplace.

Aunt Rachel could always tell a good tale, too, about Virginia or her travels. She could nurse the aged, the babies, the sick, with skill and tenderness. Days were busy; only the evenings were lonely.

When Congressman Jackson arrived in Philadelphia, the nation had just finished its first contested election. John Adams was to be the next president. People who had been shocked at the electioneering hoped that Federalists and Republicans would settle down and do the nation's business.

A depression seemed to threaten. Jackson was told that Hamilton's paper money and national bank made the trouble; this idea stayed in his mind.

During the session he made one important speech introducing a bill to pay costs of those Indian forays of 1793. James

Madison favored the bill, and it passed.

Jackson voted for building three frigates and against buying peace with Algiers. He voted against spending $14,000 to
furnish the executive mansion in the new capital and against
spending *any* money except for a definite need. His scrupulous
honesty made him careful of tax money.

At the end of President Washington's term, Jackson saw
John Adams take office. Then he left for home. He could
hardly wait to see Rachel and sit by his own fireside.

On his first ride into Nashville, he was surprised and
pleased to learn that the new postal service had delivered copies
of his speech. News of this, and that his bill had passed,
delighted home voters. Men rushed to shake his hand. He
was vastly popular.

"You're the man to represent us, Jackson! You know what
we want," several told him.

"The country's doin' something for you in this mail service,"
Jackson pointed out. "Reading about the goings-on at the
capital will help us understand government."

Jackson soon learned that popularity brings obligation.
He was elected to fill a vacancy in the United States Senate,
and had to go to Philadelphia again. Because of his respect
for the Senate he ordered a coat with a velvet collar and
breeches and vest of brown cloth to wear there.

But he found the Senate tedious. Endless days passed
awaiting news about England and France and their war intentions. Andrew hated such delay. After a time he asked leave
to go home. His first act in Nashville was to resign his office.

"I hope I am never elected to anything again," he told his wife. "Sittin' around arguing is not to my taste." Rachel smiled, hoping he would stay at home.

Nashville men understood his feeling; he was more popular than ever. He even acquired a few enemies, as successful men often do. He heard so much talk about his "wonderful service" that he tried to think what, if anything, he had done.

He had come to know Thomas Jefferson and Edward Livingston of New York. He had met and thanked Aaron Burr for his help in getting the statehood of Tennessee put through. But these men were eight hundred miles away; it was unlikely that he would ever see them again. He felt he had done little.

An appointment as judge pleased Jackson much more than federal office. He had the Superior Court, popularly called the Supreme Court, of Tennessee — a post of honor. Jackson liked the bench. He felt proud of the people's trust, and he served with honor and integrity.

Life was full and rewarding. Thirty-three year old Andrew Jackson looked back, at the turn of the century, on the twelve years he had been in Tennessee. He had been prosecuting attorney for the western district and for the United States; he had been congressman, senator, judge. He was a successful planter, lawyer, and school trustee; and he owned the finest house in the community.

Then, all of a sudden, it seemed to people in Nashville, a terrible depression settled like a blight over the nation. The business structure collapsed. Andrew Jackson, like thousands of others, faced failure.

Honest Jackson

THE DEPRESSION surprised and shocked Americans. Many successful men, like Andrew Jackson, had little cash on hand. Jackson had bought goods in Philadelphia and paid with the notes from Allison. Now Allison did not keep his promise. Jackson lacked cash, too, but he felt honor bound to keep his promises.

"We have to get the cotton off for New Orleans," he told his overseer. Jackson helped to load the flatboat himself. As it pushed off he said, "Get the best you can, in cash."

"That I'll do, Sir," the man promised.

But New Orleans wharves were piled with unsold cotton. Jackson's fine crop brought only a fraction of the former price. To raise cash enough to pay the merchants, Jackson had to sell land and horses at a loss.

"I surely hope the next note will be easier to pay," he said to Rachel. But he saw difficulties ahead before he could be free of debt. And he hated debt.

Jackson was still struggling with money matters when an unexpected honor came to him. John Sevier had been Tennessee's governor for three two-year terms — all the constitution allowed at one time — so he could not be re-elected in 1802.

Sevier was popular; he might be governor again later. Meanwhile, he wanted to be major-general in charge of the state militia, an honor many thought he had earned by thirty years' service.

Andrew Jackson was popular, too. His position as judge was second only to that of the governor. Field officers in the state militia had the choice of their leader; their voting ended in a tie — Sevier and Jackson. Governor Roane, a friend of Jackson, solved the tie by voting for him.

Sevier was furious. His chagrin turned to hate for Governor Roane and for Jackson. He became a powerful enemy. He campaigned against Roane and won the next election as governor. This did not end his hatred of Jackson; it seemed that he could not forget. The break between those two leaders was a tragedy for Tennessee; the state needed both men.

The unsought trouble with Sevier took away part of Jackson's joy in his new office. Finances troubled him again, too. It was hard, each year, to raise the cash to pay the Philadelphia notes.

"Now that Louisiana Territory belongs to us the President will be appointing a governor," a friend chanced to remark. "You'd be a good one, Judge."

Jackson thought a minute.

"I could sell out and start over," he said.

"You wouldn't like to leave here."

Jackson was not sure; the idea had some appeal. Saying little, he left for Washington to ask for the appointment.

But President Jefferson was not at the capital. He was

in grief-stricken retirement at Monticello because of the death of his daughter Maria.

"Stop and see him on your way home," a friend suggested.

But childless Andrew Jackson had imagination. He could understand a father's mourning.

"I would not intrude upon a man at such a time," he said, and went directly home. Later Jefferson appointed a governor. There was no evidence that he had even thought of Jackson.

When he got home, Jackson saw that he must take vigorous action about his growing debts. He strode into the house one day looking so weary that Rachel ran to him in concern.

"Might as well tell you right off, Mrs. Jackson," he cried roughly, fighting emotion. "We have to move."

"*Move!*" She was astonished. "Where?" Her brown eyes studied him anxiously. But quickly his face relaxed; his voice grew more natural now that the worst was told.

"You know that old blockhouse on the farm I named Hermitage? I'd planned to use it as a store someday. Would you live there? I've found a man who will buy Hunter's Hill — right now."

"We'll move at once!" Rachel did not glance at the home she loved. Her husband meant more than a house. She moved into that crude blockhouse — a two-story building. It had one large room, about twenty-four feet square, downstairs and two rooms above. There was a big fireplace, a lean-to and nearby, three small cabins — all in poor repair.

Mrs. Jackson took with them only the things the new owner did not buy. Then she went at the great task of transforming the cabin. Hermitage land was good for cotton, and the year before, Jackson had set out hundreds of little peach and apple trees. Here the Jacksons started life anew, debts paid.

While Rachel was settling their home, Jackson resigned all his offices except the military and devoted himself to making the plantation pay.

Cotton was his main crop, and until the depression had always brought a top price in New Orleans and London. Jackson was a progressive farmer; he had the first cotton gin around Nashville, and he was alert for new ideas. Now he needed something that he could control better than a far-away marketplace. He thought over this need continually.

All his life he had loved horses. As a lad he had learned how to care for them, how to judge their quality. Horses did well for him. In years of wilderness journeys his animals were companions as well as a means of travel.

So it was no surprise to his wife and friends to have him decide to raise fine horses as a business. He had every ability for success. He was a skilled veterinarian, he had kept records of all his horses and had a personal feeling for each one.

In looking around for a stallion to head his stables, Jackson was attracted to Truxton, a Virginia horse billed to race Greyhound of Tennessee near Nashville. Jackson saw that race — Truxton lost.

"All the same," Jackson said to a friend, "I like Truxton. I think that with better care and training he could win." He saw Truxton's owner; he had heard the man was hard up.

"Suppose I train your horse," Jackson said, "and race him again with Greyhound. If he wins, I'll buy him." They agreed on a price.

Word got around that Jackson was putting Truxton through hard, daily training. Men came to watch and stayed to disapprove.

"That horse'll never run." Men who saw Truxton passed the word around. "The general's training out all his spark."

But Jackson knew horses as he knew men. He inspired those who were near him.

The race was to be run at Hartsville, and it drew men from

all over middle Tennessee. Betting was spirited; buildings, farms, horses, stock, and even cash were put up.

Truxton was a beautiful bay horse with white hind feet. He was fifteen hands, three inches high; all bone, muscle and grace. But Greyhound, the winner of many recent races, was the favorite. Jackson staked $5,000 on Truxton, though how he got the cash to put up, no one knew.

No matter. Truxton won.

From that day Jackson's fortunes improved. He bought Truxton and later, Greyhound, built stables, hired Billy Phillips as a jockey, and leased nearby Clover Bottom racetrack. He became one of the successful turfmen in Tennessee.

Soon Nashville had a distinguished visitor, Aaron Burr. People buzzed with excitement. The town had never had such an important guest as a recent vice-president of the United States.

The fact that Burr had been in retirement since he had killed Alexander Hamilton in a duel did no harm to his reputation in Tennessee. Westerners often settled disputes by fists or guns. Gentlemen did not take personal quarrels to court. A duel at dawn was the accepted code of "honor." Burr's duel might even be a point in his favor.

Aaron Burr's visit to Nashville was during his extensive journey through the west. He was especially welcome because of his helping Tennessee toward statehood in 1796. He planned to stay only a day — time enough to interest some prominent men in his ideas for westward expansion.

Reluctantly he let Andrew Jackson take him to the Her-

mitage. When he rode toward the blockhouse he was sorry
that he had bothered. A log house was beneath his dignity even
though he did need the help of the most popular man in the
state.

An hour of Rachel's hospitality changed him into a willing
guest. He settled down in one of the little cabins and spent his
days eating delicious meals, relaxing under the spell of Rachel's
music and gracious ways; visiting with neighbors who shared
these pleasures. He did some business, too.

"I hear you are going to have a boat yard of your own," he said to Jackson as leave-time came. "I want you to build boats for me — your first order." He made it very definite.

New friends crowded around Burr's horse, wishing him well as he left. An aura of success lingered after him.

"You're lucky to have such a good order right off, General," a neighbor said. "You'll make money and help the government at the same time." It was a pity no one went along with Burr. They might have noticed that he had a new story about this western plan at every place he stopped.

But neither a celebrity nor boatbuilding distracted Jackson from Truxton. Every evening he went to the stable to see his favorite settled for the night. Truxton expected him. Jackson's presence seemed to lift the horse above what he had ever done before. He became a money-maker for his owner.

All this time Jackson was pushing his store business though it was a risky understaking. When he ordered goods in Philadelphia, he never knew how much his order would cost. Transportation was a problem, too. Merchandise had to be carted to Pittsburgh. Then it went by boat to Nashville; or, sometimes to Louisville, and then by packhorse to his store near Clover Bottom. Often the final cost was so high the customer would not take the order.

But merchandise got sold; people needed things. Jackson opened a store at Lebanon and one at Gallatin. A chain of stores was an innovation, then. He sold rifles and frying pans; coffee and salt; chairs and musical instruments, calico and silk. Since little cash was in circulation he had to take all sorts of goods in payment — land, cotton, fur pelts, pork, tobacco — and sell these in New Orleans. Naturally, returns were slow. Horses made money more quickly. Gradually the boatbuilding and store came to be managed by a good friend, tall, trustworthy John Coffee, a well-known surveyor, and John Hutchings.

In the autumn of 1805, after Burr's visit, Truxton was at his best. Jackson arranged to race him against Captain Erwin's fine horse Ploughboy. This November event attracted wide attention, and betting was high. Then suddenly Ploughboy was withdrawn. Captain Erwin paid a forfeit to Jackson, in notes, not cash.

This dramatic change caused talk. Men who had counted on making big money were angry; Jackson's friends said slurring things about these forfeit notes. Jackson ignored the whole matter, but to his surprise talk did not die down. So after a while, he arranged another race, this to be in April, at Clover Bottom. Again thousands of dollars were bet on both sides.

The evening before the race, Truxton hurt his thigh; it developed a bad swelling. Jackson considered withdrawing him from the race and paying the forfeit. But during the hours of the night, as Jackson worked with the horse, the thigh got better, and Jackson decided to risk his running. Unknown to Jackson, word of Truxton's injury got around. Betting, which had been about even, turned to favor Ploughboy.

The day of the race thousands arrived at Clover Bottom. They came by carriage, by wagon, on horse and a-foot. They lined the long track and saw "crippled" Truxton win every heat.

The horse seemed inspired. He ran the last heat in a pouring rain, making a mile in three minutes and fifty-nine and a half seconds, a speed that left Ploughboy sixty yards behind.

People roared with excitement that masked the angry shouts of hundreds who lost heavily. Captain Erwin's son-in-law, Charles Dickinson, let anger loosen his tongue. Some claimed he was drunk when he raised the old story of Mrs. Jackson's lack of a divorce when she married in Natchez. Angry men lost caution as they paid their bets and went home, talking too much.

The matter did not die down. Friends of both Jackson and Dickinson kept talking until a date was set for a duel. Rachel and Mrs. Dickinson were close friends, but that fact must not interfere with their husbands' "honor."

On a brilliant day in May both men bade their wives a casual good-by. They hoped the women did not know that with supporting friends they were bound for Kentucky where duels were legal. At dawn the next morning the two principals faced each other.

Dickinson, like Jackson, was a crack shot. He had boasted that the coat button over Jackson's heart was to be his aim. But he did not notice that Jackson had chanced to unbutton his coat. Dickinson aimed at the button, but it hung an inch and a half from its place. That chance saved Jackson's life. He fired an instant after the bullet hit him — and killed Dickinson.

Jackson's friends bound up his wound and somehow got him home. Rachel nursed her husband devotedly. If she knew the reason for the duel she kept that knowledge locked in her heart. Her grief for her friend's widowhood subdued Jackson's anger at the whole sorry business. And soon he had new cause for worry.

Talk about Aaron Burr as a traitor reached Tennessee, and Jackson's loyalty to his country was questioned because he had entertained Burr and was building boats for him. Later word came that Burr was arrested for treason in attempting to set up a new government in what had been the Spanish Southwest. Men gathered on the square in Nashville each time the express came to town.

"They goin' to try him in Richmond, Virginia," one said after a glance at the newspaper.

"I can't think that man is a traitor! Remember how agreeable he was. I liked him. Why, out at the general's —"

"Jackson must a' got fooled for once," the tone was sly.

"Jackson's no traitor. He's smart, too. He'd a known if —"

"Don't you be sure. Smart men get fooled."

"Not the general. The thing's all politics."

The talk settled nothing, but coming after the duel it did not help Andrew Jackson.

Time passed. The Jacksons had hosts of loyal friends. Money troubled Jackson at times, but life at the Hermitage was good. He bought more land and improved the estate. Visitors, old and young, rich and poor, were welcomed at the blockhouse.

The Donelson family increased and prospered too, in those early years of the 1800's. Jackson was a friend to them all. When Rachel's brother Sam died, Jackson took his two sons to the Hermitage to raise and educate. John Coffee, the friend and partner, married Rachel's niece Mary and thus strengthened the tie between the two men.

Rachel and Andrew enjoyed their times alone, too. As they sat before the fire, evenings, Andrew smoked his long clay pipe and rested. Playfully he taught Rachel to smoke, too. Some women did, of course. Other times he played on his flute and Rachel accompanied on her guitar or the harpsichord. If they had had children, life would have seemed perfect.

One evening as Christmas of 1809 drew near, a horseman pulled up noisily at the door. Jackson hurried to open it.

"Oh, you're from brother Severn's!" Rachel exclaimed, peering from behind Andrew. "Is there trouble?"

"Yes, the missus is took sick. Kin you come?"

"At once!" Rachel threw a cloak around her and picked up a bag she kept packed ready for sick calls. Andrew had already gone to fetch her horse. She left at a gallop.

Some hours after she reached her brother's home, twin sons were born to Severn's wife. Their joy that all were alive was mingled with dismay — Severn now had four little children and his wife was very frail. Rachel saw his worried face.

"Oh, Brother! Will you let me take the second baby home and care for him?" She was tenderly bathing this little new-comer by the fire. "We would be very good to him."

"Will Andrew mind?" Severn asked. "You can see that we have our hands full with the one new baby we expected."

"Mind?" Rachel's tone was sure. "Mr. Jackson will be happy. He loves Sam's boys. But this baby we shall have for

our very own. It is settled then, Severn? I may take the baby?"
Severn smiled his assent.

Later, when the new mother was comfortable and the household in good order, Rachel wrapped that second baby in a blanket and cradled him in her arms as she rode home.

"Mr. Jackson!" she cried. And as he came to greet her, she said quickly, "We have a baby! At last we have a baby of our own." She told him all that had happened.

"We'll call him Andrew Jackson, junior," he decided happily. "He shall be my dear son and my heir. Come morning, I shall make a cradle for him with my own hands. Oh, Rachel, God is good to us!"

"Old Hickory"

THOSE WERE happy days at the Hermitage. Andrew, junior, was a healthy, attractive baby. When the legal steps of adoption were complete, Andrew Jackson began calling his wife "Mother," a word sacred to him because of love for his own mother.

Sam's boys, John and Andrew Donelson, were older; they stayed at the Hermitage when they were not away at school. Jackson liked to have children around his home. Often in the evening, he took Andrew, junior, on his knee and told stories. John and Andy Donelson stretched out on the floor, close by.

"Now tell us when you rode through the wilderness," Andy said.

"Tell the one with the two men," John added. "Andrew likes it, too."

"You know that one," Jackson smiled.

"That's why we want it," Andy grinned back.

"I was ridin' along toward Nashville," Jackson began. With these words the boys relaxed, content. Rachel watched the firelight play on their faces — wonderful children, she thought.

"One morning a couple of rough-looking fellows waylaid

me. I knew right off they meant to rob. But seemed they meant
to have some fun first. One aimed a clumsy gun at me.

"'Git off'n that horse,' he yelled, ''n dance. *Dance*, I say!'"

"I pretended to be the stupid country bumpkin."

"You can do it, Uncle Jackson," John giggled. "I've seen
you."

"Let him tell it," Andy cried. "So've we all."

Jackson chuckled. So they had caught on that he could
put on an act. Well, sometimes it helped in a pinch. He talked
on.

"'I kint dance without m' slippers,'" I told them, "'they're
in m' saddlebag, back thar.'"

"'Then get'm fast n' dance,' the tall one ordered.

"Actin' like I was mighty scairt, I swung down n' went to my spare horse, opened the right hand saddlebag and reached inside. So far, I was movin' slow and awkwardlike. Suddenly, like a flash, I had a pistol in each hand and was aiming steady at those two men.

" 'Dance yourselves!' I yelled at 'em. 'Dance, or I'll fire!' "

"And they did," little Andrew said.

"Indeed they did." Jackson drew the boy close. "I kept those men dancing till they fell down from weariness. Then I mounted and rode off. They made no move to follow."

"Now show us how they danced," Andy commanded. This was the climax of that tale. Uncle Jackson set the boys down and capered for them until the boys rolled on the floor in glee. The next night they had him tell another from his endless store.

So months passed by.

In the summer of 1812, a discordant note sounded on the Cumberland. Jackson was in town, near the square, when sound of a galloping horse caught people's attention.

"It's Billy Phillips!" a man shouted. "Your Billy, General." The jockey had gone to Washington and got a place as express rider for the government. His arrival meant news.

Billy drew up, waving his cap; his horse was covered with foam and dust.

"War!" he shouted. "The United States Congress has declared war on England!"

"Glory be!" an old man shouted. "I never thought they'd buck up 'n do that!" People crowded around Billy to hear more as he handed over his dispatches.

"Make way! Make way!" a voice behind shouted. "Don't crowd! I've his relay! You can't stop the United States Express!"

Billy grinned as he slid from his weary horse, flung himself on the fresh mount and took food handed him.

"I'm nine days from Washington," he laughed when people gasped. "I'm making a record to New Orleans!" He was off like a streak, his heels digging into his horse's flanks. The crowd stared till he was out of sight. Then they looked at each other uneasily.

War. What might that mean?

"The government will call General Jackson right off."

"Nobody in the country better." People had begun to talk.

"But maybe the President is peeved. Everyone knows the general likes Monroe better'n Madison."

"And there's that Aaron Burr business. I heard folks in Washington City hold that against the general."

Some shook their heads, puzzled and doubtful.

No doubts occurred to Andrew Jackson. He sent word to Tennessee's Governor Blount, volunteering his service and that of twenty-five hundred men in the state militia. The governor forwarded the offer to Washington and added that the troops were ready. They could leave at once for Canada.

The Secretary of War replied with appreciation. No doubt it was agreeable to have such prompt support when many Americans were heartily against this war. But no orders came to Nashville. Major-general Jackson and his men fretted. But actually, this delay was not surprising.

The announced purpose in making war on Britain was to stop affronts to American sailors and commerce; and to end Britain's help in Indian war. But underneath these stated reasons lay a real purpose. Groups of men called War Hawks, led by John C. Calhoun of South Carolina and Henry Clay of Kentucky, wanted war so that Canada and the Floridas could be annexed. New England, a section that needed commerce, feared that if the south and west became safe from Indians, more people might move away. Led by Daniel Webster, they blocked war. In this confusion of purpose, it is no wonder that orders were not sent to Nashville.

Autumn had come by the time President Madison was heard from. Jackson and fifteen hundred men were to go to Natchez.

Nashville was wild with excitement.

"I'll bet we're to go straight to Florida and straighten out the mess there," a lieutenant said as he collected weapons.

"Florida! I heard we were to go to New Orleans."

"The general would hate that. He's no faith in that Wilkinson down there — none. But he'll do what he's ordered."

"Anyway, we're goin' somewhere," a neighbor gloated. The men had been as impatient as the general to test their military skill.

The order said that troops were to "equip themselves . . . with ammunition, camp equipment, blankets . . ." The government would pay in the usual way. Jackson decided that meant when Congress got around to it. He signed personal notes for what was needed. For economy, he suggested that the men wear hunting coats or shirts of blue or brown homespun, dark socks and pants.

More than 2,000 men came to Nashville on the appointed day, December 10th. These volunteers were mostly sons or grandsons of Revolutionary War soldiers. They were eager to fight this second war of independence.

That very day the weather turned bitterly cold. The worst storm in the memory of settlers blew icy blasts over the town. No one had expected so many would come; there was no possibility of shelter for all. Wood had been gathered in case a few fires would be needed.

"Use it up!" the general ordered. "We'll get more tomorrow!" The whole lot was gone before morning.

General Jackson and Quartermaster William Lewis spent the night going from company to company.

"Keep moving, men! Use the wood while it lasts," Jackson said, over and over.

"If you lie down, keep your feet to the fire," Lewis added.

"Weather can't get us down, eh, men?" Jackson's grin always got an answering smile.

As day broke, the two officers went to an inn for coffee. A guest there quickly berated them for poor planning.

"You should have had inside quarters ready," he scolded.

"That from you, who slept in a bed!" Jackson roared. "Let me hear one more word from you, and I'll run a red hot poker down your noisy throat!" The complainer shrank back, silenced.

When they were outside, Lewis eyed Jackson slyly. "You

put on quite an act in there, General," he said.

Jackson laughed. "My boys tell me I do sometimes. I couldn't let talk like that in there get started."

Luckily the storm and cold did not last. Soon the cavalry, 670 mounted men under Colonel John Coffee, set off through the forest for Natchez. On the seventh of January the infantry held a dress parade in Nashville Square and then embarked on flatboats for the thousand-mile river trip to join them.

While the men were embarking, Jackson was writing to the secretary of war. ". . . . 2,070 volunteers, the choicest of our citizens, go at the call of their country to execute the will of their government . . . They will rejoice at the opportunity of placing the American eagle on the ramparts of Mobile, Pensacola and Fort Augustine, effectually banishing from the southern coasts all British influence." He wrote vigorously. Andy Jackson of the Waxhaws was at last going to lick the British.

The men kept in fine spirits on the journey in spite of cold wind and rain. When they came to Natchez on the 15th of February they found Colonel Coffee and his men had arrived. Immediately Jackson was handed a puzzling dispatch from General Wilkinson of New Orleans — an order to stay in Natchez. Jackson shrugged off dismay. A short delay might be good for travel-weary troops.

"We'll set up camp a few miles out," he told his aide, "and hold daily drills to keep the men fit."

As for the general, he spent hours writing letters. Many were for Rachel; he sent these on whenever he could get a messenger. These were tender letters, beginning, "My dear love." And he wrote official reports to Nashville and to Washington. One express said that if he and his men were not needed in the south they would eagerly serve in Canada. A reply did not come till March.

The general tore the message open and scanned it quickly. Then he flushed red. His hair seemed actually to rise in wrath. His strong hands shook. His voice trembled with anger.

"I am ordered to dismiss my men — *here* — *now*. We're not needed! The President thanks us — he does indeed!"

His aides stared at him, incredulous.

"How can I dismiss my men without pay?" Jackson shouted as he raged back and forth in his tent.

"How shall they get hundreds of miles without food? I am responsible for my troops. They are the flower of Tennessee. I shall get them safely home." He glared at his aides, then dashed off to find his quartermaster.

The aides looked at each other uneasily.

"Is this Wilkinson's trick to make us join up with *him*?"

"His recruiting officer was around here yesterday till the general sent him packing. I wouldn't blame a man for joining — it will not be easy to go home — unwanted."

"Harder for the general than for us. Me, I stay with him."

"So do we all. He'll find what to do. I trust him."

The general set out for the city. He had no funds, but he signed personal notes for needed food and supplies. Natchez merchants backed him loyally.

"I am grateful to you," the general told one merchant who had practically emptied his store.

"Your note is good as cash, General. We've no worry."

Jackson smiled. He knew that if the government did not repay him for these purchases he would be ruined.

On the journey home the general walked with his men the whole way. The sick rode in rented wagons; the ailing rode horses — three of which belonged personally to Andrew Jackson.

The Natchez Trace followed the old Chickasaw trail and was not yet a good road. The going was bad for men, worse for wagons. Trees had to be cut down, canebrake hacked away, bridges contrived. Cold rain washed out fires and turned the trail to mud. Day after day the general strode along, now stepping up his pace, now dropping far behind, to be with different companies. The men watched him with growing affection.

"He's a tough one, our general," a private said.

"I never knew him till now. He's sound as hickory." The man held up a hickory cane he had cut as an aid on the march.

"Sound as hickory!" passed down the line. Most of them had hickory canes; the wood was strong, the phrase had vivid meaning.

Before the march ended the word "Old" was added in affection.

"Old Hickory's lookin' after us best he kin."

"I'd go anywhere for him!" Men who until now had known only Jackson's temper and drive saw his kind thoughtfulness.

Days passed. As they neared home, sick men perked up. The troops marched proudly into town — in spite of sore feet and gaunt faces. People rushed to cheer.

Almost immediately the general volunteered for service in Canada where the war was not going well. He held the troops together until the answer came, "not needed."

As the men scattered to their homes they spread the general's fame. *The Nashville Whig* carried an article about him:

"Long will their general live in the memory of the volunteers of Tennessee for the humane and fatherly treatment of his soldiers; if gratitude and love can reward him, General Jackson has them . . . We fondly hope that his merit will not be overlooked by his government."

"Those are fine words, General," Rachel's eyes glowed with pride as she called him that name for the first time.

"Fine words, yes." But Washington was far away. Congress had no praise. Jackson's accounts were disputed, and no move was made to pay Natchez expenses. Jackson was close to ruin as he gathered cash to send to the trusting merchants.

One of Jackson's friends, Colonel Thomas Hart Benton, was leaving for Washington when he heard Jackson's situation.

"I'll take it up with congressmen I know," he promised.

He kept his word. He told men there, "Your administration can't be re-elected if you disdain Tennessee. You *must* pay the expenses of those volunteers you ordered out."

So Congress voted the money. The President, through Benton, sent good wishes. And Benton wrote, explaining that the order to return home was caused by confusion in Washington, not by lack of regard for Jackson and Tennessee's troops.

Difficulties seemed passed. Jackson was held in high esteem — when suddenly a painful incident developed.

On the march from Natchez, Major Carroll had quarreled with Jesse Benton, brother of Thomas Hart. No one was surprised; petty differences were common. At home they could hardly recall what started their disputes. But they still argued. They planned a duel, and Carroll asked Jackson to serve as his second.

"Man! I can't!" Jackson exclaimed. "I'm too old." He was forty-six, old for dueling. "Anyway, you two have no business to fight. Get together and talk your differences out."

This good advice made matters worse.

"The general scorns me and my cause!" the major complained.

Jesse Benton talked about "getting Jackson." He said he'd been belittled. They held the duel, but it was a sham that settled nothing. People took sides. Colonel Benton's return from Washington did not check the reckless talk.

Tempers were hot on an early September day when Jackson and Coffee rode to town on business. Men eyed them. Jackson scented trouble.

As the two men passed an inn, Colonel Benton stepped out and by chance or intent, put his hand into his pocket. Jackson thought he meant to shoot. He whipped out his own pistol and covering Benton forced him into the inn and through the hall. They were at the back door when Jesse Benton arrived, running.

Seeing Jackson's weapon, he fired, at close range. Jackson fell, sprawled out, his left shoulder shattered by two bullets and some metal Jesse had loaded with them.

Coffee, who had hurried through the hall, sprang at Jesse with a knife. Lives were saved only by quick, brave action of bystanders who dragged the men apart.

Blood was spurting from a great wound in Jackson's chest. His face was gray. He seemed to be dying. Somehow they carried him across lots to the other inn and got a doctor.

Suddenly the summer-long quarrel evaporated. The general was dying — the man who had had no part in the dispute.

"Kin they save him?" someone asked. Tears fell unheeded.

"'Taint likely. His blood's soaked two mattresses a'ready, I heard." Crowds stood about uneasily, almost guiltily. Someone came out and galloped off on Jackson's horse.

"He's fetching Mrs. Jackson," the word went around. "But the doctor doubts the general will live to see her."

Jackson lived through the night. Through the next day — and the next. After a time he was taken home, on a bed in a wagon, weak, shaken; his shoulder with one bullet left in it would never serve him well again. That they knew.

Silent, Jackson lay on his bed at home. Every move was agony, every thought a regret. Days passed.

On the 18th he heard new sounds — a galloping horse, a scattering of gravel, an urgent voice and Rachel's "Hu-u-sh!"

"That's John Coffee!" Jackson whispered, wondering. "Something's up, and I lie abed like a woman." He turned his face to the wall in bitter shame. Then he half raised himself, listening. Yes. It *was* Coffee, very insistent.

"Come up, John!" Jackson called, and gritted his teeth with the pain of speaking. "You've news! Out with it!"

Creeks
on the Warpath

Mrs. Jackson and Colonel Coffee heard the general call. She put her arm across the door and shook her head.

"But Aunt Rachel," he whispered, "I *have* to see him. There's a Creek uprising, a massacre near Mobile Bay."

"And he is needed? He cannot even rise from bed!"

"John! Come up!" the voice from upstairs called. Coffee pushed past Rachel and ran up the stairs as she followed.

The sight of the sick man shocked Coffee. Jackson's face was haggard from pain and loss of blood. His shoulder was swathed in bandages. Dare this man hear bad news? But it must be told. Jackson was head of Tennessee troops.

"You'll hardly believe it, Andy," Coffee began. "Two messengers got in today. Red Eagle led a massacre on Fort Mims — looks like only those two escaped. Nashville is scared the uprising will spread. They shout for troops — and you."

Jackson groaned.

"John! Can't you see? You must leave!" Rachel cried.

"Go on, John," Jackson laid his good hand on Rachel's.

"People aren't as heartless as it might seem," Coffee said. "They know you've been near death. But their fear is spreading like a fog. They want protection. Maybe the British are

back of this. I heard a score of men say today that they'd join up if you'd lead."

Rachel's eyes were intent upon her husband. A minute more and she'd send John packing. Jackson was silent.

"The governor will have to order out troops," Coffee went on. "Folks want you. The Benton shooting's past. Don't worry about that now. Point is, who can lead?"

Jackson flushed. He moved his good hand across his face as though to brush away weakness — and chanced to touch that old scar. In a flash the scene at his cousin's house in the Waxhaws was vivid before him — the arrogant British officer, the gleaming sword, the spurting blood, the pain. That lad, though hurt, had courage to plan for neighbor Thompson; strength to carry on and then walk forty miles to Camden. Could a man do less?

Instantly a miracle seemed to take place, there on the bed. Listless eyes sparkled deep blue. Gaunt cheeks flushed. The sick man half raised himself on his right arm.

"Who can lead them? Tell them their general is coming! Let the governor call the troops." His voice was steady now. "When they are ready I shall be there, though I am carried on a stretcher. Think I would let a shoulder keep me from serving my country?"

"But, General," Coffee began to protest.

"No buts. You see, I pay for my folly. It is not easy to lie here when I long to ride back with you. However, I have learned a lesson. That is something. Hurry, John. Tell the governor I'll be ready soon. We'll avenge this frightful wrong."

John Coffee ran down the stairs, mounted and was off. Only Rachel saw her husband fall back, fainting, on his pillow.

"You can't go, General," she said, her heart sad for him. "You know that yourself."

"Save your words, Mother," he struggled to speak firmly. "I'll have a little time. I'll *make* myself well." He raised up, turned slowly and swung his feet toward the floor.

"This is going to take more doing than I thought," he admitted as he fell back weakly. "But I'll get there." Presently he added, "I can be making plans. I'll need all the facts."

That day and through the night Jackson pieced together the facts, as he knew them, that must have led to the massacre at Fort Mims. Likely Tecumseh was back of it, he decided.

This Shawnee chief in the North was one of the greatest of American Indians. Tecumseh, with his brother, a mystic called The Prophet, got the idea of uniting all tribes of American Indians from the Great Lakes to the Gulf into a mighty force that would push the whites from the continent forever. His project got underway after a decade of relative peace between the races.

The United States government had developed a fairly good policy about Indians. They sent white men as Indian agents to act as go-betweens, protecting rights of both settlers and Indians.

Many of these agents were dedicated men. They taught agriculture and marketing; their wives taught sewing, cooking, baby care, and English. One fine couple was responsible for the good feeling between Tennessee settlers and the Cherokees. An era of peace seemed coming when Tecumseh visited the south to rouse converts to his project.

In the year 1811, the chieftain was having some success till General William Henry Harrison defeated the forces of The Prophet at Tippecanoe. The glamor of the brothers vanished like a puff of smoke. The struggle between the races might have gradually died out — but for the British.

They offered Tecumseh arms, whisky, and partnership. And also what he cared for most, a chance to save face by joining with them. When some hesitant warriors suggested that they fight alone or be neutral, Tecumseh was firm.

"No," he told them, "I have taken the side of the King,

my white father. I will suffer my bones to be bleached on the sands before I join any neutral council." The first British victories were due to his efforts.

These encouraged him to go south again and talk with the Creeks. Their chieftains were cool. They liked their agent. But the magic of Tecumseh's oratory won a small group; they named themselves Red Sticks. Encouraged, Tecumseh traveled from village to village. His voice held Creeks spellbound as he recounted old wrongs — all true enough — and urged revenge. He traveled in secret. Neither the settlers nor the United States government suspected what he was doing.

Then Tecumseh won a man worthy of his great scheme, Red Eagle. This new disciple was a handsome, brilliant man of mixed descent. His ancestors were French, Scottish, English, with about one-eighth Creek. He might have lived as a white, as his brother did; but he chose the Creeks. He liked the name Red Eagle better than William Weathersford to which he was born. He had land, slaves, and was content — but for one thing. In giving Indians benevolence the United States withheld equality.

Red Eagle knew that British ships were in the Gulf; he knew of British successes in the north. And Tecumseh told him that Britain promised equality to their red brother. Red Eagle gave himself wholeheartedly to the British cause.

On the 30th of August, five days before the shooting affair that nearly killed Andrew Jackson, Red Eagle led the surprise attack on Fort Mims. The slaughter was appalling. The seven-eighths white in Red Eagle's nature was sickened;

he sternly reproved his warriors. But he kept his earned leadership — and the Creek War was launched.

Governor Claiborne of New Orleans rushed news of Fort
Mims to New York, but the journey took thirty-one days. When
the messenger arrived, newspapers were full of Commodore
Perry's success on Lake Erie. The massacre was hardly noticed.
It didn't seem important; the south was far away.

At Nashville the feeling was different. Tennesseans knew
what Indian war meant. While terror swept the town, a sick
man in an upper room pondered over history and laid his plans.

Military aides and personal friends responded to Jackson's
call and came to his bedside. Some friends were Indians he
trusted; they brought word of Creek strength. He sent them
out to Choctaws, Cherokees, and even to Creeks to enlist such
help as they could get. And daily, as he worked with his mind,
he struggled to strengthen his body for heavy work ahead.

The governor called out troops. Townspeople demanded
revenge. The general would get it for them, they said.

But in the general's mind a wider service than revenge
was taking shape. He planned to enter the Creek region in
Alabama and push south to Mobile Bay; to make a wide, safe
pathway through the Creek land to the Gulf. This route would
assure supplies in an invasion of Florida. He determined to
crush Spain, Britain's ally; perhaps troops in Georgia and
Mississippi Territory would help. But mostly he must rely on
men from Tennessee. Shrewdly he kept this plan to himself.

On the first of October, General Jackson walked a little,
his arm in a sling. On the seventh, he took command of his

troops in Fayetteville, eighty miles away. Colonel Coffee and his cavalry were already in the Creek area; the two forces were to meet on a certain high bluff overlooking a southern loop of the Tennessee River.

Jackson's infantry had hardly got marching when word came that the cavalry was threatened with attack. In literally one minute after that message arrived, the infantry began moving in double quick time. They marched thirty-two miles in nine hours — a record for movement of infantry.

That effort of a sick general against a cruel foe ended in frustration. Word arrived that the first messenger was mistaken. Coffee was not in danger.

"Don't let this new word get around," Jackson cautioned his aides. "Now the men are inspired. We must join up with Coffee while this mood is on." He got them marching again.

On that high bluff the joined forces built a stockade. Jackson called it Fort Deposit. Then Coffee's troops moved on across the rugged Raccoon Mountains and built a second stockade they named Fort Strother. Now they had two forts — and almost no supplies to put in them. Jackson's peculiar military gift was in the quick movement of infantry. But to move, he must have food for his men. None came. He decided to take a risk.

"We shall cross those mountains if we have to eat acorns," he announced. And gave the order to march.

They crossed and found that Colonel Coffee had foraged and got some Indian corn. Thus encouraged, the infantry attacked and wiped out two Indian villages. The men were jubilant. But without food, Jackson dared not push on south to the Creek stronghold on the Alabama River. Coffee's corn was gone.

Twenty-five hundred men needed a thousand bushels of grain and twenty tons of meat each week — and he had almost nothing. He had given his personal supplies to the sick. He ate acorns — there was nothing else. Creek spies knew his plight; he could feel their eyes on him, gloating in his humiliation.

While waiting, Jackson inspected the conquered villages. The warriors had been killed or had crept away. Only pitiful groups of women and children remained — all hungry. The general ordered them fed from a small cache of corn just discovered. As he turned away he noticed a small boy, alone.

"Take him and feed him," he said in Creek to a squaw.

"He should be killed," she retorted bitterly. "His father is dead. His mother is dead. What good is his life? Kill him!"

Shocked, the general picked up the child and carried him to camp. He found a bit of brown sugar, mixed it with water and crumbs and fed the child. It was plain that he was famished. When the little one slept, Jackson called an aide.

"Take him to the rear, to Huntsville if you can," he ordered. "As soon as possible, get him to Nashville. Mrs. Jackson will care for him."

"Do you know his name, General?" the aide asked, hiding his surprise. Jackson eyed the sleeping child thoughtfully.

"I call him Lincoyer," he said. "Guard him well." Then he turned to his letters.

The general's work included the writing of many letters; George Washington had found that a large part of his work during the Revolution. Jackson's daylight hours were for training troops, interviewing spies, checking reports. Nights were for letters: to the President, the secretary of war, the governor of Tennessee, the quartermaster, the merchants. And last, but never neglected, to Rachel.

A small tin writing kit was a part of the general's field equipment. Its black-painted sides opened to reveal a bottle of ink and one of sand with glass stoppers. A small holder kept

a candle erect, and sealing wax in a narrow groove was at hand for closing a folded paper.

Jackson whittled his own pens from quills picked up on the ground. He set this simple kit on a table or on his left knee as he wrote. His left shoulder was hunched oddly to ease pain.

An express rider to Nashville always carried a letter to Mrs. Jackson. The general wrote of his problems, his deep religious faith, his love for her. She cherished each one, reading and rereading them daily. When small Lincoyer arrived, she loved him and cared for him devotedly.

Now three small boys played together around the Hermitage. There was Andrew, junior, the son and heir, the brown-skinned lad, Alfred, who belonged to the plantation and was Andrew's body-servant, and the copper-skinned Indian refugee, Lincoyer. They shared their bumps and their pleasures and were conscious of no difference in Mrs. Jackson's care.

Letters from the Hermitage told of simple things, the children, the relatives, the plantation. Rachel wrote of her faith in God, her prayers for her husband's safe return, and always of her love. Her father had taught her good penmanship. Spelling was phonetic, of course, but Jackson had no trouble understanding. He spelled that way himself; most people did.

Weeks passed. Supplies did not come. Jackson did not know until later that drought and low water in rivers far away caused as much of the delay as lack of money. He welcomed food friendly Creeks and Cherokees brought when they came to volunteer. In accepting their service he insisted that they wear white feathers and white deer tails to distinguish them in battle. These and most of the soldiers were loyal and patient. Young Davy Crockett was a help to morale; he told endless amusing tales that intrigued idle men.

And so they waited — the enemy in front, famine crowding behind. Naturally some soldiers fretted.

"Where's Red Eagle — that's what I'd like to know."

"What we staying *here* for? Me, I'm a-goin' home."

"I'm with you. Let's go!" Instead of two, half a dozen men slipped off through the trees to the trail north.

Word of this came to the general. He leaped to his horse and galloped off in a roundabout route to their trail. There he pulled across the path just as they came near — it was quite a group, now, that had joined together.

"Halt!" he shouted. "Turn back!"

"We got a right to go home, General," a man near said. "We're hungry."

"I know. We all are," Jackson admitted. "But that gives you no right to desert." They made no move to turn.

"*Back*, I say! Or I shoot you as deserters!" As he faced them across their path a stream of words flowed over them and held them spellbound. The general might be short on spelling, but his speech was remarkable. Silent, terrified, they listened. Then, meek as boys caught stealing jam, they hurried back to camp. The mutiny was ended — for that time.

The next day food came, but not nearly enough.

Soon it was December. Men who had enlisted for Natchez clamored to go home.

"Our year of service is over, General," one soldier said.

"Over!" Jackson dashed from his tent to face them. "*Over!* You had six months of that year at home!" He scolded them like a father; stormed at them like an outraged officer; and held them. But he dared not push south with a hungry, halfhearted army.

For ten miserable weeks the general endured humiliation, despair, and physical suffering. But he was never tempted to give up. Fortunately the Creeks were to the south, building up a stronghold they thought impregnable. Then one good day a spy brought word that the stronghold was taken. Soldiers crowded around to hear the news after the formal report had been made.

General Claiborne, brother of the governor of Mississippi, had attacked. Indians had fought hard, but in vain. Many had died. Others had escaped.

"Did they get Red Eagle?" that question was important.

"No. You'd never believe how he got away," the spy said. "He was the last to quit the stronghold. He mounted his gray horse — right in sight of us — and dashed for the bluffs along the Alabama River. We saw him ride at full speed along the rim. Claiborne's men were behind and across the river, watching.

"He dashed on, spurring his horse. Then suddenly he turned, and man and horse leaped into the water. For seconds they disappeared. Then we saw them, swimming for dear life."

"They got away?" Listeners could not believe.

"Not a shot was fired. I reckon every man was dumb-founded at such courage. Never have I seen the like."

So Red Eagle, a brave enemy and a dangerous foe, was still free. Jackson dared wait no longer. Nine hundred new recruits arrived. He must hold their interest, use their loyalty, and find Red Eagle.

"We march at once," he ordered. Men eagerly obeyed.

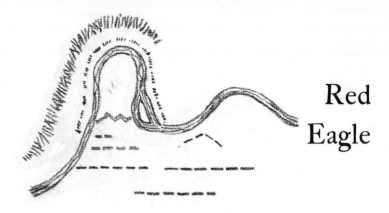

Red Eagle

A CURIOUS BEND in the Tallapoosa River was General Jackson's first objective. Here the river made a great loop forming a peninsula which the Indians called Tohopeka and the whites, Horseshoe Bend. The Indians had camped on this hundred-acre tract; with water on three sides and wilderness behind they felt safe. But to make doubly sure, they built log barricades across the neck of land, making a breastwork that looked more British than Indian. And they made canoes, kept tied under the river bank in case they needed to escape.

All this Jackson knew from his spies. He knew, too, that some eight hundred warriors and three hundred women and children were in this camp — under Red Eagle.

In the evening of January 21st, the general and his troops camped a few miles north of Horseshoe Bend.

"The Creeks know we are here," Jackson told his aides. "Haven't you felt their eyes upon you as we marched? They'll attack at dawn — and we'll be waiting. Tell the men to fire at gun flashes if it's too dark to see men."

The Indians did attack. But it was they who were surprised, not Jackson's troops. Before the warriors could recover, Jackson had pushed them back, back, to their camp.

Alas! The general dared not follow and strike a final blow. He lacked guns, ammunition, food. To win in the end, he must retreat now. He ordered dead horses skinned to make litters for moving the wounded and quickly marched north. The Creeks had suffered heavy losses and did not follow.

During the next weeks, Jackson's troops fought in the region south of Fort Strother. His method was to throw troops in a line halfway around a village. Then, in the center of that crescent, he created an incident to attract Indians in defense. As warriors were busy fighting, Jackson completed the circle. Thus surrounded, the Indians had to surrender or be killed. In this way several villages were taken.

Some supplies and new recruits arrived. But new men would not accept discipline. In spite of modest successes, mutiny threatened — shocking, with victory so near.

In an effort to stiffen morale, Jackson refused to intervene when a court-martial of officers sentenced young John Woods to death for insubordination. Jackson wrote sadly to Rachel about the youth. But he thought his duty was clear.

During these same weeks of minor attacks, messengers rode north with news of the battle at dawn and the taking of villages. Newspapers printed glowing tales. Even small victories delighted readers since news from Canada was disheartening. This period was the first time that Andrew Jackson's name was prominent in the nation's press.

People in Tennessee were proud. Enlistments soared. Soon Jackson had hundreds of new recruits and the help of the 39th United States Infantry — five thousand men in all. He marched south, built stockades and had men enough to leave a force to guard his supplies. At long last he was ready.

A quick march with two thousand men brought him to the higher land back of Horseshoe Bend, on a day late in March. He fired a few warning volleys. Then he waited till nearly noon to allow women and children to leave the fort by canoe.

General Coffee and his troops had crossed the river. He reported on the evacuation. And he sent swimmers across to destroy the few canoes that were not used.

Then Jackson attacked with fury. Young Sam Houston was one of the first to mount the breastworks, defying arrows

and musket fire. An arrow caught in his hip, but he dashed on, leading the charge. Only after several wounds did he fall fainting, to the ground.

The Creeks had but eight hundred warriors; the battle became a massacre. In mid-afternoon Jackson offered a chance to surrender, but the Creeks merely laughed. They fought on, to death. As troops took over the fort they counted 557 dead. Coffee's men counted about two hundred dead floating down the river. Jackson had forty-seven killed and about 150 wounded.

On this day, March 27th, 1814, the power of the Creek nation was broken. The massacre of Fort Mims was avenged.

But the glow of satisfaction was short-lived. In the evening Jackson learned that Red Eagle had by chance been away from the fort. This was a serious flaw in the victory. Jackson raged, vastly disturbed.

"Think how great the victory, though, General," one of the officers consoled him.

"Victory!" Jackson's tone was scornful. "In war, nothing is done until all is done! Red Eagle may even now be gathering forces to attack us. We must move on quickly."

They marched south to the point where two rivers meet to form the mighty Alabama. The Creeks thought that land by the meeting of the waters was holy land. They believed that no white man could step on it and live.

When Jackson's soldiers came — and lived, Creeks fled. The United States flag was raised, and the fort was given a new name, Fort Jackson.

But where was Red Eagle? Safe under the flag of Spain? Jackson did not know.

A while later, the general sat in his tent writing his reports. Suddenly he heard loud talk — what was this?

"Now we've got you, Bill Weathersford!"

"Don't touch me!" a well controlled voice replied. "I'll shoot you right in front of your own general's tent!"

Jackson sent writing things flying and sprang to open his tent-flap. There, a dozen feet away, stood Red Eagle. His handsome face was gaunt. Ribs showed skeletonlike; he was bare to the waist. He held a bridle, and beside him was the horse that had taken that mighty leap, now gaunt as his owner. Across the saddle sprawled a newly killed deer. The man's worn, desolate look caught Jackson's attention even as he aimed his pistol.

"You dare come to my tent?" Jackson shouted. "You, who led the massacre at Fort Mims?"

"I dare." Red Eagle lifted his head proudly. "I am not afraid of you, General Jackson. Kill me if you like." His tone was casual, indifferent. No man moved.

"I want nothing for myself," Red Eagle went on. "I have come to beg you to feed our women and children. They are starving in the wood, back of you. Your soldiers have taken their cribs of corn. Have destroyed their fields. Will you feed them?"

A flush reddened Jackson's long, lean face.

"You ask this?" he cried. "You who killed our people?"

"I tried to stop that massacre at Fort Mims," Red Eagle

admitted frankly. "I wanted a military victory, not murder. But my warriors would not heed me." He paused. Jackson did not speak.

"Now I am done with fighting. Once I had a hope — but it is gone. Once I could call upon my braves — now they lie dead. The dead do not hear; they cannot answer. I wish to avoid more misery.

"Will you be fair, General? Will you tell our nation where to go that they may live in peace? I throw them and myself upon your mercy. Speak. And we shall listen."

As he stopped the spell was broken. Men shouted.

"Kill him! Kill him!"

"Silence!" The general roared. "He who would shoot this brave man is a coward!" He turned to Red Eagle.

"I meant to hunt until I found you. I would have killed you or dragged you here in chains. But you have come alone, on an errand of mercy. You are hungry? Come, we'll eat together."

Red Eagle's face lit up with a smile. He turned and lifted the carcass off his saddle.

"I brought you a gift, General Jackson," he said. "Will you accept it?" Jackson assented, and the two great leaders went into the tent.

After a while Red Eagle came out, mounted his horse and rode away. His dignity, his daring, even more than his words, had convinced Jackson that he spoke the truth. He was done with fighting. And so they let him go. He had Jackson's promise to feed the women and children.

The Creek War was over. The power of that mighty nation was so broken that it never again was a serious threat to peace. The starving were fed and then moved north, nearer supplies. The United States fed them until the next harvest.

Troops of the regular army under General Pinckney came from Georgia and took over Fort Jackson. The volunteers were free to go home.

Jackson had sent reports to his government. He had included among these a sketch of the environs of Horseshoe Bend so that the battle could be understood. Now he wrote to his wife. He asked her to bring five-year-old Andrew, junior, and meet him at Fayetteville.

The triumphant march must have pleased her though it could not make up for hours of anxiety. Rachel Jackson was never moved by fame.

Four miles from Nashville, the road became lined with admiring throngs. People shouted wildly.

"Welcome home, victor!"

"We knew Old Hickory'd lick 'em!"

At a ceremony on the Court House steps, several men made speeches and a handsome sword was presented to the general. In the evening he was feted at a banquet. The town had stopped work to honor its famous citizen.

Through this acclaim, Andrew Jackson conducted himself with quiet dignity. The man who accepted that sword was quite a different man from the one who shared a foolish quarrel a few months before. This man's character had strengthened; though physically he was weaker. Poor diet, sleepless nights, great burdens, and his shattered shoulder had broken his health, perhaps beyond repair.

John Overton, who stood near him, whispered a friendly warning in the midst of the plaudits.

"Your standing today is as high as any man in America. But watch your step, General. Many envy you."

Among those Overton meant were the men who had mutinied and got away. Had they stayed, they might have shared this day's honor — instead they were scorned. That infuriated them.

News of this welcome home, as well as the battle of Horseshoe Bend, was widely printed. The name, Andrew Jackson, was beginning to be familiar; people enjoyed the drama of his success.

The United States government seemed to appreciate Jackson, too. They made him a major-general of the U. S. Army. This office was higher than the same rank in the state militia. He was assigned to the seventh district, an area that included Louisville and Mobile.

Jackson's first act in his new work was to go south and work out peace terms with the Creeks. Chiefs of the warring tribes, and of friendly tribes, too, came to Fort Jackson to hear the terms. Stoical by training, they listened; their faces were without expression as the hard words were read.

It seemed a cruel peace. Friend was treated hardly better than foe. Twenty-three million acres — part of Georgia and about half of Alabama — were to be given to the United States. This was more than half of the land the Creeks called home.

The chiefs temporized. They tried to delay signing.

"The cost of the war you began must be paid," Jackson told them. "Creeks must be separated from Spain and Britain, and your allies against us." No need to put in words what all knew. That the United States would gain a way to the Gulf and choice land for settlers.

"There is more," Jackson added solemnly. "The path that Tecumseh trod must end. Until it does, your nation cannot expect happiness nor ours security."

The chiefs signed. They had no choice.

Many Americans who rejoiced in the victory thought that the peace terms were too harsh. Others, like Andrew Jackson, who knew the Creeks, saw that the white man was steadily overcoming the savage. Perhaps a stern victory that insured peace would in the long run prove to be the kindest.

The treaty was sent on to Washington, where it was finally ratified, eighteen years later. Jackson pushed south to investigate tales brought to him about acts of Britain and Spain along the Gulf and in Florida. Back in his mind he felt that the real gainers from the Creek War were the British, who by handing out powder and shot through their Spanish allies to three or four thousand Creeks had kept four times that number of American regulars and volunteers engaged in war. He meant to watch the British.

This same summer the war between England and France ended and England was free to send ten thousand men to fight on the Eastern Coast. Maine was taken; Boston Harbor was occupied; Long Island Sound was invaded. New England prepared to meet at Hartford and secede from the United States.

About this same time, the British threatened from Quebec in a successful effort to divert attention to the north. Then they sailed up Chesapeake Bay and marched on to Washington. The capital city could have been ably defended; but army and government, alike, vanished for safer parts. Only a few brave souls stayed behind to see the Capitol, the White House, and other buildings burned. This in the very month the stern peace treaty with the Creeks was being signed.

Then came news from the navy. The Americans had defeated the British and sent the fleet sailing off into the blue. When General Jackson heard that, he began watching for the fleet off Pensacola.

The general had come down to Mobile and there he found new problems. He talked frankly with spies and his officers.

"Spain has stopped trying to look neutral," one said.

"I think they will plan to make Pensacola their base." The general felt certain of this. "They'll march on across to Baton Rouge and cut off New Orleans."

"They could take the city easily, then."

"They'll count on using slaves from Jamaica and New Orleans — I hear that," a spy asserted.

"Does Claiborne know all this?" Jackson asked. No one was sure. So Jackson sent a rider to ask the governor to make the city ready for attack.

Claiborne glanced at the message; gave a casual order to be prepared. He had little interest in Jackson's ideas. He was engrossed with a problem of his own just then.

The Battle of New Orleans

Governor Claiborne's problem was how to collect taxes on luxurious goods illegally sold in New Orleans. These things were brought in by a group of men called "Baratarians" and Claiborne was determined to catch their leader and stop the whole business.

This man, Jean Lafitte, had gathered around him a group of daring men who raided and robbed ships on the Caribbean Sea. The stolen goods were brought to a hide-out by Lake Barataria, southwest of New Orleans, and from there sold to merchants in the city.

Claiborne's task was hard, for people liked the finery Lafitte had to sell and they liked the price he could ask, without duty. Lafitte had become so bold that he had engaged Edward Livingston, once a United States Congressman, and friend of Andrew Jackson, to act as his attorney; there was not a keener lawyer to be had.

Mr. Livingston found plenty to do. Pierre, Jean's brother, had been arrested and held in prison that fall of 1814. A large cash award was offered for Jean himself — offered, but not claimed, though Jean walked boldly on the city streets, unchallenged. Rumors about the Lafittes had reached the British, and since they expected to take and hold the city, they thought they might as well have Jean on their side. So a naval officer was sent to Barataria to see him.

Bravely the British officer left his long boat and walked from the beach toward the pirate hideout.

"I wish to see M. Jean Lafitte," he told the guard.

"Your papers, Sir?" The guard recognized the British uniform, but with the state of affairs in the city he was cautious.

"Take this to him." The officer handed over a letter.

Immediately a handsome man came from the house. He carried the letter, apparently unopened.

"Welcome to Barataria," he said. "The time is right for dining. Please come in and eat before we talk business."

He led the way into a good-sized, rather plain building. But to the officer's surprise, the inside of the house was handsomely furnished. The two sat at a damask-covered table and from fine china and crystal ate an excellent meal.

After dinner Lafitte opened the letter. It offered him a captaincy in the British navy and delicately hinted at financial advantages. Talk led to mention of the attack on New Orleans, coming soon.

"I shall have to give thought to this honor." Lafitte appeared pleased. "I will let you know —" he named a day. Before the lieutenant quite knew how it happened, he was in his long boat, returning to his ship.

Lafitte sent that letter to Governor Claiborne and with it an offer of his services to repel the attack. Lafitte's way of earning a living was illegal, but he considered himself loyal to the United States.

Claiborne showed the letter to a friend.

"It's forgery — plain forgery," the friend exclaimed. "Don't have any dealings with a pirate! Don't even reply."

Claiborne did not answer the letter. But to protect himself he sent it on to General Jackson at Mobile.

A day or two later, one of Lafitte's men intercepted a British letter which confirmed news of the impending attack. That night Pierre mysteriously escaped from prison and the brothers, with their associates, left Barataria. They went around the city, to the north, to be ready for action against the British. Because of this sudden move they were away when the governor's guard attacked and practically destroyed the Barataria hide-out. The governor thought he had cleaned the pirates out. He would have been astonished to know that all the while they waited to the north, ready to help the city.

The Baratarians were just making camp when General Jackson received the British letter to Jean that the governor had sent.

"So the British are enlisting pirates, are they?" was his only comment as he turned to other business.

For many weeks the general and his troops had been busy on the Gulf Coast. He took the fort thirty miles below Mobile, an act that made the British move to Spanish Pensacola. On his own responsibility, Jackson captured that place, too. When he returned to Pensacola he found orders not to attack Pensacola; it was neutral.

"Neutral! With the British there!" Then he chuckled. "It's lucky that order didn't come sooner!"

In this same lot of mail he found a letter from Edward Livingston, telling of threats to New Orleans. Jackson trusted his old friend. He, too, feared for the city. He sent express messengers to Washington beseeching the government to send guns and ammunition and troops to save New Orleans. He well knew the government's ignorance and indifference about the south but surely they would not let their gateway city be taken by the British!

The secretary of war, William H. Crawford, was not concerned. Secretary of State James Monroe understood the need and did his best. But the credit of the United States had fallen so low that he could not borrow money — not even for guns.

"Perhaps you will have confidence in *my* honor since you have none in your country's," Monroe told the lenders, bitterly. He pledged his fortune to raise the cash he needed. Guns and ammunition were started down the river — he hoped they would get there in time.

Meanwhile word of the impending attack seeped into New Orleans. People were desperately frightened. Jackson got letters from Livingston and others begging for help.

"Come and save us!" "We need you now!"

Jackson saw he must act.

General Jackson was an infantry man. Likely that was a reason for his feeling that the British would land troops somewhere along the coast and march upon the city. Why should they attempt the long, winding hundred miles up the river? So he left a large part of his army at Mobile with orders to be ready for quick movement.

Then he sent John Coffee, now General Coffee, and his men across to Baton Rouge. There Coffee was to wait for reinforcements expected from Kentucky and Tennessee — and the guns Jackson hoped were on the way downstream.

On November 21st Jackson left Mobile for New Orleans.

"I shall travel slowly," he wrote to Washington, "searching along the shores for a point where the British are likely to land." He did not mention that he was so stricken with fever that he could hardly sign that letter.

A few days later, in distant Jamaica, a fleet of fifty British ships, carrying ten thousand soldiers, sailed for American shores. Many of the troops were veterans who had served under the

Duke of Wellington. General Pakenham, the Duke's brother-in-law, was in command. Jackson did not know any of this; neither did people in the city.

In due time, Jackson got to New Orleans. The city was truly in desperate need. Weapons and money were scarce. Jackson made a stirring speech in the square; Livingston stood by him and translated it into French so all could understand. People shouted approval. They thought Jackson would save them. But when he commandeered all weapons and put the city under martial law they were displeased.

Jackson paid no attention to praise or blame. He knew he must make the gay city into an armed camp if he was to save it — and all the river shipping.

His problem was made harder by the many routes into the city. Would the British come by the river on the south or by the two lakes on the north? Would they come by the bayous and swamps on the southwest — Barataria lay that way — or by the sugar plantations to the east? All these must be studied and watched.

Then, one day, the confident British fleet arrived northeast of the city; the ships scattered among small islands and cast anchor. A few boats were sent out to reconnoiter.

"We saw a small fort by the channel," was the report. "And on the inland lake a few sailing ships."

The admiral promptly gave orders that these were to be attacked.

Lieutenant Jones, of the American navy, saw the ships coming. Cleverly he enticed them into Lake Borgne where they quickly went aground. He had guessed that they were not built for shallow water.

But his small ships and the little fort could not stand against a large fleet. A bravely fought, hour-long battle ended in defeat, and the northern route to the city was wide open.

New Orleans was stricken with panic. Jackson kept his dismay to himself. The British had *not* come by land where he could outflank them. They were now at the very door! He had literally no time for new preparations.

Moreover, the general's fever was complicated by a hemorrhage of the lungs. He was so ill he could not stand; he could hardly sit up. But from a couch in the Livingston home he mapped the campaign. He had the citizens called to the square where Livingston addressed them. Bands played martial music, people with frightened faces cheered — and at once work began on fortifications. They had no time to lose.

On cold, rainy December 22nd, a British vanguard left their ships. By small boats and then by forced march through swamps they came to a sugar plantation southeast of New Orleans. Two of the plantation men escaped and brought this news to Jackson. He was still ill. His army was widely scattered, watching approaches. But he hesitated not a second.

"Gentlemen! The British have come. We fight tonight."

That was at noon. By four o'clock, two thousand men with Jackson at their head were marching to that sugar plantation. And an armed sailing ship was getting into position to give help from the river.

British soldiers had a formal code of warfare. It did not

include night fighting. So they were completely surprised by an attack after dark. They rallied and fought bravely, but they could not stand against troops trained in Indian fighting. While Americans fought, the ship poured shot and shell until a fog spread, blotting out everything, friend as well as foe. Jackson recalled his victorious men.

The victory was not decisive. But it left the British discouraged and bewildered by their losses.

Jackson withdrew to the Rodriguez Canal, a ditch that connected the river with a swamp east of the city. The canal was dry at the moment and there he set up fortifications. His plan was simple. He counted on the swamp at his left; the river with an armed ship at his right. The new, strong breastwork was in front and reinforcements behind. All this ready, he watched and waited, often studying the scene through a spyglass from a third floor window of a house near by.

General Pakenham attacked two days before Christmas. He destroyed the ship that had done the earlier damage. But Jackson had a second ship, armed and ready. It swung into position and opened fire on British troops as they marched forward.

Jackson swept the terrain with field glasses. He saw the oncoming British. He turned north and looked intently. A motley crowd of men were running swiftly toward the American line of fire.

"Who are they?" he asked his aide sharply.

"Baratarians, Sir. We had word that they had left their hiding place. We heard they were coming to join up."

Jackson's keen eyes observed that each man had a gun and knew how to carry it. They were running hard. His notion about accepting help from pirates changed.

"Assign them to a place in line — there!" he pointed. The aide dashed off. Reinforcements would help.

Pakenham soon found that the double bombardment, from ship and infantry, was too much. He withdrew — giving Jackson a chance to finish his fortifications.

The Americans did not stop fighting. Every night they

fired occasional shots upon the enemy camp. The harassing continued through the first week of January.

On January eighth, Jackson was ready.

Because of a heavy fog, Pakenham chose that morning to attack. His red-coated soldiers marched in close formation straight forward to the American lines.

Jackson had seen activity in the British camp. He had personally inspected all his line, regulars, free men of color, troops from Tennessee and Kentucky, Baratarians — a varied

lot; three thousand men against ten thousand British regulars; three thousand held together only by patriotism and the general's will to win.

Suddenly the fog Pakenham counted on lifted. Americans need not fire blindly; their cannon literally mowed down the advancing charge. With amazing courage, British ranks closed, only to be cut down again — and again. On the redcoats marched, into range of small weapons and the deadly Kentucky rifle in the hands of skilled hunters.

Jackson was here, there, everywhere, directing the battle. He saw that his troops did well; the marksmanship was superb.

Pakenham was killed early. Many of his officers were killed or wounded; the carnage was shocking. The courage of both British and Americans was thrilling. Soon the British saw that they had no choice but to retreat; human flesh could not stand against that deadly fire.

As the enemy fell back, the American band played, "Hail Columbia," and Jackson and his aides rode along the line of cheering men. In that moment the scars inflicted by the British officer's cruel sword on Andy of the Waxhaws was avenged. The general would carry the physical blemish, but the hurt to his spirit was healed.

Only later did he learn the frightful cost to Britain: 2,600 casualties, nearly a third of these dead on the field. Americans had eight killed and thirteen wounded.

This battle of January the eighth, 1815, was decisive. The enemy soon left the American shores.

The general went back to the city that evening and found

it wild with rejoicing. War was over. People thought that easy days had returned. Merchants called early the next morning to collect pay for food, arms, and ammunition.

Jackson had no money and no such plans. He continued martial law, though he had trouble with civil authorities about his decision. His days were spent with angry merchants, his evenings being gratefully entertained like royalty — a combination hard to find in any other city.

Meanwhile, Rachel Jackson had despaired of getting word about when she was to visit her husband. So she took the matter into her own hands. With Mrs. Overton and two other friends, and Andrew, junior, she came to New Orleans by boat. The little party arrived a few weeks after the victory.

The general was thrilled to see Rachel. Others might observe that she was travel worn and weary from responsibilities of a large plantation; to him she was perfect. Proudly he introduced her to the elite. His delight was moving.

Mrs. Livingston, an heiress and a beauty, welcomed Mrs. Jackson into her home. She saw that the traveler had rest. She called her own skilled dressmaker and ordered gowns for the parties ahead — a New Orleans woman could do no more! Rachel bloomed under her kindness.

The general's health improved, and he lost the look of strain. On Washington's birthday a great ball was given for the Jacksons, and there a handsome topaz necklace was presented to the general's lady. Jackson watched proudly, and his eyes gleamed as she gracefully accepted. Later she wore the jewels when he had her portrait painted.

The final incident of the battle of New Orleans was, perhaps, the most amazing of its whole story. Word came that peace was agreed with England *before* that battle was fought. No one in the western hemisphere had suspected that peace was near.

It was known, of course, that Russia, needing trade, had called envoys from warring countries to make a peace. It was no secret, either, that English and American delegates labored for months in the Flemish city of Ghent and made little progress toward agreement.

War with France had ended; the British no longer needed to impress American sailors. The Indians were conquered; that issue was gone, too. The aims of expansion were tactfully referred to a commission — nothing was left to be settled.

And so, the "Peace at Ghent" was signed on December 24th, just as the fog revealed the British retreat at the end of that first of three engagements near New Orleans.

Americans had gained nothing by this war that had been so costly in men, money, and prestige. Jackson's victories were the only land successes that mattered. But his victories, even the one that came too late, affected history. Europe learned at New Orleans that Americans knew how to fight; the lesson would be long remembered.

Jackson's countrymen had learned something, too. Patriots who were ashamed when their government fled the capital, letting the enemy invade and burn, could now hold their heads high. They could be proud of America, proud of a general.

Andrew Jackson was the man of the hour.

Man of
the Hour

The United States of America needed a hero as 1815 began. The people had endured a troubled winter. The capital city was still blackened from enemy burnings. News of the British armada sailing from Jamaica seemed to mean that New Orleans would be taken.

Congress was frustrated and ill-tempered. New England newspapers had announced, "no more taxes shall be paid until we have peace." On the ninth of January eastern papers printed word of the arrival of the British fleet off the Mississippi coast. The final blow was thought to be only a matter of days.

To add to the gloom, a three-day snowstorm blanketed the entire northeastern part of the nation. People sat at home, sad and discouraged. Was this the end of the republic?

It was not until the fourth of February that good news got through. People could hardly believe the headlines!

"ALMOST INCREDIBLE VICTORY!!!"

The whole front page of more than one newspaper was used to tell of Jackson's victory on the eighth of January.

As the news spread, eastern cities ordered public illumination. People made "transparencies" for their front windows; some of these were clever pictures of General Jackson on horseback. Men, women, and children paraded the streets enjoying the display, applauding the man who won the victory. Many had not even noticed his name before.

"Who is this Jackson?" was a common question.

"Where does he live?" "Why haven't we had him sooner?" People exchanged news about him — often items were fiction!

In a few days, formal reports prepared by Jackson and Livingston began to arrive in Washington. It was a relief to know that early reports were true. The British, though numbering more than three to one, had been soundly beaten.

But that was not all of the story. Nine days later, express riders from New York brought to the capital a new word — peace. Not only were the British defeated and gone, but peace was now official. The nation rejoiced, and a new era of prosperity began.

All the while these events were happening in the faraway East, General Jackson worked long hours of each day. He had no suspicion of his growing fame; he was absorbed by the problems and frictions of victory. A warm and sunny April was spreading beauty over New Orleans, hiding the blood-soaked field with bloom before his business was done.

On the sixth of that month the general and his party left by boat for Natchez. The brief rest of the voyage barely gave him strength for the parade and banquet in that city. People were determined to give him many honors.

From Natchez the party went by land to Nashville. Young

Andrew delighted his father by riding the entire distance on his pony.

"He does well for a six-year-old," Jackson was assured.

"Naturally," the general agreed, with a broad smile.

People ran from isolated cabins on the way, to shout a greeting. Tiny settlements hailed the gaunt rider with joy.

"You kept the river open for us, General!" men cried.

Riders brought word of his approach to Nashville, and a great crowd of citizens gathered on the square to greet him. Town officials and students from the academy made speeches of welcome. The general replied with modest dignity and with appreciation.

This acclaim was pleasant, but a strain on a sick man. When at last he got to the Hermitage and had seen his neighbors

gathered there, he climbed the stairs and collapsed on his own bed, grateful for its comfort. Mrs. Jackson conferred with the welcoming committee and tactfully persuaded them to postpone until late May the ball they planned.

The log home seemed beautiful; Rachel had made it so comfortable. For four months Rachel nursed her husband, never once relaxing her vigil. At first she had to feed and care for him like a baby. He could eat very little, but she saw to it that the little was nourishing. She served his meals on her pretty dishes. She kept his room gay with flowers — he thought it was wonderful to be at home.

Kinfolk and friends came to see him, but she held them off. "No, you can't see your uncle this morning," he often heard her say. "Stop by the garden and get some beans for your mother, Nephew. And come again next week."

Or if the visitor was a man from town she would say, "I'm afraid the general is not up to talk this morning. That hemorrhage of his lung takes a time to heal. Come again soon."

Gradually, under this nursing, he gained strength to sit up, to come downstairs, to rest in the garden. But it was slow.

"Where's that letter from Washington?" he remembered to ask one day.

"You're not to think of letters!" Rachel was startled.

He waited until the next day, then read the letter. It was an invitation, practically an order, to come to Washington as soon as he was able.

"That's now," he said, getting up with a show of vigor.

"Not yet, General. Not yet." So he waited awhile.

October days were cool and pleasant and the trees glorious red, copper and gold, when he got away. Rachel and Andrew junior went with him in a handsome coach with four fine horses and a spare led behind. Mrs. Jackson determined to care for him properly on this journey.

One of their stops was at Lynchburg, Virginia, where a great banquet had been arranged in his honor. Seventy-two year old Thomas Jefferson came from Monticello to attend. He proposed the first toast:

"Honor and gratitude to those who have filled the measure of their country's honor." He used the word "those" to include James Monroe, a friend of both Jefferson and Jackson and a likely candidate for the next president.

Jackson promptly replied with a toast to Monroe.

The Jacksons got to Washington in mid-November and were royally entertained. The general was given an ovation wherever he appeared. Those who had expected to see a rough frontiersman marveled at his dignity and fine manners.

The general's great height, the more marked because he was still thin, his long face and Roman nose, his up-standing hair, gray but thick, and his keen blue eyes, made him noticed even in the city of distinguished men. Newspapers recorded his every move and noted that he seemed unspoiled by fame.

The travelers returned home in January. Roads were hard and frozen ruts made it rough going, but they got home safely.

A letter from General Coffee awaited Jackson. The Creeks were restless with their unconfirmed treaty. The matter seemed so important that Jackson went to New Orleans to get facts and hold conferences. It seemed clear that the south would never be secure while Florida was in Spanish hands. The British had given Florida to Spain in 1783. Someday, surely soon, the general thought, this area must belong to the United States.

On his way home from New Orleans, Jackson stopped for conferences at several Indian villages. He listened patiently to Cherokees, Chickasaws, Choctaws and agreed to requests to the limit of his authority. His tact greatly increased good feeling.

His success in these conferences was certainly one reason for his appointment that summer on a commission to make a peace treaty with the Cherokees. This treaty brought about their final transfer to land west of the Mississippi.

He had hardly reached home when a minor personal event saddened him. Truxton, all this time Jackson's favorite horse, had to be retired to pasture. The general hated to see him go.

For nearly a year, Jackson was at home, managing his plantation, raising fine horses, enjoying his home and friends. In the late summer of 1817, he was called to a dying friend, John Hutchings, now living near Huntsville. He went at once.

Hutchings had been associated with Coffee and with Jackson in the store and boatbuilding businesses. He had married a niece of Rachel, so he was relative and friend. The dying man's great longing was to have Jackson take his motherless son, who at the age of six was about to become an orphan. Jackson readily accepted and was made the lad's guardian.

While he was in Huntsville word came to him from the new president, James Monroe. Jackson was ordered to take charge at the "border" — the line between Georgia and Florida.

This was action Jackson very much approved. As Hutchings had died, the general took his new ward, Andrew Jackson Hutchings, home to Rachel and immediately raised troops for the Florida expedition.

His orders were curious. He was to pursue the enemy but not to attack a Spanish fort. Did this mean that he was to enter Florida ready to occupy it for the United States? He wrote frankly to President Monroe: "Let it be known that possession of the Floridas would be desirable to the United States and in sixty days it shall be accomplished." He got no answer.

So when Jackson and his troops entered Florida he thought he had tacit orders to take it for his country. In his vigorous way, he took Indian villages and the Spanish fort, St. Mark. He ordered shot or hung men he considered traitors. He marched on Pensacola — and the Spaniards fled. In five months he brought peace on the border, ousted hostile Indians, and conquered Florida — all with little bloodshed.

Most Americans were thrilled by this action. Those living in the south and west were especially pleased as life would be safer now, they thought. New lands would be opened. Jackson, the people's hero, had done this for them.

But the government at Washington was dismayed. They had not meant to *conquer* Florida; they meant to buy it. Henry Clay attacked Jackson for his violation of neutrality. Congress tried to censure General Jackson, but did not quite dare — the nation approved what he had done. Criticism might mean political ruin for the administration.

In the midst of this, Jackson went to Washington and to his friends' surprise, stayed quietly in his hotel until Congress, its galleries crowded, vindicated him. When he did appear, he was widely entertained. The controversy seemed to make, rather than lose, friends and supporters for him.

During various journeys Jackson had seen many fine homes, in New Orleans, Philadelphia, Baltimore, and other cities. One day the thought occurred to him that his Rachel should have a fine house, too. They loved the blockhouse and were content there. But once the idea of a new house came, Jackson could not forget it. He could well afford anything he wanted; the plantation was very successful, and he had no debts. The whole nation was prosperous. Corn and cotton sold high. He decided to build.

The Jacksons often walked at the close of the day, enjoying a breeze and the garden. That was the time to speak.

"Suppose we built a new house, Mother?" he asked. "Where would you like to have it?"

"A new house!" She was astonished. "Why take on more work for yourself?" That was a fair question. He was better, but he used a cane for walking.

"Oh, I'd not hurry. Build next year maybe. But where?"

She glanced around. They chanced to be standing on a bit of level land a few hundred feet from their home. He could walk easily here. The land lay well for a garden, too.

"This might be a good place," she said. "Do you like it?"

"Here it shall be. I'll draw some plans. We can make bricks on the place. Maybe we could start building next spring. You begin to think what you want to put in the house."

As he had time between his work for his country, Jackson drew plans for a house such as was popular in the south. He sketched a wide center hall with two rooms on each side; a gracious staircase and hall and four rooms on the second floor. All rooms were medium-size with high ceilings.

Like Washington and Jefferson, Jackson intended to build the house himself, with a builder to oversee the work when he had to be away. But his work must begin long before actual building.

Tennessee limestone for the foundation was quarried on his own land, by his own people. A bed of good clay was found and the top soil taken off. Then for many days a boy on a mule treaded lime into the clay, preparing the material to make bricks.

Other plantation hands built kilns and molded the clay with hand molds and attended to the baking. Poplar trees furnished timbers, and red cedars were cut for the flooring.

By spring of 1819 everything was ready to build. One evening the general was walking over the site with his friend William Lewis.

"But why do you build *here?*" Lewis asked, puzzled. "It's so flat. Why not choose higher land with a view?"

"Mrs. Jackson chose this spot," the general said. "I am building the house for her. You can see that I shall never live in it — not with this trouble in my lung."

Major Lewis looked sadly at his friend and could offer no comfort. That Dickinson bullet was a serious handicap.

But slowly, along with the building of the house, the general seemed to grow stronger. Presently new furniture, china, crystal, mirrors, books, and pictures arrived and were put in place. Choice possessions were carried over from the block-house. Rachel had especially prized a set of serving dishes, creamy white with green trim. These had a place of honor in the new cupboards. Her harpsichord and guitar, her volumes of English poetry, were brought over and put in place.

In April of 1819 Jackson had engaged an English gardener who laid out grounds and gardens. At the end there was quite a flurry in the settling, for word came that President Monroe was coming to Nashville for a brief visit. He would like to stay with his friend, Andrew Jackson.

Years before, George Washington had made a personal visit in the South. Monroe now decided to tour the West. With that part of the nation growing rapidly, he felt he must see it.

Monroe left Washington in April, bound for Atlanta, stopping at towns and villages on the way. He came to Nashville by steamboat in June. To townspeople's surprise the boat did not dock at Nashville but continued upriver to the Hermitage dock. Jackson hurried to meet the party there and drove them the short distance to his new house.

The Nashville celebration came a few days later. On this day a Nashville committee met the President four miles from the city. Escorted by Tennessee volunteers, the President passed by cheering throngs to a banquet at the Nashville Inn. The next day there was a reception at the Nashville Female Academy, followed by a ball.

The *Nashville Whig* had a long article about all this. Of the ball, it said: "We have never seen more taste and beauty than was displayed in arranging the room or a more numerous and brilliant assembling of ladies." The editor spoke of a new portrait of General Jackson which was a center of the decoration and of the many transparencies showing the Battle of New Orleans. The city welcomed President Monroe — but it did not overlook its own hero.

When the President's party left, Jackson went along with them to do the honors for the West.

As townspeople talked over the splendors of the visit, some mentioned that it was a good omen to have a president as the first guest in the general's new house.

"I was right glad to see a real president," one man remarked. "But after all, Monroe can't hold a candle to our general. Why can't Andy Jackson be president?"

"Well, why not?" men asked each other, pleased.

"The general would be good at it. He gets things done."

Changing
Times

GENERAL JACKSON came back from his trip with the President confident that the matter of the Floridas was practically settled. He had not reckoned with the Spaniards. Sailing ships were slow mailcarriers; official letters were polite but vague. Two years passed before a treaty was signed and five million dollars purchase money paid to Spain. Only then did Spain deliver her lands in Florida to the United States.

These years were pleasant ones at the Hermitage. The relative quiet gave Jackson the best chance in years to regain health. The Dickinson bullet still gave trouble, but his shoulder and his general health improved.

The English gardener had made charming gardens and had laid out a stately driveway in the form of a guitar, Rachel's favorite instrument. The upper part of the guitar formed the single drive in from the road. Halfway to the house, this separated and curved gracefully in the shape of the instrument, with the rounded base at the main entrance of the house. Plantings of cedar trees edged the drive and enhanced its beauty.

Rachel and Andrew took great pleasure in their home.

The guest room was usually occupied and, as in the blockhouse, rich and poor were welcome.

One of the early guests was an itinerant portrait painter named Ralph Earl. Rachel took a fancy to his work and invited him to stay to paint the general. Shortly, he married one of her nieces and later, after his wife's death, came back to the Hermitage and made his home there.

Several times during the period of waiting for the Spanish treaty, there had been talk about Jackson being governor of the Florida territory. He did not want the position; he disliked administrative work. He wrote to Monroe asking that his name not be considered. Rachel was relieved when she saw that letter sent to the Nashville postoffice.

But before the semi-weekly post left for Washington, a letter from the President came. Jackson was appointed.

"But you'll not take it!" Rachel exclaimed.

"Mother, you know my rule. Never seek an office nor decline if you are chosen."

"But, Mr. Jackson . . ." she was so dismayed that the old name slipped out. "You wrote . . ."

"Oh, I know I did. I've sent a boy to town. I hope to get my letter back. I'd not like to disturb the President. And I have a reason for accepting. Perhaps I have a duty to friends who are poor since the depression. I can appoint officials down there. We can have a congenial group and serve our country well. And you and Andrew can go with me." His quick mind had arranged the whole business.

The trip by steamboat to New Orleans took eight days —

it seemed incredible speed. On the boat with them was their newly painted coach. Its curtained windows and seats upholstered in leather made it a proper vehicle for a governor and his family. New, well-made harness and four handsome horses assured speed on the journey from New Orleans to Pensacola.

Honors in the city pleased the general, but he was impatient to get to his work. So very soon they drove along the Gulf toward the new territory.

They had left the Hermitage in April when the place looked beautiful. The contrast along the Gulf was dismaying. Rains and mud bogged the coach. Mosquitoes in clouds were a horror. And Spanish codes of etiquette almost daunted the general. Almost, but not quite, for he finally disregarded codes.

On the 17th of July he officially entered the city. American soldiers stood at attention on the square while the Star Spangled Banner was sung. The Spanish flag was hauled down and the Stars and Stripes run up amid cheers.

Alas! His first official mail was distressing. The President had made important appointments — to the very offices the new governor had expected to fill. Not one was Jackson's choice. The dream of congenial associates vanished, even while he tackled administrative work he hated.

His thoughts turned to the Hermitage — why had he left it? He wrote a letter to his brother-in-law, John, hinting at his return. And he added:

"We are happy to learn that little Andrew Hutchings is contented. Say to him that his cousin Andrew junior will bring him a pretty when he returns, and I will buy him a pony."

Rachel wrote letters home, too. In one to a friend she wrote, "The general wants to get home as much as I do."

She was not surprised when he resigned in October, and they left for Tennessee.

However, there were compensations. She was with her husband, this time. And they stopped in New Orleans to shop for the new house. Boxes and bales went with them, by boat.

It was November when they got home. Fires blazed in every room, and Rachel found her household well. Unpacking the purchases was a joyful task. There was a beautiful mahogany bed with fluted trim, very elegant, and a French mattress to put on it. There was a handsome sideboard and dishes,

crystal — oh, many choice things. She bustled about making the house ready for the visitors who were sure to come.

Andrew Jackson was happy, too. In June, from the South, he had resigned his commission in the army. Now, since resigning the governorship, he had no civil or military office. He was free for the first time in thirty-two years, a refreshing thought.

"I'm fifty-four years old," he remarked to Overton, "and I'm through with traveling about the country. My health is always better at home; I mean to stay here."

Overton looked at him affectionately. He did look well.

"Come on to the office, John." Jackson led the way. "I've subscribed to twenty newspapers around the country. Some ought to be there now." The two friends went to the small building Jackson used for his business.

When Overton left, Jackson called the children in for lessons. Part of the home task he assigned to himself was the education of the lads under his care. The oldest, Andrew Donelson, had graduated from West Point a year before and was now in the army. Other nephews were away at school. But Jackson spent many hours teaching those who were at home, along with Andrew, junior, and his ward, Andrew Hutchings.

Old friends took to dropping in, to read Jackson's papers and to talk. There was quite a group, including Overton and John Coffee.

John Henry Eaton was United States Senator from nearby Franklin, Tennessee. Eaton was a lawyer and had written a biography of Jackson through the victory at New Orleans.

William Lewis, the quartermaster of war days, was a

wealthy planter, who liked people and had ample leisure. Lewis had a flare for politics, all the more marked because he wanted nothing for himself.

Together these and other friends talked over the affairs of their state and the nation. They agreed that times were changing fast.

Only a little more than thirty years had passed since George Washington selected his first cabinet. His distrust of political parties made him choose men of different ways of thinking, as Alexander Hamilton and Thomas Jefferson, for his advisers.

President Adams was a Federalist, like Washington, which at first only meant that he favored the federal Constitution. In Adams' term, people began to fear too much federal power as interpreted by Hamilton. Americans saw they did not want a big debt; they did want more state control of matters not covered by the Constitution.

Thomas Jefferson had been educating citizens along this way of thinking, and he was twice elected president. James Madison, who came after him, thought much as Jefferson did. They called themselves Republicans because they thought of government as a republic, a union of states, not a great federal power. Jefferson plainly said, "The best government is that which governs least."

By 1816, the end of Madison's terms, the Federal party had died out and the Republican party was changing. Some had not wanted the war. Others did not want a national bank lest it get too powerful. James Monroe was elected twice, but in the second campaign (1820) it became plain that things

were different. Men talking in Jackson's study wondered what
might happen next.

"People are tired of getting their presidents from one state,"
Eaton said one day. "Virginia has furnished eight terms out
of nine. A man the other day spoke of the 'Virginia dynasty'
as though it was a royal family."

"You can't break away from a kind of succession as long
as we have the caucus system for nominating."

"True," Lewis agreed. "But why have a caucus? Why
let a small group of party leaders in Congress call themselves
a caucus and nominate a president? They practically elect
him. The nation has grown. Other sections have good men."

"Why shouldn't states nominate? Why not let state legislatures put up a man for the nation to vote on?"

"While you're asking questions," Lewis remarked, "I'll ask,
too. Why not nominate Andy for president?"

"*Me!*" Jackson was shocked.

"The idea is not new to you, surely."

"Oh, I've heard it mentioned," the general admitted. "But
to have *you* say it — that's different. I know my limitations, and
so do you. I can lead in battle. But I hate running civilian
affairs. I want no office."

"The American people may decide differently," Lewis said.
But he did not argue — not then. Quietly he began to work.

The Tennessee legislature met in July of 1822 and nominated Andrew Jackson for the office of president of the United
States. This was the first time that a state had placed a candidate in nomination; it set a new precedent.

Jackson was pleased but calm. He thought it was a pleasant gesture to a native son. Rachel was startled.

She went to her room where Hannah, the faithful serving woman, was waiting to brush her hair, a daily custom. But this day, Mrs. Jackson was silent, almost sad. Hannah brushed — fifty strokes on this side, fifty strokes on the other. The silky, dark hair slipped over Rachel's shoulders.

"Something troubles you," Hannah remarked.

"Yes, it does, Hannah. I've just heard that the legislature wants the general to run for president."

"Presidents come from Virginny," Hannah said, to comfort.

"They don't *have* to, Hannah, though usually they have. But times are changing. Mr. Lewis says that every time he is here. And you never can tell what the general may do if people push on. *He* doesn't want any office. But William talks of his duty." Hannah brushed, silently.

"The general is needed here," Mrs. Jackson said thought-

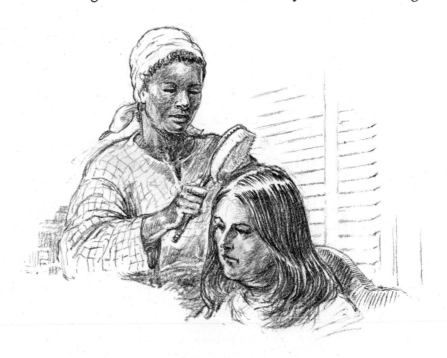

fully. "The boys need him. Andrew, junior, is so dear, but he is headstrong. I pray every day. We'll have to wait and see."

During the waiting time two events happened in the Jackson family, the first very personal.

Both Mr. and Mrs. Jackson had been Christian believers from childhood; they were brought up in the Presbyterian faith. During the period following the Battle of New Orleans, Peter Cartwright, the greatest evangelist of his time, preached in Tennessee. From there he went on into Kentucky and to Indiana. For twenty years he was a religious influence in the West.

The Jacksons and their neighbors heard Cartwright. He moved them to a more active practice of their faith. Grace before meals became the family custom; Rachel and Andrew read their Bibles often, not just at bedtime as before. And Rachel began to wish for a church near home because sometimes it was not possible to get in to Nashville for services.

Her wish was granted that year, 1823. The general, with the help of a few neighbors, built a small church across the road from the Hermitage. The simple building was attractive. Both the exterior and the floor inside were of red brick and the pews were painted white. Two great fireplaces made the room comfortable in winter. Such a sturdy building would last many years. Rachel was proud and pleased.

The second event of 1823 was not so agreeable. Jackson was again elected to the United States Senate.

"Get Andy to Washington," Lewis had said to John Eaton. "Let people know him. That ought to do it!" They were heart and soul in the work they had undertaken.

During the fall session of the Congress Jackson stayed in the Franklin House, a popular hotel between Washington and Georgetown. John Eaton had rooms there; the proprietor, William O'Neale, liked to have government officials as his guests. The oldest daughter, Margaret O'Neale Timberlake, called Peggy, was married and had two small, attractive children. She won the general's friendship with her children, her cheerfulness, and her music.

Rachel went to Washington with her husband for a later session. She felt she knew Peggy, for the general had written and talked of her earlier. Rachel approved of the young woman and was a kind friend.

General Jackson attended the Senate meetings faithfully. He talked little, but when he did speak he favored better roads, for their military value, and some armament. He heard Clay make a brilliant speech on protection. Clay said that it was the "American System" to charge a high duty on imports so as to protect and develop home industries. Jackson also heard Daniel Webster speak effectively against some of Clay's ideas.

It was plain that such topics made good speech material but there were only two real issues as the campaign neared; how to nominate a president and which to elect among many candidates.

The caucus had nominated William Crawford of Georgia in spite of the fact that he was seriously ill. Various states nominated Henry Clay of Kentucky. And of course the President, John Quincy Adams, hoped to be re-elected. Calhoun was Secretary of War, Clay was Speaker of the House of Representatives. All three would use their official positions toward winning.

Until he realized how the candidates were lined up, Andrew Jackson had been indifferent about running for office. "I'm not a politician," he told his friends.

But when he read that list of candidates his fighting blood was aroused. Here was a battle with votes, not guns. From that moment he was out against the whole field.

In March, with the campaign of 1824 growing, Pennsylvania endorsed Jackson. Other states followed this lead. Most of the speeches were about the caucus.

"Will you let a few men make your president?"

"It's the people against the politicians!"

"Down with the caucus!"

The exciting campaign among four men, all in one party, drew toward its end. The general had been home for some weeks. With Mrs. Jackson, he set out for Washington, and duties in the Senate, just as soon as he had voted. On the way he heard talk both for and against himself.

"That's a fine coach the general's got! Bet he rides straight to the White House."

"Thomas Jefferson rode a horse when *he* went to Washington," a disgruntled bystander remarked. Jackson did not appear to notice.

"Had any election news yet?" he asked the innkeeper.

"Not yet! Bet you'd like to hear, too."

Indeed, the general would like to know. He could hardly wait to get to Washington.

A Tragic Campaign

By THE TIME the Jacksons got to Washington, the general knew the unofficial result of the election. Official announcement must wait until Congress got the count in February. He had 99 electoral votes, Adams 84, Crawford 41, and Clay 37. No one was elected; the choice must be made by the representatives in Congress.

This recalled to many men the long days of voting when Thomas Jefferson was first elected president. As then, people stopped on the streets to discuss the election. In taverns, at parties, around home firesides it was the favored topic.

"Crawford is out. I hear he is still not able to speak."

"But he can throw his votes . . ."

"Can he? My guess is, those votes will be split up. It takes a strong man on the job to hold votes in a time like this."

"Clay holds the deciding power. I'm astonished he didn't make a better showing, himself. But he can make the next president — and he will."

"As fourth man he's out of the running, isn't he?"

"Yes, the top three are up for voting. But Clay will never turn his votes to Jackson — his rival from the West!"

"The man Clay favors will be the next president!"

Many agreed on this, though new rumors turned up every day. About the last of January there was a persistent tale that Clay had agreed to throw his votes to Adams and in return Adams promised to make Clay his secretary of state. This would put Clay in line for the presidency.

"Clay's a smart man. He's got a deal — you'll see." This was conceded by many astute observers.

Jackson and his followers heard the story. But they could not quite believe that the final election would go against a popular choice. Jackson's popular vote was 152,901 while Adams had 114,023 votes. The city buzzed with talk.

On the ninth of February the House of Representatives convened; ballot boxes were opened and votes counted. The next business was to vote on the three highest names. There were twenty-four states; a candidate must get thirteen votes to win.

A hush fell over the assembly as Daniel Webster rose to announce the result of the first ballot; Adams thirteen, Jackson seven, Crawford four. John Quincy Adams was to be the next president. The quick ending astonished most men.

That evening at a party the defeated General Jackson happened to meet Mr. Adams face to face. Onlookers stared, expecting a scene. There was none.

"How do you do, Mr. Adams?" Jackson said graciously.

"Very well, sir." Adams' tone was frigid. "I hope General Jackson is well." The two bowed and passed on. Startled observers thought the general was the calmer of the two.

On the fourth of March, Adams took the oath of office.

Three days later he announced Henry Clay as his secretary of state.

President Adams' term of office was a troubled time. He was a minority president; put in office against an expressed preference of the voters. He was surrounded by men who angled to succeed him. It was a pity, for Adams was an able man, and loyal. At another time, he might have been been one of the great American presidents.

People quickly took it for granted that Andrew Jackson would run again. The general, so lukewarm at first, now determined to give the people the man they wanted. On his way home from Washington that spring of 1825, towns and villages turned out to greet him. His campaign really began then.

At home, neighbors flocked to encourage him. Plans were made; newspapers, committees, regions, were organized. The keynote was the "people" against the "politicians"; which should run the country?

Meanwhile there was an agreeable interruption.

One of the pleasures of the Washington season had been the Jacksons' friendship with General Marquis de Lafayette. The French officer was visiting in America and while at the capital naturally met the American hero-general. The men developed a real friendship. When, later, Lafayette went south to visit Jefferson, he traveled to Nashville to see General Jackson at the Hermitage. Nashville people got up a parade, and a great banquet as the happy days of visiting drew toward a close.

Then people's thoughts turned back to politics. Jackson

resigned from the Senate to be free. In October the legislature of Tennessee nominated him for president. Now he had official backing — three years ahead. As Jackson and his intimate friends looked over the scene they saw that his supporters were as mixed a group as his soldiers had been at the battle of New Orleans, ten years earlier.

The West was largely for Jackson: he stood for them as no public man before had done. Followers of Calhoun and Crawford joined, though they would have preferred a southern man. Later Martin Van Buren, who had tried his lot with Adams, came to see Jackson and to offer his services. Van Buren was a New Yorker and a master of politics. Andrew Donelson, who had resigned from the army and turned to law, came to stay near Jackson and act as his secretary. He was called "Jack" because there were so many Andrews; he would be a great help. Jackson's forces seemed strong and vigorous.

But there was strength against them. The campaign was already personal. Men spoke of themselves as Jackson men or Adams men. The division in the party was shown by new words, Democrat-Republicans, Jackson; and National-Republicans, Adams. The personal nature of the contest was a handicap to President Adams. Jackson men held a majority in Congress and they blocked many of Adams' moves, adding to his difficulties.

In the midst of all this, sadness came to the Jacksons. The Creek Indian lad, Lincoyer, had grown to be a tall, intelligent youth, much loved at the Hermitage. In his late teens he contracted tuberculosis and died — the first death in Jackson's family. He was buried with honor in the family lot in the garden. Grief over his death made the political strife seem harder to bear.

It soon was evident that the campaign, already personal, would be violent, even bitter. Jackson had known that in 1824 the old story of Robards' belated divorce was printed here and there. He had tried to keep the matter from Rachel. Now he saw that protection was going to be impossible.

"It is better that you know from me, Mother," he told her frankly. "William is going down to Natchez to search for legal papers on our marriage there. John Overton promises to write out a true account that we hope will stop the talk." Tears rolled down her face, but she said nothing. What were words? Couldn't people see their thirty-five years together?

William Lewis stayed in Natchez for some weeks searching. He confirmed Rachel's visit with the Greens and her marriage in their home. But he found no legal records. He hurried home to confer with John Overton.

"Could it be that the Spanish governor, when he left, took all papers with him?" Overton wondered. "Perhaps he sent them to Spain? That would be a logical act."

"But not one that will help us now," Lewis said. "Well, do your best, John. If I had thought this would come up. . . ."

"Mrs. Jackson has courage," Overton said.

"Looks as though she will need it." Sick at heart, Lewis picked up a paper from the east; false tales and hard words were spread over the front page — adulteress, bigamist, loose woman. Jackson himself was called atheist, liar, wife stealer.

John Overton wrote a straightforward account of the facts of the divorce and marriage. He had it sworn and sent out copies. But the attacks continued. Many of the worst seemed to come from Kentucky. Tales of the Dickinson duel, the Benton quarrel, John Woods' sentence and the execution of mutineers in the Indian wars were spread across the pages in lurid, exaggerated phrases.

In the midst of this, Harry Lee, son of the Virginia Lee family, George Washington's friends, came to Nashville and offered his help with the heavy correspondence. He stayed with William Lewis, though most of the time they worked at the Hermitage. Jackson was grateful for his help.

As the day of voting drew near, people around Jackson began to feel that, in spite of the evil talk, he would win. Eastern politicians had overlooked the changed requirements for voting. The constitutions of many new states gave a man without property the right to vote. Thousands of these new voters were westerners — and for Jackson. Beginning with 1824, and increasing each year, these new voters were a powerful influence in local, state, and national elections.

Tennessee returns came in first; in many towns and villages every vote cast was for Jackson. In the whole state, only three thousand were against him. Soon it was found that he had carried every Southern state and every state west of the Alleghenies. In the final count he had 178 electoral votes to Adams' 83 — an amazing victory.

Friends congratulated Rachel as well as the general.

"Isn't it thrilling!" they said.

"For Mr. Jackson's sake, I am glad he got it. For myself, I never wanted it." Those were true words.

Even before the unofficial returns were all in, men rushed to Nashville to congratulate the general and offer their services. Visitors found that townspeople were planning a great banquet on December 23rd, the anniversary of the first of the battles at New Orleans. Rachel, as first lady, was to share honors in the feast of food and oratory.

The general was happy with victory, but the edge of joy was dulled by anxiety for his wife. She had not been well; the strain of the campaign had made her worse. The doctor ordered quiet — but how could one be quiet in that overflowing house? Or now, with clothes to prepare and the long journey to Washington ahead.

On the 17th of December Mrs. Jackson had a severe heart attack. The general was panic-stricken. He stayed close by her bed as Hannah nursed his wife tenderly; he would see no one, not even the nieces.

Downstairs the kinfolk wondered.

"I heard she was in town; did she hear more of that campaign trash?"

"Hannah says she was not away from home all day."

"What's the difference? She's ill. Poor Uncle Jackson!"

Hours later, when she could speak, Rachel urged the general to rest. "People are counting on you for the banquet." But what was a banquet compared with Rachel. He stayed by her side.

In a few days she seemed to rally. On the 22nd she insisted that the general go in the next room and rest. He had just dropped to sleep when a scream awakened him.

"Come quick! She gone! Miss Rachel done gone!" Hannah was holding her, and sobbing.

"She isn't! She can't be!" The general shouted for help. He sent for the doctor; he tried to revive Rachel. But she was gone.

News came to the city as early morning workers were decorating the banquet room. They took down gay streamers and hung black crepe. The whole town mourned. The next afternoon, the eve of Christmas, everyone who could ride, drive, or walk went to the Hermitage. Silent and weeping, they saw Rachel Jackson buried in the garden she loved. She had been

"Aunt Rachel" to scores, a friend to hundreds who felt a personal loss.

Those who were near enough to see the general were shocked.

"He's aged twenty years since last week," many observed.

"Being president will mean nothing to him, now."

Andrew Jackson stood silent as the grave was filled. Then friends got the broken-hearted man into the house.

These friends had a task before them. In three weeks he must leave for Washington — the city of gaiety and intrigue. Letters came by the sackful. Visitors by the score crowded the house of a man whose only thought was his grief.

"It's lucky that the general had his inauguration address outlined. We had talked it over." Jack Donelson and Lewis conferred on work to be done.

"Yes, we both knew just what he meant to say: economy in government; no debt. Guard the public funds. Proper regard for the rights of the states and a fair Indian policy."

"You write it out, Jack; I'll go over it; and Harry can polish it up — he's the writer of this crowd." Later, Jackson made only slight changes in the paper they had ready for him.

In mid-January the President-elect and his party boarded a steamboat for part of the journey to Washington. As Jackson drove to the dock, William Lewis eyed him anxiously.

"This is not easy, General," he said. "But you can do it."

Jackson lifted his head, proudly. "I know my duty."

It was a comfort to have these friends with him. Jack Donelson was to be his personal secretary; Jack's wife, Emily, one of Rachel's nieces, would be mistress of the White House.

The river trip was one long ovation. At villages and towns people crowded the dock, begging for a glimpse of the general. Between towns, farm folk sat on the banks for hours, in spite of the cold, hoping to see their new President.

In a month Jackson was settled in Washington. He saw all who came to his hotel rooms — but he told nothing. The capital city fairly buzzed with guessing.

Daniel Webster wrote home: "Nobody knows what he will do. Efforts are making to put him up to a general sweep of all offices." But Webster was guessing like the rest. Later, he wrote:

"A great multitude, too many to be fed without a miracle, are already in the city, hungry for office." Then he listed a few men who had worked for Jackson, "Our friend Isaac Hill from New Hampshire, Mr. Noah from New York, Mr. Kendall from Kentucky. But nothing is settled . . . Probably General Jackson will make some removals but I think not too many immediately."

The fourth of March was a beautiful sunny day, a blessing for the crowds. The whole nation seemed to be there, massed in a sea of faces before the east side of the Capitol. Carriages could hardly move through the crowded streets. Jackson walked.

President Monroe was the first to hold the inauguration ceremony on the east portico. This beautiful day was perfect for continuing the custom.

When Andrew Jackson and a few others stepped into sight, a mighty roar of greeting reverberated against the building.

Then there was a hush as the chief justice administered the oath of office. People tried to see the central figure — they studied him eagerly, lovingly, curiously, according to their reason for being there.

They saw a tall, lean man who would be sixty-two in a few days, but who looked very much older. The high hat he had laid aside was banded in crepe; a ten-inch band of crepe was stitched to his left sleeve. For a moment, as he spoke the solemn words, his eyes sparkled. Was he thinking of his mother? She had sensed something special in her youngest son; she would have been proud today.

The President began to read his address, his voice clear and steady. Grief had bowed him; nothing could defeat him. Underneath his coat he could feel the touch of Rachel's miniature, which he wore daily; it made her seem nearer. Her Bible, read night and morning, gave him strength for today.

Only a few of that vast crowd could possibly hear his voice. But all stood politely quiet. They applauded at the end — and dashed for the White House and the promised reception. A horse awaited the President, but he had difficulty going even at a slow pace.

The people had taken possession of the White House. Friends locked arms to get the President inside; then they pushed him on through and out the back door. He went to bed, exhausted, in his room at the hotel.

The crush was frightful. Furniture, dishes, pictures were broken. Servants carried a great bowl full of punch out onto the lawn. Crowds broke windows in the rush to reach it. "The people" had taken over with vigor. As hours passed and the crowd drifted away, thoughtful citizens began to wonder anxiously.

Was this mass today, representative of America? What did they really want of their President? And what would he do? No one in all the nation could answer that.

The Challenge of Office

Washington buzzed with talk when the President's cabinet appointments were announced. Men gathered to discuss his choices.

"Van Buren might make a good Secretary of State but he is governor of New York. Think he will resign that?"

"Oh, yes. He would rather be in Washington."

"Eaton is a natural choice for Secretary of War — he's an old friend of General Jackson. But the others — Ingham, Branch, and Berrien — are southerners. They're for Calhoun. John wouldn't want his vice-presidency again, except that it puts him in line for president in 1832. These cabinet appointments ought to make that certain."

"Look! Every man on this list is against Clay. The general does not forgive him for the campaign slander, does he?"

"Clay has repeatedly denied that he started the talk against Mrs. Jackson. Doesn't the general believe him?"

"Frankly, no," a close friend of Jackson replied. "The worst of the slander came from Kentucky. If Clay didn't start it, at least he didn't stop it, and he might have — he is powerful. The general is usually generous, but he will never forgive that wrong to his wife."

"Clay has presidential ambitions as well as Calhoun. This cabinet will not help *him!*" Men smiled knowingly.

William Lewis's name was not on the list; he had declined any appointment.

"I'm off for home soon, General," Lewis said after the inauguration. "You know that was my plan."

"But I need you, William!" Jackson looked so stricken that Lewis accepted a minor post and stayed on at the White House. Soon he gathered around him a group of able men who came to be known as the Kitchen Cabinet — back door men, who had influence in promoting the President's ideas of government.

They were interesting men. One of the strongest was Amos Kendall, once a timid New England farm boy. He graduated from Dartmouth and went out to Kentucky where he tutored Henry Clay's children, studied law, and became a newspaper editor. Through these experiences he lost timidity, gained ease and knowledge, and became a powerful political influence. He broke with Clay before 1828 and came to Washington to support Jackson's campaign.

Another Kitchen Cabinet man was Isaac Hill — "Ike" — of New Hampshire. As a lad he was poor, and a cripple; he learned to write while working in a small print shop. By the time he came to Washington, he was writing brilliant newspaper articles which sparkled with wit and sarcasm — against Andrew Jackson's opponents.

Along with him was Francis Blair, whose indifference to dress tended to obscure his fine education and background. On Blair's first call at the White House, Jackson enjoyed him so much he asked him to stay to dinner. Supposing they were to be alone, Blair accepted — only to discover that he was attending an affair of state, where his shabby clothes were in marked contrast to the official regalia around him. Jackson put Blair beside him, and they talked all through dinner.

Blair quickly became a devoted worker for the President's policies. His fine literary style was evident in many state papers and newspaper articles. These three men made the administration newspaper, *The Globe,* a powerful national influence in support of the President. With Lewis, Jack Donelson, and occasionally others, they worked hard for Jackson.

The machinery of the new administration had hardly got underway before the problem of patronage threatened its success. This idea, that government office should be given to a man who worked during a campaign, was not new. New York and Pennsylvania had been through many a patronage battle in state affairs. The phrase, "To the victors belong the spoils," which many people heard now and thought referred to Jackson's actions, originated earlier in New York state.

Jackson's problem was really acute. Many of the crowds who came to Washington for the inauguration stayed on, expecting to get government jobs. Jackson, when he promised to improve civil service, had not expected so many applicants or so much pressure on him to make appointments. Finally he conferred anxiously with Lewis.

"Men talk to me, too," Lewis said. "They remind me that they voted for a change and they expect to get it. They say some government employees have held the same job for sixteen

years. You're to turn these out and appoint new men who have
a right to work under you."

"No man has a *right* to a public office," Jackson said sternly.
"Neither the man who is in, nor the man who is out. Working
for the government is not the only way of earning a living."

But after this talk, the President began a rather reckless
course of firing and hiring. Some of his changes improved the
service; others did not. Because of this recklessness, many
thought Jackson made more changes than he actually did make.
In his entire presidency he changed only about twenty per cent
of officeholders — not a startling record.

The patronage storm was still raging when a small cloud
appeared on the Washington social scene.

In January, before the inauguration, John Eaton, with the
President's approval, had married the widow, Peggy O'Neale
Timberlake. The Jacksons had known and liked Peggy when
they lived at the O'Neale hotel during Jackson's time in the
senate. Since then Jackson had heard gossip about Eaton's lik-
ing for Peggy, and he was pleased when the two were married.

At the time, the newlyweds were hardly noticed in Washington. But after Eaton, as Secretary of War, became a cabinet
member, Washington social leaders raised their hands in horror.

"We can't associate with an innkeeper's daughter!"

"She's been talked about, too!" others said with lifted
brows. Social queens snubbed Peggy at the early parties in the
fall of 1829, and the matter came to the President's attention.

"Another fine woman slandered!" Jackson roared, roused
as he had not been since Rachel's death. He gave no thought to

the difference in background or the ideas involved; it was, he said, a repetition of the slanderous talk about Rachel. "Those women *shall* receive Peggy Eaton. Mrs. Jackson loved her. I know she is good."

He gave orders which he supposed settled the matter. Certainly he had other business needing his attention.

In December, Jackson's first message to Congress was due. He had a very personal way of getting that message written. In general he knew what he wanted to say to the lawmakers. So when he had a thought, he wrote it down on any handy bit of paper and tossed it into his tall hat on his desk. Some notes

were on the margin of a newspaper; some covered two or three pages of legal paper. As the time for delivering the message drew near, Jackson dumped out the contents of his hat and called Jack Donelson.

"There it is, Jack," he said. "Make it sound right." Jack had quite a job, but he did it well.

When the work was ready, the President read the result. He crossed out bits, he added here and there, using the steel pen he had accepted when he came to Washington. Then the message was discussed with his cabinet and the Kitchen Cabinet. And finally Jackson went over it again.

Congress and the nation eagerly awaited this first official word from their leader. In voting for him they hardly cared what he was "for"; they only knew they wanted him.

The tone of the message was calm and deliberate. It told of national strength and prosperity; of the intelligence and well-being of the people. Then it turned to foreign affairs.

"We shall cause all our rights to be respected," Jackson had written. And then it mentioned Britain: "Everything in the condition and history of the two nations is calculated to inspire sentiments of mutual respect, and to carry convictions in the minds of both that it is their policy to preserve the most cordial relations."

Many people were astonished at these mild words about Jackson's once bitter enemy. Those who knew him well, understood; they had heard him say that Britain was respectful to the United States only after the defeat at New Orleans. It was a victor's hand that was now extended in friendship.

The message asked for an improved manner of nominating a president. The caucus was dead; even the word began to have an unpleasant flavor. He asked that the office be limited to four or six years, as long tenure in public office was corrupting.

Tariff was mentioned vaguely; the nation's finances were reported in good order. The debt would be paid soon. After that, surplus revenues could be used for making roads and canals. The message suggested distrust of the United States Bank and that perhaps the government should have its own bank for its own business.

Jackson kept a watchful eye on Congress as it went to work. A trusted friend was present at every session, and the President never went to bed until he had listened to the report of the day's actions. So he was not really surprised when the third knotty problem of his first year, tariff, came up.

That word, tariff, was often seen in newspapers. But many citizens hardly bothered to understand that it meant a tax, a "duty," to be paid on goods brought into the country. Tariff had been started to get money for the young nation, and to make foreign goods more expensive than American. This was called "protecting home industry." But *how much tariff to charge* was a problem. A man's opinion depended on where he lived.

New Englanders wanted a high tariff.

"How can factories make money when Americans buy cheap things made abroad?" they asked. "Slap on a high duty and make people buy home goods."

Southerners disagreed. They sold cotton abroad, bought

things overseas, and had very few factories of their own.

"Why should we pay a high duty?" they asked. "We don't need protection. We like to trade abroad. Let northerners do as they please. We'll have low tariff."

Unexpectedly a bill about another matter brought the dispute into the open. This bill proposed to restrict land sales in the west. Westerners wanted settlers so they were against the bill. Easterners wanted to check the movement west so they were for the bill.

Southerners were quick to turn the dispute to help their tariff problem. They declared that no section of the nation had a right to impose a law on any other section. A state could do as it wanted, regardless of federal law, because a state was more important than the union of states.

Daniel Webster, a New Englander, chanced to enter the Senate chamber one day as Senator Hayne of South Carolina was ending an impassioned speech favoring states' rights. On the spur of the moment, Webster took the floor and talked on the side of a strong union — and against states' rights. He closed with a bold challenge to Hayne to reply.

Hayne accepted the chance to explain his views.

On the 21st of January, 1830, the galleries were packed when the southern orator rose to make his speech. Many people had not studied the arguments for and against states' rights. Now they listened intently as Hayne spoke of the problems of Kentucky and Virginia in 1798, and of New England, when she opposed the war of 1812 and threatened to secede.

"Is the federal government to be the judge of its own

power?" Hayne cried. "Is it without limitations? High tariff would ruin the South!" He spoke well as he answered his own questions.

Five days later the chamber was again crowded when Daniel Webster replied. He was a tall, dignified man; many thought him the most brilliant speaker of his time.

"The Constitution by the will of the people is the supreme law of the land," he said. "It must be obeyed or changed." His audience listened intently through the four hours of his speech. He told them that the Constitution had created a supreme court; it, not the states, should decide what laws were right. He ended with ringing words:

"While the union lasts we have high prospects spread out before us, for us and for our children. Hayne says, 'liberty first and union afterward.' I speak another sentiment, dear to every American heart — Liberty AND union, now and forever."

These brilliant speeches were reprinted in newspapers over the nation. They were read by thousands — in colleges, in homes, in debating societies, in rural groups, in Gentry's crossroads store in southern Indiana. There a young clerk, Abe Lincoln, got the paper and read both addresses to men gathered to hear the news. Like many other Americans, they had given no thought to the meaning of their Constitution; now they did.

Through February and March of 1830, talk surged in Washington. The mystery no one could solve was, "Which side is the President on?" Both factions claimed him.

April came, and with it the annual birthday banquet in honor of Thomas Jefferson, who had died four years earlier. South Carolinians had charge of the affair, and they made sure that the speakers and the toasts would favor states' rights. Of course the President would speak.

"Have you any idea what he'll say?" a committeeman asked.

"Oh, he's for states' rights. I know that."

"But do you? Can you prove it? Can you quote him?"

No, they could not prove; they could not quote. The President had been very silent.

The guests gathered. The dinner was served. The time for speeches and toasts drew near. The President sat silent and thoughtful. He was introduced — and for seconds he did not move. Then he unfolded his long legs and stood tall before them. He lifted a glass, cast his eyes around the hushed room — and spoke fateful words:

"Our federal union, it must be preserved!"

Men gasped. Calhoun leaped to his feet and tried to save the day for the south.

"The union, next to liberty, most dear! May we all remember that it can only be preserved by respecting the rights of states and distributing equally the benefits and burthens of the union."

Calhoun meant well. But too many words are never as effective as a few. Jackson's terse, honest statement would be remembered; in it he had boldly dealt states' rights a staggering blow.

President
Jackson

AFFAIRS OF STATE were engrossing during Andrew Jackson's years in Washington, but he found time for his beloved family. The Hermitage had been a center for young people. Now, undeterred by the long journey, they came to see Uncle Jackson.

Emily, Mrs. Jack Donelson, made a charming hostess; and usually one or more of Rachel's nieces were with her. William Lewis and John Coffee both had beautiful daughters who joined the White House circle. Army and Navy officers, and government young men, flocked around them like bees over a clover field. Jackson loved the gaiety and romance they brought to the big house.

In the late summer of 1830, the President and quite a party of friends went home for a visit. Mary Ann Lewis was left as temporary mistress of the executive mansion.

An unusual happening prompted Jackson to take this trip. Choctaw chiefs had invited him to come to a treaty-making council, and he had accepted. This would be the first time that a president had met with an Indian tribe. Jackson felt that too many treaties with the United States had been heedlessly broken, and he hoped by his presence to make this one impressive to his fellow-countrymen.

In the years since 1814 when the power of the Creeks ended, several tribes of Creeks, Choctaws, Cherokees, and Chickasaws had kept their promises and had learned to live the white man's way. A Cherokee, Sequoya, had invented and taught a system of writing; that tribe was the most advanced of all the southern Indians, in arts, and in practical living.

But, strangely, the Indian growth toward civilization disturbed the whites. They feared the Indians might keep lands the whites wanted for themselves. Georgia broke several treaties in defiance of federal law, just as South Carolina proposed to defy federal law about tariff. Jackson had come to think that the only solution was for the Indians to move west, and that he was the one to persuade them. Perhaps this move, like the hard treaty with the Creeks, might, in the end, prove wise.

At the council, the President addressed the Indians through an interpreter.

"Old men," he told them, "lead your children to a land of peace before you die! Young men, preserve your people and your nation! This chance may not come again. State the terms you consider just."

Reluctantly they agreed. In a few years practically all Indians were moved westward. There seemed nothing else for them to do.

While the President was at home he gave thought to enlarging his home. The Hermitage had been crowded with visitors; more room was needed. So he made plans for remodeling which were carried out in the spring of 1831. The change included two new wings; one adding a gracious dining room, pantries, and covered passage to the kitchen; the other, on the garden side, a bedroom and study. A spacious two-story veranda was built on, too. Rachel would have enjoyed these changes, he thought.

From the peace of the Hermitage, Jackson returned to Washington to find trouble about Peggy Eaton. As the social season opened, the cabinet ladies, led by Mrs. Calhoun, wife of the vice-president, presented a solid front against Mrs. Eaton. They would not invite her to their parties; they would not at-

tend social affairs where she was invited — Peggy always accepted.

The President stormed; but the ladies defied him. John Calhoun saw his chance to be president was hurt. To make matters worse, the President discovered that it was Calhoun, not William Crawford as he had supposed, who had made trouble for Jackson in the Indian War days. The Calhouns were so out of favor at the White House that he began to consider resigning his office.

Secretary of State Van Buren was a widower so not concerned with feminine quarrels. He attended every party and was gracious to Mrs. Eaton. This pleased Jackson, and Van Buren's prestige went up as Calhoun's declined. But the social situation was so intense that Van Buren saw it should not continue. One day, while riding with the President, he made a bold move.

"I'm resigning from the cabinet," he remarked.

"You can't!" Jackson was shocked. "I won't let you."

"You can't let this dissension in your cabinet continue," Van Buren said. "If I resign, I can persuade the others to do the same. Surely you do not want three Calhoun men *now!* You must see, General, that I speak for the good of the country." He talked so persuasively that Jackson agreed.

Other cabinet men did not wish to resign, but Van Buren persuaded each one. Calhoun's lost prestige made people say that Van Buren should run with Jackson in the next election.

Jackson appointed a new cabinet which included three men of great distinction. Edward Livingston, the patriot of New

Orleans' days, was made secretary of state; happily his brilliant, kindly wife could be counted on to restore peace in Washington. General Cass, the hero of the Michigan frontier, was appointed secretary of war. And Roger Taney, one of the great lawyers of his time, was attorney general until two years later Jackson made him Chief Justice of the Supreme Court.

John Eaton was appointed governor of Florida and later minister to Spain, where the pretty Peggy was a great success. But the cabinet ladies won their point; she left Washington.

In the midst of state affairs, the President often wondered whether the White House, with its gaiety and flattery was a good place for young Andrew. The youth was popular and charming, but very susceptible to romance with its delights and miseries. Andrew was visiting in Philadelphia when he chanced to see Sarah York and to fall violently in love with her. Sarah was a Quaker, an orphan, and a very fine young woman. They were married and drove to the White House where the President gave a large reception in their honor.

From his first glimpse of Sarah, Jackson loved her; she returned his affection. In a few days she began dropping in at night to see that he was comfortable. She was the only one who read to him from Rachel's Bible; who knew that Rachel's miniature was propped on the bedside table each night so he could see it on wakening. Sarah was one of his few real joys after Rachel died.

The year 1832 was to be a memorable one for the United States. It was a campaign year for the election of a president. South Carolina and the tariff would be debated. The charter for the United States bank was pending and other matters important to the nation were coming up. Andrew Jackson would have many hard problems to solve.

The President's friends were much concerned about his health. They had understood his grief and hoped that time would ease his heartache. But time only made worse his trouble from the bullets he carried. In January, he was persuaded to let a Philadelphia surgeon examine him.

"I dare not touch that bullet near your heart," the surgeon told him. This was the Dickinson bullet. "But the one in your shoulder could be removed."

"Then out with it!" Jackson ordered. He bared his shoulder and took a firm grip on his walking stick. The surgeon cut — and in a few minutes the Benton bullet was out.

Sarah wanted to nurse the President, but he would not be pampered. He attended a state dinner that evening and soon could use his shoulder without pain. His health improved; his springy step returned. Friends rejoiced to see him so well for the campaign ahead.

Before long Sarah and Andrew left to live at the remodeled Hermitage where he was to be the manager. The plantation missed Mrs. Jackson's skilled direction; the income had declined since the President could give it little attention. He hoped Andrew would do well with it; all his own thought was for his country.

When Jackson became president, he already distrusted the United States Bank. This was a private bank, that got its name and semi-official backing because the government was part-owner and kept public money on deposit. The bank had a twenty-year charter from 1816, and wealthy Nicholas Biddle was its president. Jackson talked the problem over with friends.

"I'm no economist, you know," he said. "But I think the bank and Biddle have too much power. I distrust money power."

"The banks works out well," Lewis remarked.

"Maybe now," Jackson granted. "But Biddle could wreck the country if he took a notion. He is not accountable to the voters. They can't put him out. The bank should not be re-chartered."

"What would you do with government funds, General?"

"Keep them in state banks until we find a better way."

Word got around — and Biddle flooded Washington with propaganda leaflets. So great was his influence that in July both houses of Congress passed a new charter and sent it up for the President's signature.

While friends gave Biddle a victory dinner, Jackson called a cabinet meeting.

"I shall veto that bill," he told them sternly.

There was a storm of protest.

"You dare not, Mr. President! This is a campaign year!"

"If you must veto," someone suggested, "send a mild veto message so another bill can be introduced."

But the President refused advice. His veto message was a vigorous statement of his views; the bank was unconstitutional, it gave too much power into private hands. The message expressed concern for all men, not just the more favored, more gifted. These were entitled to protection for their labors, but the law must also protect those who lacked advantages. Law must give protection to all. "In the act before me," the message ended, "there seems to be a wide departure from these just principles."

Congress did not pass the bank charter over the veto, so the question was laid before the voters.

In the midst of the bank fight, Jackson was warned that the problem of states' rights was stirring again.

"We ought to go down to South Carolina and settle that idea once and for all!" the President exclaimed angrily.

Many citizens certainly would have approved drastic action when it was known that the South Carolina delegation in Congress voted to nullify federal tariff laws. The President was furious.

"This act doesn't prove anything!" Men in a conference group at the White House tried to sooth him. "It's just talk."

"Oh, I let them talk and threaten to their hearts' content," the President retorted. "Free speech is not treason.

"But just let them *act!*" he roared angrily. "If *one* drop of blood is shed in defiance of the law of the United States, I'll hang the first man of them that I can get my hands on to the first tree that I can find!"

His vivid warning was not heeded — and the campaign went on, a bitter battle with words for American votes.

Henry Clay was the presidential nominee of the National-Republican party — which soon was to be known as the Whigs. Van Buren was running with Jackson for the Democrat-Republicans; in this campaign the double word was dropped, and Jackson followers called themselves Democrats. Americans watched the contest anxiously as the fateful day of voting drew near.

The voters spoke decisively; they gave Jackson 219 electoral votes and Clay 49.

The President took this as a mandate to act. He ordered General Scott to keep southern forts ready, and he had the navy stay near Charleston. Then he waited.

During this period, General Sam Dale, who had been a messenger at the battle of New Orleans, called on the President. They talked of those exciting days and then of present trouble and the danger to the union. As the President walked to the door with his guest, puffing vigorously at his pipe, he said:

"If this thing goes on, Sam, our country will be like a bag of meal tied in the middle and open at both ends. Pick it up, and the meal will run out. I have to tie up that bag!"

"I hope things will turn out all right, Sir," Dale said.

"They shall go right!" Jackson shouted, and threw his pipe onto the floor violently. The clay pipe broke with a shattering sound. Dale grinned. The general hadn't changed.

After the election South Carolina passed a Nullification Act. Jackson decided to write a proclamation to the people, giving his ideas about this dangerous act. He had faith that they would understand him and make known their approval. He drove his pen furiously over the paper, tossing sheet after sheet to the floor to dry.

"There! That's what I want to say! The Union must be

preserved, without blood if this is possible," he told Livingston. "Fix it so it sounds right."

Livingston gave the paper a bit of scholarly polish, but the ideas were Jackson's. The most vital phrases were put in bold type when editors printed it in the news:

"I consider the power to annul a law of the United States, assumed by one state, incompatible with the existence of the Union, contradicted expressly by the letter of the Constitution, unauthorized by the spirit, inconsistent with every principle on which it was founded, and destructive of the great objective for which it was formed."

A large part of the nation was aroused by this defense of its pride and dignity. Excitement was so great that Jackson's second inauguration was hardly noticed. There was no reception; he was far too busy. As the speeches and newspaper articles continued, both sides hoped that the dispute would be resolved by a conclusive victory.

In Jackson's next message to Congress he reported the excellent state of the nation's finances. Debt was about paid; soon the treasury would have a surplus. High tariff was not needed for income, and so duties were gradually reduced.

Peace came with this compromise. Both sides claimed victory. The North said that federal law was upheld. The South said that tariff had come down; they would obey federal law — but on their own terms. The final solution was shrouded in the mists of the future.

Through all this time the bank matter was unsettled. Jackson felt that his majority of votes showed the people were

against the bank, as he was. But he knew sudden changes were
bad for business. So in the fall of 1833 he very gradually began
to deposit public money in various state banks. Clay, Biddle,
and their like, set up a great cry — "Jackson is ruining the
nation!" Biddle stopped loans to business. Factories closed.
Men were thrown out of work; many were hungry.

"See what your President has done!" Biddle said. "This
is what happens when he plays with public money."

Delegations came to Washington to beg Jackson to put
public funds back into Biddle's bank. At first the President saw
them and politely tried to explain.

"Biddle is proving the very thing I said! He has the power
to wreck business — now he is doing it."

More visitors came — and listened indifferently. Jackson

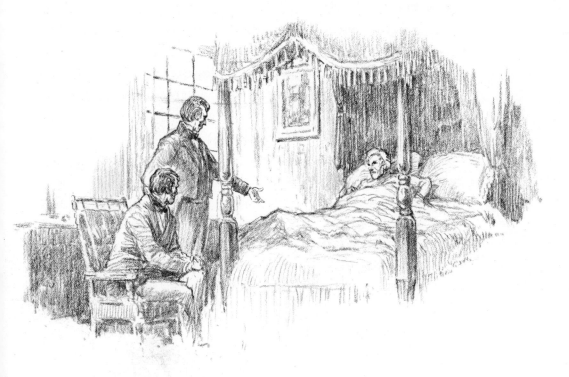

suspected that Biddle or his cohorts sent them. The delegations went to Congress, too; the Capitol was in an uproar day after day.

In the midst of it, Jackson had another lung hemorrhage and had to take to his bed. He received the delegations just the same — but not quietly, as before.

"You ask me for money?" he shouted at them. "Go to Biddle! *He* has money, plenty of it! The bank is a monster, sucking the country dry."

The visitors hurried away, startled, thoughtful.

"You put on quite an act, General," Lewis remarked as Jackson sank back on his pillow.

"Have to do it, sometimes," the President admitted. "I wish Congress would come calling. I have plenty I could say to *them!*"

Congress did not come. Instead the Senate passed a bill censuring the President for exceeding his authority in the new way of handling of public money. Jackson men were furious at this affront to their hero — but the resolution of censure stood.

Biddle, bankers, Whig leaders, Congress, begged and threatened, but the sick man in the White House held firmly to his course. His body was weak; but his will was iron.

By the spring of 1834 the worst was over. Biddle, not the President, gave way. Money became easier. Business picked up. The first difficult step was taken. Someday this would lead to a sound system for managing the people's money.

Almost at the end of Jackson's second term, a dramatic

scene was enacted in the Senate. The old record book was brought into the chamber. Before the eyes of all, the secretary drew black lines on a page and wrote across the resolution of censure: *"Expunged by order of the Senate."*

Jackson men relaxed. Their hero was vindicated.

Andrew Jackson was pleased to turn the country over to his chosen successor, Martin Van Buren. He declined the new President's invitation to stay on — he was needed at home.

On the 22nd of February, before Van Buren's inauguration, President Jackson gave a party at the White House. He was still frail, so he received his hundreds of guests seated in a big chair. Emily stood by him, gracious and charming; and nearby was Van Buren, smiling and happy.

An admirer had sent Jackson a huge cheese, even larger than the one given Thomas Jefferson on a similar occasion. The President wished to share it with his guests — many had heard this and came prepared with napkins for wrappings. Others took pieces, cut with a saw-knife, and went around the White House rooms munching happily. The party was a success.

But the final leave-taking was for intimate friends. Jackson's last afternoon in Washington was spent in Blair House, across from the Executive Mansion. The general might have talked of his exciting past; instead, his thoughts were on the future. With Hill, Blair, Lewis, Kendall, he discussed Texas, and the dispute with England on the Oregon border.

They followed him to his coach, reluctant to see him go. The door slammed. He smiled at them and called:

"Keep your eyes on Texas, boys, and never give up 54-40!"

Days later as he turned in at the Hermitage gate; the view pleased him. There had been a bad fire in 1834, and in rebuilding there were changes. Now he liked the handsome new ver-

anda with its white columns. He glanced toward the garden and approved the new tomb over Rachel's grave.

Sarah, Andrew, and their children ran to greet him. Hannah, Alfred, and others of his people gave him welcome. "It is good to be at home," he told them.

Andrew Jackson, just turned seventy, was thin, white-haired and tall, without a hint of a stoop. His health was good except for occasional lung hemorrhages from that bullet. He admired the house, the beautiful French wallpaper in the hall, his own rooms, the gardens, and most of all his fine grand-children.

Then he went to work on plantation figures. They were not encouraging. He had always made his land pay. Qualities of common sense, shrewd judgment, and sterling honesty had served him well. Andrew, junior, though charming and de-voted to his father, was weak and careless. People knew it and took advantage of him. Jackson had guessed this. Now he knew.

Neighbors had known for some time.

"Young Andy's no business man," was freely spoken.

"You can't blame the man, really," some defended. "The general always spoiled him. It ain't as though the Jacksons had a lot of children, like some of us. Young Andy's his nearest; seems like the general never could say no to him."

Jackson got matters into his own hands. The plantation began to prosper. But over it hung the shadow of unexpected debts, contracted by Andrew. The general had to sell more than one cherished piece of land to keep his credit clear. But Jackson did not complain; he willed the whole estate to his son, hoping that somehow he would care for Sarah and the children.

During the eight remaining years at the Hermitage, Jackson had many pleasures. People from all over the country and from abroad came to see him. He joined the church; his faith was the same as always, but joining seemed to bring Rachel nearer.

In June of 1845, the general had a severe lung hemorrhage. As he grew worse, the wails of the plantation people reached his room.

"Tell them not to cry," he said kindly. "We shall all meet in heaven." Soon he was gone.

They laid him to rest in the garden, by his Rachel.

The tall, slender figure of Andrew Jackson, with brushed-up white hair and brilliant eyes, is familiar to Americans; but his personality has often been obscured by partisanship. He was a very human man, neither saint nor tyrant, interested in affairs of home and farm. He had a quick temper and moods of violence; but he had, too, honesty and steadfastness, imagination and tenderness, integrity and loyalty to friend and country.

It was given to him, Andy of the Waxhaws, to show for the first time that a poor boy, lacking family prestige, money, and formal education, could become president of the United States. To gain this high place, he broke the control of politicians who, from the birth of the nation, had kept the highest office in the hands of educated, well-born, devoted men deemed well qualified for the duties laid upon them. His achievement was a unique step forward in man's climb toward equality of opportunity for all men.

In a very special way, Andrew Jackson was the product of the new world. Born in the year when men's longing for freedom was striving for expression, he grew up under the dual influence of the Declaration of Independence and the war to make its ideal a reality in the lives of the people. The tragic loss of parents and brothers forced him to fend for himself and led him to go west, where others like him were striving and achieving.

On the frontier, men were judged by what they were — not by family or education. On the frontier, other poor boys were getting rich, entering professions, making names for themselves — Andrew Jackson was quite at home in such company. Years on wilderness trails and in army service gave him opportunity to observe men, to learn their ambitions and their dreams, as they faced life in the developing west. He was not a philosopher; at the time, this knowledge was simply stored in his mind and heart.

Andrew Jackson had never sought office, but his defeat in the 1824 presidential election angered him. He thought his large popular vote showed that the people wanted him and that in electing Adams the politicians in the House of Representatives had robbed the people of their choice. This anger carried Jackson through a three-year campaign that in 1828 brought victory.

But the man who entered the White House in 1829 was changed, molded by deep grief into a different person. This new man wanted nothing for himself; his thought was for his country. His knowledge of people flowered into rare understanding of the average American and what he wanted from his government. He remembered the farmer and the merchant — he had worked, even as they. Surely the dreams of a factory hand were much the same.

He roused himself from mourning and worked to make the United States the kind of country where such people could prosper. Men sensed this and gave him an almost fanatical devotion.

Andrew Jackson believed in the rule of the majority. He believed in the union of states, with each state enjoying as much freedom and responsibility as was possible within the framework of the Constitution. He believed in strong bonds of fellowship and vigorous leadership in the service of all the people. His faith made the ideal of democracy have new and vivid meaning and gave courage to all who work for the freedom and betterment of mankind.

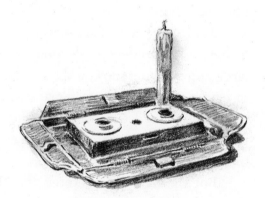